# MEDICAL NUTRITION THERAPY

## A Case Study Approach

MARCIA NAHIKIAN NELMS, PhD, RD, LD

**Southeast Missouri State University**

SARA LONG ANDERSON, PhD, RD, LD

**Southern Illinois University, Carbondale**

**WADSWORTH**

**THOMSON LEARNING**

Australia • Canada • Mexico • Singapore • Spain
United Kingdom • United States

Publisher: Peter Marshall
Development Editor: Elizabeth Howe
Assistant Editor: John Boyd
Editorial Assistant: Andrea Kesterke
Marketing Manager: Jennifer Somerville
Advertising Project Manager: Stacey Purviance
Project Manager, Editorial Production: Sandra Craig
Print/Media Buyer: Robert King
Permissions Editor: Joohee Lee

Production Service: Matrix Productions Inc.
Text Designer: Cynthia Bassett
Copy Editor: Linda Purrington
Cover Designer: Ross Carron
Cover Image: PhotoDisc
Cover Printer: Phoenix Color Corp.
Compositor: G&S Typesetters
Printer: Quebecor/World, Dubuque

Printed in the United States of America
2  3  4  5  6  7  05  04  03  02

For more information about our products, contact us at:
Thomson Learning Academic Resource Center
1-800-423-0563
For permission to use material from this text, contact us by:
Phone: 1-800-730-2214
Fax: 1-800-730-2215
Web: http://www.thomsonrights.com

ISBN: 0-534-52410-9

**Wadsworth/Thomson Learning**
10 Davis Drive
Belmont, CA 94002-3098
USA

**Asia**
Thomson Learning
60 Albert Street, #15-01
Albert Complex
Singapore 189969

**Australia**
Nelson Thomson Learning
102 Dodds Street
South Melbourne, Victoria 3205
Australia

**Canada**
Nelson Thomson Learning
1120 Birchmount Road
Toronto, Ontario M1K 5G4
Canada

**Europe/Middle East/Africa**
Thomson Learning
Berkshire House
168-173 High Holborn
London WC1 V7AA
United Kingdom

**Latin America**
Thomson Learning
Seneca, 53
Colonia Polanco
11560 Mexico D.F.
Mexico

**Spain**
Paraninfo Thomson Learning
Calle/Magallanes, 25
28015 Madrid
Spain

## Dedication

For the men in my life: Jerry, Taylor, and Emory

*—Marcia Nahikian Nelms*

Writing a book dedication is a lot like the acceptance speeches for the Oscars (except this won't go on and on and on, it will just seem like it)—there are people to thank who have made contributions to your success. This is my humble attempt to thank those special people in my life:

To my husband, who lets me be me
To my daughter, who taught me what real courage is
To my mother, who taught me unconditional love
To my father, who taught me how to have a sense of humor
To my grandfather, who taught me how to play
To my grandmother, who taught me how to be strong
To my son-in-law, who taught me to be convivial (and who shares my taste in pizza)
To my dear friends Georganna and Connie, who taught me what true friendship means
To my mentors, Ann Knewitz, Trish Welch, and Elena Sleipsevich, who taught me how to be a professional
To my co-author and friend, Marcia, who taught me to stop dreaming about this book and just do it
And to our publisher, Pete Marshall, for believing this whole thing would work.

*—Sara Long Anderson*

# CONTENTS

## Unit 4

# MEDICAL NUTRITION THERAPY FOR PANCREATIC AND LIVER DISORDERS  195

## Unit 5

# MEDICAL NUTRITION THERAPY FOR NEUROLOGICAL AND PSYCHIATRIC DISORDERS  235

## Unit 6

# MEDICAL NUTRITION THERAPY FOR PULMONARY DISORDERS  257

## Unit 7

# MEDICAL NUTRITION THERAPY FOR ENDOCRINE DISORDERS  287

## Unit 8

# MEDICAL NUTRITION THERAPY FOR RENAL DISEASE 345

## Unit 9

# MEDICAL NUTRITION THERAPY FOR HYPERMETABOLISM, INFECTION, AND TRAUMA 375

## Unit 10

# MEDICAL NUTRITION THERAPY FOR HEMATOLOGY– ONCOLOGY 409

## Appendices

# PREFACE

The idea for this book actually began more than 10 years ago as we began teaching medical nutrition therapy for dietetic students. Entering the classroom after being clinicians for many years, we knew we wanted our students to experience nutritional care as realistically as possible. We wanted classroom teaching to serve as a bridge between the textbook and the clinical setting. We relied heavily on our clinical experience to develop realistic clinical applications.

Using clinical applications or case studies is not new. Case studies are often used in nutrition, medicine, nursing, and many other allied health fields. The case study forces the student to integrate knowledge from many sources, supports the use of previously learned information, puts the student in decision-making role, and nurtures critical thinking. This text differs from a simple collection of case studies in the pedagogy used.

Each case study in this book uses the medical record as its structure. The student is asked to seek information for treating the case by using the exact tools that he or she will employ in the clinical setting. As the student moves from the admission form to the physician's history and physical to laboratory data and documentation of daily care, he or she will need to select relevant information from the medical record. This "real world" approach helps prepare the student for progressing from classroom to professional setting.

Each section in this book is introduced by a brief summary of the diagnoses the student will encounter. Each case identifies student objectives. The case is presented within the context of a physician's history and physical examination. Appropriate laboratory data and other components of the medical record that would document patient care accompany each case. Finally, case questions and applications focus on pathophysiology; assessment; clinical, nutritional and behavioral outcomes; interventions; and appropriate patient follow-up.

To be consistent with the philosophy of the text, each case requires that the student seek information from multiple resources to complete the case. With each case there is an extensive reference section that should provide an adequate framework from which the student may begin research.

We have chosen cases to represent the most common diagnoses that rely on nutrition therapy as an essential component of the medical care. Thus these cases represent the types of patients the student most likely will encounter. The concepts presented apply as well to many other medical conditions not presented here. The cases represent both introductory and advanced practice, so instructors can choose cases that match students' level of expertise.

The cases we have developed here cross the life span, allowing the student to see the practice of medical nutrition therapy during pregnancy, childhood, adolescence, adulthood, and senior years. We have tried to represent those people and patients we encounter in practice today. Placing medical nutrition therapy and nutrition education within the appropriate cultural context is crucial.

The cases in this book lend themselves to use in several important ways. They fit easily into a problem-based learning curriculum. They also can be used as summaries for classroom teaching about the pathophysiology and medical nutrition therapy for each diagnosis. The cases can be integrated into the appropriate rotation for dietetic internship and medical or nursing school. We have built objectives for student learning within each case around the competencies for dietetic education both in the didactic portion and the supervised practice component. This construction allows an additional way for nutrition and dietetic faculty to document student performance as part of program assessment.

We have created this text to provide a learning environment that will support students as they integrate knowledge and develop critical thinking skills. We have developed these "laboratories" and real-world situations to foster curiosity about new ideas, enthusiasm for learning, tolerance for the unfamiliar, and the ability to critically evaluate new ideas.

# Acknowledgments

We first need to thank our publisher, Peter Marshall. He has believed in this book from the onset, and because of him, it has happened. We have had excellent editorial guidance—we cannot begin to thank Beth Howe enough. We would like to extend a special thanks to the following reviewers: Elaine Blyler, California State University, Northridge; Gilbert A. Boissonneault, University of Kentucky; Mallory Boylan, Texas Tech University; Nancy Cotugna, University of Delaware; Susan Dahlheimer, Indiana University of Pennsylvania; Lynette M. Karls, University of Wisconsin, Madison; Patricia Plavcan, consultant; Valentina M. Remig, University of Akron; and Christine Rosenbloom, Georgia State University.

We had expert contributions to the book from our guest authors. Anne Marietta, PhD, RD, LD; Joe Pujol, PhD, FACSM; Jeremy Barnes, PhD; Mark Langenfeld, PhD, FACSM; Deb Cohen, MMSc, RD, LD, CNSD; Laurie Bernstein, MS, RD, FADA; and Sandra Dunning, MS, RD, LD, have shared their expertise in developing their cases in a way we could never have done without them. Sara Lopinski, MS, RD, and the dietetic staff at St. John's Hospital in Springfield, Illinois, were crucial in providing support for the development of several cases. We also thank William Marietta, MD, for his medical review of the cases.

My family and friends provide the solid foundation on which I have been able to write this book. My friends and colleagues—especially Cathy, Miriam, and Georganne—have nurtured me through this past year as I have come to always count on them. I thank my sister Marta and sister-in-law Pat for periodically coming to my home and taking over so that I could write. The men in my life—my sons, Emory and Taylor; my father, Bob; and my wonderful husband, Jerry—have always been there for me. Thank you for everything.

—*Marcia Nahikian Nelms, PhD, RD*

# ABOUT THE AUTHORS

## Marcia Nahikian Nelms, PhD, RD

*Associate Professor*
*Director, Didactic Program in Dietetics*
*Southeast Missouri State University*

Dr. Nahikian Nelms is an associate professor and director of the Didactic Program in Dietetics at Southeast Missouri State University. Prior to teaching, Dr. Nahikian Nelms practiced in both public health and clinical settings for 12 years, specializing in the care of patients in hematology, oncology, and bone marrow transplant. Today Dr. Nahikian Nelms continues to consult for agencies providing medical care for HIV and AIDS patients.

Dr. Nahikian Nelms has published numerous peer-reviewed journal articles and has contributed to four textbooks in the past. Her research has focused primarily in the areas of nutrition and cancer and early childhood nutrition. Her academic awards include Outstanding Teacher and Service Provider for the College of Health and Human Services at Southeast Missouri State University. In addition, she was named Outstanding Dietetic Educator in a Didactic Program in Dietetics for the state of Missouri.

Dr. Nahikian Nelms has actively participated in local, state, and national activities associated with the American Dietetic Association. She has most recently served on the Commission on Accreditation for Dietetic Education.

## Sara Long Anderson, PhD, RD

*Associate Professor/Director, Didactic Program in Dietetics*
*Animal Science, Food and Nutrition*
*Southern Illinois University, Carbondale*

Dr. Anderson is an associate professor and director of the Didactic Program in Dietetics in the Department of Animal Science, Food and Nutrition at Southern Illinois University, Carbondale. Prior to obtaining her PhD in health education, she practiced as a clinical dietitian for 11 years. Dr. Anderson has been the nutrition education/counseling consultant for Carbondale Family Medicine since 1986. She is an active leader in national, state, and district dietetic associations, where she has served in numerous elected and appointed positions. Dr. Anderson served as president of the Illinois Dietetic Association from 1993–1994, and has been an invited speaker at over 30 professional meetings.

Dr. Anderson is a co-author of *Foundations and Clinical Applications of Nutrition: A Nursing Approach.* Dr. Anderson is author or co-author of 13 peer-reviewed journal articles, 8 co-authored articles, 3 professional newsletter articles, 25 peer-reviewed abstracts (published in professional journals), and 5 sole-authored book reviews.

Dr. Anderson has received various awards and honors for teaching, including Outstanding Dietetic Educator (ADA), Outstanding Educator for the College of Agriculture, College of Agriculture Fellowship to attend Faculty Summer Institute on Learning Technologies, and a nominee for the Food and Agricultural Sciences Excellence in College and University Teaching Awards Program.

# ABOUT THE CONTRIBUTORS

Laurie Bernstein, MS, RD, FADA, is a metabolic nutritionist at the Children's Hospital in Denver, Colorado, and a Senior Instructor/University Colorado Health Science Center/School of Medicine–Department of Pediatrics. She is a specialist in providing nutrition care for those with inherited metabolic disease. Most recently she has been involved in the design and development of education materials for patients with inherited metabolic disease, specifically *Eat Right, Stay Bright Guide for Hyperphenylalaninemia*.

Jeremy T. Barnes, PhD, is an assistant professor in the Department of Health and Leisure at Southeast Missouri State University, where he teaches a variety of graduate and undergraduate health promotion and wellness classes. Dr. Barnes holds a PhD in health education from the University of Missouri, Columbia. He is a National Commission for Health Education Credentialing Inc. (NCHEC) certified health education specialist (CHES), an American College of Sports Medicine (ACSM) health/fitness instructor (HFI), and a National Strength and Conditioning Association (NSCA) certified strength and conditioning specialist (CSCS).

Deborah Cohen, MMSc, RD, LD, CNSD, is a clinical faculty associate at Southeast Missouri State University in the dietetic internship and the Department of Human Environmental Studies. Previously she worked as a nutrition support specialist with the bone marrow transplant team and in oncology/hematology at the Clinical Cancer Center at the University of Nebraska Medical Center in Omaha, Nebraska, for seven years. She brings to this project extensive experience with both autologous and allogeneic bone marrow and stem cell transplant patients.

Sandra Dunning, MS, RD, LD, is the registered dietitian at Fresenius Medical Care in Carbondale, Illinois. She brings to this project over 15 years experience in renal nutrition.

Mark Langenfeld, PhD, FACSM, currently is the chairperson of the Department of Health and Leisure and a professor in both the Department of Biology and the Department of Health and Leisure. He is an exercise physiologist with a longtime interest in nutritional aspects related to physical performance, including ultraendurance events. A Fellow of the American College of Sports Medicine, he enthusiastically promotes physically active lifestyles and lives that ethic by commuting to work by bicycle.

Anne Marietta, PhD, RD, LD, is an assistant professor at Southeast Missouri State University, where she teaches and directs the dietetic internship. Dr. Marietta received her PhD in health education from Southern Illinois University. She has been a practicing registered dietitian for over twenty years both in Pittsburgh, Pennsylvania, and in southeast Missouri.

Thomas Joseph Pujol, EdD, FACSM, is an associate professor of exercise physiology in the Department of Health and Leisure at Southeast Missouri State University. Dr. Pujol earned his EdD from the University of Alabama in 1991. His BS and MEd were awarded by the University of Louisiana at Monroe. Dr. Pujol is a Fellow of the American College of Sports Medicine. His research interests include subjective responses to external stimuli during exercise and exercise testing.

# Unit One

# MEDICAL NUTRITION THERAPY FOR LIFE CYCLE CONDITIONS

## Introduction

This section introduces medical conditions commonly found in specific life cycle groups. Each life cycle group has its own unique nutritional concerns related to factors such as age, pregnancy, or growth. You will apply your knowledge of nutrition assessment, physiology, and normal nutrition to address the medical problem. The life cycle cases also require knowledge of the unique psychosocial concerns for each group.

The admitting diagnoses are not the primary health concern for these cases, but significant health problems are uncovered as a result of the hospital admission. These cases could also be encountered in outpatient clinics, private physician offices, or public health clinics.

The first case highlights nutritional concerns of pregnancy. Adequate and appropriate nutrition is the foundation of a healthy pregnancy. This case focuses on one of the most common nutritional interventions during pregnancy: prevention and/or treatment of iron deficiency. Hemoglobin levels below the normal range during pregnancy have been linked to increased risk of poor pregnancy outcomes, including low birth weight, preterm delivery, and perinatal mortality.

The second case uses the diagnosis of rheumatoid arthritis as the context for applying principles of nutrition assessment in adults. This case also highlights physical symptoms that interfere with adequate oral intake. These symptoms, common in arthritis, appear in numerous medical conditions. This case also addresses difficulties with supplementation and drug–nutrient interaction. The goals of medical nutrition therapy coincide with treatment goals for the disease: Maximize nutritional status, minimize side effects of disease and treatment, and enhance quality of life. You can apply these same goals to almost any diagnosis.

Seniors are the fastest growing segment of the U.S. population. Nutrition plays a large role in the health status of this group. The older adult population may be at nutritional risk because of the physiological effects of aging, complex medical problems, and potential psychosocial and economic concerns. The third case in this section lets you address these issues within the context of polypharmacy, a common problem for the elderly.

The importance of nutrition in the performance of sports activities, whether competitive or recreational, is generally accepted. The fourth case challenges you to apply normal nutrition for the adolescent in the context of support for young athletes. Concepts of hydration, overtraining, and general sports nutrition will be used throughout.

The last case in the section introduces an adolescent athlete at risk for an eating disorder. This advanced case lets the student explore the assessment, pathophysiology, and treatment options for anorexia athletica. A number of risk factors for the disorder have been identified, as you will identify throughout the case.

# Pregnancy with Iron Deficiency Anemia

*Introductory Level*

## Objectives

After completing this case, the student will be able to:

1. Identify the physiological changes in pregnancy and apply those principles in nutrition assessment.
2. Define the pathophysiology of iron deficiency anemia.
3. Evaluate and interpret laboratory indices used to diagnose iron deficiency anemia.
4. Determine the risk factors for iron deficiency anemia.
5. Define the role of iron nutriture in pregnancy.
6. Evaluate dietary intake for adequacy during pregnancy.
7. Identify possible drug–nutrient interactions in iron supplementation.

Sarah Henley, a 31-year-old pregnant woman, is admitted to University Hospital after a fall on the ice. She is admitted to rule out premature labor, but because of her low hemoglobin levels, a complete hematological workup is done. Mrs. Henley is diagnosed with hypochromic, microcytic anemia.

 UNIVERSITY HOSPITAL

# ADMISSION DATABASE

Name: Sarah Henley
DOB: 10/7  age 31
Physician: F. Bowman, MD

| BED# 1 | DATE: 1/17 | TIME: 0300 | TRIAGE STATUS (ER ONLY): ☐ Red ☒ Yellow ☐ Green ☐ White |
|---|---|---|---|

**Initial Vital Signs**

| TEMP: 98.6 | RESP: 15 | SAO2: |
|---|---|---|

PRIMARY PERSON TO CONTACT:
Name: Michael Henley
Home #: 201-245-3321
Work #: 201-255-2890

| HT: 5'5" | WT (lb): 145 | B/P: 128/72 | PULSE: 88 |
|---|---|---|---|

ORIENTATION TO UNIT: ☒ Call light  ☒ Television/telephone
☒ Bathroom  ☒ Visiting  ☒ Smoking  ☒ Meals
☒ Patient rights/responsibilities

| LAST TETANUS unknown | LAST ATE 12 noon | LAST DRANK 12 noon |
|---|---|---|

**CHIEF COMPLAINT/HX OF PRESENT ILLNESS**

Went out to pick up mail at end of driveway–slipped and fell on ice.

Began to have small amount of bleeding and some abdominal pain.

Questioned if she was beginning premature labor.

PERSONAL ARTICLES: (Check if retained/describe)
☐ Contacts ☐ R ☐ L          ☐ Dentures ☐ Upper ☐ Lower
☒ Jewelry: wedding band
☒ Other: glasses

**ALLERGIES: Meds, Food, IVP Dye, Seafood: Type of Reaction**

VALUABLES ENVELOPE: no
☐ Valuables instructions

**PREVIOUS HOSPITALIZATIONS/SURGERIES**

Two previous pregnancies–one vaginal delivery 3 years ago and

one cesarean 18 months ago

INFORMATION OBTAINED FROM:
☒ Patient          ☐ Previous record
☒ Family          ☐ Responsible party

Signature  *Sarah Henley*

**Home Medications (including OTC)**   Codes: A=Sent home   B=Sent to pharmacy   C=Not brought in

| Medication | Dose | Frequency | Time of Last Dose | Code | Patient Understanding of Drug |
|---|---|---|---|---|---|
| prenatal vitamins | 1 | every am | several days ago | c | no |
|  |  |  |  |  |  |
|  |  |  |  |  |  |
|  |  |  |  |  |  |
|  |  |  |  |  |  |
|  |  |  |  |  |  |
|  |  |  |  |  |  |
|  |  |  |  |  |  |
|  |  |  |  |  |  |
|  |  |  |  |  |  |
|  |  |  |  |  |  |

Do you take all medications as prescribed?  ☐ Yes  ☒ No   If no, why? The prenatal vitamins make her nauseous.

**PATIENT/FAMILY HISTORY**

☐ Cold in past two weeks
☐ Hay fever
☐ Emphysema/lung problems
☐ TB disease/positive TB skin test
☒ Cancer Mother
☐ Stroke/past paralysis
☐ Heart attack
☐ Angina/chest pain
☒ Heart problems Father

☒ High blood pressure Father
☒ Arthritis Maternal grandmother
☐ Claustrophobia
☐ Circulation problems
☐ Easy bleeding/bruising/anemia
☐ Sickle cell disease
☐ Liver disease/jaundice
☐ Thyroid disease
☐ Diabetes

☐ Kidney/urinary problems
☐ Gastric/abdominal pain/heartburn
☐ Hearing problems
☐ Glaucoma/eye problems
☐ Back pain
☐ Seizures
☐ Other

**RISK SCREENING**

Have you had a blood transfusion?  ☐ Yes  ☒ No
Do you smoke?  ☒ Yes  ☐ No
If yes, how many pack(s) .5/day for 15 years
Does anyone in your household smoke?  ☒ Yes  ☐ No
Do you drink alcohol?  ☐ Yes  ☒ No
If yes, how often?_____  How much?_____
When was your last drink?_____/_____
Do you take any recreational drugs?  ☐ Yes  ☒ No
If yes, type:_____  Route
Frequency:_____  Date last used:_____/_____/_____

**FOR WOMEN Ages 12–52**

Is there any chance you could be pregnant?  ☒ Yes  ☐ No
If yes, expected date (EDC): 5/15/this year
Gravida/Para: 2/2

**ALL WOMEN**

Date of last Pap smear: 10/10 last year
Do you perform regular breast self-exams?  ☐ Yes  ☒ No

**ALL MEN**

Do you perform regular testicular exams?  ☐ Yes  ☐ No

Additional comments:

✗ *Bea Masters, RN, FNP*
Signature/Title

**Client name:** Sarah Henley
**DOB:** 10/7
**Age:** 31
**Sex:** Female
**Education:** High school diploma
**Occupation:** Stay-at-home mother
**Household members:** Sons ages 12 months and 3 years, husband age 35—all in good health
**Ethnic background:** Caucasian
**Religious affiliation:** None
**Referring physician:** Frieda Bowman, MD

## Chief complaint:

"I went out to get the mail and slipped on the ice. After I got back in I noticed a small amount of bleeding when I went to the bathroom. Over the next hour, I had some abdominal pain. I was afraid something might have happened to the baby."

## Patient history:

*Onset of disease:* Pt is a 31-year-old white female gravida 2/para 2 who presented to the ER in her 23rd week of gestation. She has experienced vaginal spotting and some abdominal pain. She is now admitted for observation to R/O premature labor secondary to her fall.

*Type of Tx:* Routine prenatal care. Received PTA.

*PMH:* Two previous pregnancies delivered at 38 and 37 weeks respectively. No other contributory history. Patient states that she is much more tired with this pregnancy but has related it to having two small children. She also describes being short of breath, which is common with her pregnancies, but states it feels like it has started earlier with this pregnancy.

*Meds:* Prenatal vitamins

*Smoker:* Yes

*Family Hx: What?* CAD  *Who?* Father

## Physical exam:

*General appearance:* 31-year-old pregnant female, pale, in no acute distress

*Vitals:* Temperature 98.6°F, BP 118/72, HR 88 bpm, RR 19 bpm

*Heart:* Regular rate and rhythm, heart sounds normal

*HEENT:*

   *Eyes:* Sclera pale; PERRLA, fundi without lesions

   *Ears:* Clear

   *Nose:* Clear

   *Throat:* Pharynx clear without postnasal drainage

*Genitalia:* Normal

*Neurologic:* Alert and oriented ×3

*Extremities:* No edema; DTR 2+ and symmetrical throughout

*Skin:* Skin pale without rash

*Chest/lungs:* Clear to auscultation and percussion

*Abdomen:* Bowel sounds present

**Nutrition Hx:**

*General:* Patient states that appetite is good right now. She suffered a lot of morning sickness during her first trimester but is better now.

*Usual dietary intake:*

AM:      Coffee, cold cereal, occasionally toast

Lunch:   Sandwich or soup

Dinner:  Casserole such as Hamburger Helper, hot dogs, soup; sometimes she cooks a full meal with meat and vegetables. Her husband works nights so she doesn't always cook except on his days off.

*24-hr recall (PTA):*

AM:      2 c Frosted Flakes, ½ c whole milk, black coffee

Lunch:   Hot dog on bun, ½ c macaroni and cheese

Dinner:  3 oz Salisbury steak, 1 c green beans, 1 roll, 1 c black coffee

*Food allergies/intolerances/aversions:* Patient says there are a lot of foods she doesn't like. Describes herself as a picky eater.

*Previous MNT?* Yes  *If yes, when:* Has had nutrition information during first pregnancy. *Where?* WIC program during first pregnancy.

*Food purchase/preparation:* Self

*Vit/min intake:* Prenatal vitamins but states she doesn't take them every day because they make her nauseous.

*Anthropometric data:* Ht 5′5″, Wt 145#,  *Prepregnancy wt:* 135#.  *Previous pregnancies:* Gained approximately 15–18 lbs with each

**Tx plan:**

*Activity:* Bed rest  *Diet:* NPO  *Lab:* CBC, RPR, Chem16  IV: Lactated ringers @ 8-hour rate, monitor fetal heart tones and contractions; I&O q shift; routine vital signs

**Hospital course:**

Ultrasound, fetal heart tones were all WNL for 23-week gestation. No more contractions occurred. Due to low Hgb found on routine admit lab work, additional hematological workup completed. *Dx:* microcytic, hypochromic anemia 2° to iron deficiency. Patient was discharged the following day on 40 mg ferrous sulfate TID. Nutrition consult ordered.

# UH UNIVERSITY HOSPITAL

NAME: Sarah Henley                              DOB: 10/7
AGE: 31                                          SEX: F
PHYSICIAN: F. Bowman, MD

**************************************CHEMISTRY**************************************

DAY:                                             1
DATE:                                            1/17
TIME:
LOCATION:

| | NORMAL | | UNITS |
|---|---|---|---|
| Albumin | 3.6–5 | 3.9 | g/dL |
| Total protein | 6–8 | 6.2 | g/dL |
| Prealbumin | 19–43 | 25 | mg/dL |
| Transferrin | 200–400 | 390 | mg/dL |
| Sodium | 135–155 | 142 | mmol/L |
| Potassium | 3.5–5.5 | 3.8 | mmol/L |
| Chloride | 98–108 | 105 | mmol/L |
| $PO_4$ | 2.5–4.5 | 3.2 | mmol/L |
| Magnesium | 1.6–2.6 | 1.6 | mmol/L |
| Osmolality | 275–295 | 292 | mmol/kg $H_2O$ |
| Total $CO_2$ | 24–30 | 26 | mmol/L |
| Glucose | 70–120 | 105 | mg/dL |
| BUN | 8–26 | 8 | mg/dL |
| Creatinine | 0.6–1.3 | 0.7 | mg/dL |
| Uric acid | 2.6–6 (women) | 2.7 | mg/dL |
| | 3.5–7.2 (men) | | |
| Calcium | 8.7–10.2 | 8.8 | mg/dL |
| Bilirubin | 0.2–1.3 | 0.4 | mg/dL |
| Ammonia ($NH_3$) | 9–33 | 9 | $\mu$mol/L |
| SGPT (ALT) | 10–60 | 12 | U/L |
| SGOT (AST) | 5–40 | 8 | U/L |
| Alk phos | 98–251 | 99 | U/L |
| CPK | 26–140 (women) | | U/L |
| | 38–174 (men) | | |
| LDH | 313–618 | | U/L |
| CHOL | 140–199 | 165 | mg/dL |
| HDL-C | 40–85 (women) | 55 | mg/dL |
| | 37–70 (men) | | |
| VLDL | | | mg/dL |
| LDL | < 130 | 125 | mg/dL |
| LDL/HDL ratio | < 3.22 (women) | 2.27 | |
| | < 3.55 (men) | | |
| Apo A | 101–199 (women) | | mg/dL |
| | 94–178 (men) | | |
| Apo B | 60–126 (women) | | mg/dL |
| | 63–133 (men) | | |
| TG | 35–160 | 120 | mg/dL |
| $T_4$ | 5.4–11.5 | | $\mu$g/dL |
| $T_3$ | 80–200 | | ng/dL |
| $HbA_{1c}$ | 4.8–7.8 | | % |

# U<sub>H</sub> UNIVERSITY HOSPITAL

NAME: Sarah Henley                    DOB: 10/7
AGE: 31                               SEX: F
PHYSICIAN: F. Bowman, MD

\*\*\*\*\*\*\*\*\*\*\*\*\*\*\*\*\*\*\*\*\*\*\*\*\*\*\*\*\*\*\*\*\*\*\*\*\*\*\*\*\*HEMATOLOGY\*\*\*\*\*\*\*\*\*\*\*\*\*\*\*\*\*\*\*\*\*\*\*\*\*\*\*\*\*\*\*\*\*\*\*\*\*\*\*\*

| | NORMAL | 1 | UNITS |
|---|---|---|---|
| DAY: | | 1 | |
| DATE: | | 1/17 | |
| TIME: | | | |
| LOCATION: | | | |
| WBC | 4.3–10 | 7.2 | $\times 10^3/mm^3$ |
| RBC | 4–5 (women) | 3.8 | $\times 10^6/mm^3$ |
| | 4.5–5.5 (men) | | |
| HGB | 12–16 (women) | 9.1 | g/dL |
| | 13.5–17.5 (men) | | |
| HCT | 37–47 (women) | 33 | % |
| | 40–54 (men) | | |
| MCV | 84–96 | 72 | $\mu^3$ |
| RETIC | 0.8–2.8 | 0.2 | % |
| MCH | 27–31 | 23 | pg |
| MCHC | 31.5–36 | 28 | g/dL |
| RDW | 11.6–16.5 | 22 | % |
| Plt Ct | 140–440 | 282 | $\times 10^3$ |
| Diff TYPE | | | |
| % GRANS | 34.6–79.2 | 36.2 | % |
| % LYM | 19.6–52.7 | 41.3 | % |
| SEGS | 50–62 | 52 | % |
| BANDS | 3–6 | 4 | % |
| LYMPHS | 25–40 | 31 | % |
| MONOS | 3–7 | 3 | % |
| EOS | 0–3 | 0 | % |
| TIBC | 65–165 (women) | 172 | $\mu$g/dL |
| | 75–175 (men) | | |
| Ferritin | 18–160 (women) | 10 | $\mu$g/dL |
| | 18–270 (men) | | |
| ZPP | 30–80 | 18 | $\mu$mmol/mol |
| Vitamin B$_{12}$ | 100–700 | 250 | pg/mL |
| Folate | 2–20 | 2 | ng/mL |
| Total T cells | 812–2318 | | $mm^3$ |
| T-helper cells | 589–1505 | | $mm^3$ |
| T-suppressor cells | 325–997 | | $mm^3$ |
| PT | 11–13 | 12 | sec |

## Case Questions

1.  Evaluate the patient's physical examination. Is any sign or symptom abnormal? Are any signs or symptoms consistent with her diagnosis of iron deficiency anemia? Why or why not?

2.  In reading the patient's medical record, you note that she is "gravida 2 para 3." What does this mean?

3.  Mrs. Henley's physician indicated the need for additional lab work when her admitting CBC revealed low hemoglobin. Why is this a concern? Are normal changes in hemoglobin associated with pregnancy? If so, what? Do any other hematological values normally change in pregnancy? Explain.

4.  Discuss each of the following laboratory tests, and explain what they measure. Determine the normal values. Then evaluate Mrs. Henley's lab results and compare.

| Test | Function | Normal value | Mrs. Henley's value |
|------|----------|--------------|---------------------|
| Hemoglobin | | | |
| Hematocrit | | | |
| MCV | | | |
| MCH | | | |
| MCHC | | | |
| RDW | | | |
| Reticulocyte count | | | |
| Ferritin | | | |
| Transferrin saturation | | | |
| Total iron binding capacity | | | |
| Zinc protoporphyrin (ZPP) | | | |

5.  There are several classifications of anemia, including megaloblastic anemia, pernicious anemia, normocytic anemia, microcytic anemia, sickle cell anemia, and hemolytic anemia. Define each.

6.  Several stages of iron deficiency actually precede iron deficiency anemia. Discuss these stages—including the symptoms—and identify laboratory values that might be affected.

7.  What other nutritional data available from this medical record can be used in your nutrition assessment?

8.  Check Mrs. Henley's prepregnancy weight. How much weight has she gained? Plot her weight gain on the maternal weight gain curve. Is her gain adequate? How does her weight gain compare to current recommendations? Was the weight gained during her previous pregnancies within normal limits?

9.  What factors in her pregnancy histories indicate any additional risk factors for the development of iron deficiency anemia? Were her other pregnancies normal?

10. Could anything else in her medical, nutritional, or lifestyle history affect her pregnancy outcome? Explain.

11. Calculate Mrs. Henley's energy and protein requirements. What standards did you use and why?

12. Assess Mrs. Henley's 24-hour recall. How does it compare with the requirements you calculated in Question 11?

13. Assess this patient's average daily iron intake. How does this compare to the RDI for iron during pregnancy? Are there any other nutrients you should be concerned about? Explain.

14. What are good dietary sources of iron? Is the absorption of iron affected by any other conditions? Explain.

**15.** Mrs. Henley was discharged on 40 mg ferrous sulfate TID. Are there potential side effects from this medication? Are there any drug–nutrient interactions? (Explain.) What instructions might you give her to maximize the benefit of her iron supplementation?

**16.** As you discuss the medical nutrition therapy for iron deficiency anemia with Mrs. Henley, what other nutrition problems might you discuss regarding her prenatal nutritional status?

**17.** You note in Mrs. Henley's history that she has received nutrition counseling from the WIC program. What is WIC? Would you refer her back to that program? If so, how would you make this referral?

# Bibliography

Allen LH. Anemia and iron deficiency: effects on pregnancy outcome. *Am J Clin Nutr.* 2000;71(5 suppl):1280S–1284S.

American Dietetic Association. Pregnancy. In *Manual of Clinical Dietetics,* 6th ed. Chicago: American Dietetic Association, 2000:109–127.

Blackburn MW, Calloway DH. Energy expenditure and consumption of mature, pregnant and lactating women. *J Am Diet Assoc.* 1976;69:29–37.

Centers for Disease Control. CDC criteria for anemia in children and childbearing-aged women. *MMWR.* 1989;38:400–404.

Commission on Accreditation for Dietetics Education. Knowledge, skills, and competencies for dietitians. *Accreditation Manual for Dietetics Education Programs,* Revised 4th ed. Chicago: American Dietetic Association, 2000.

Garn SM, Ridela SA, Petzoid AS, et al. Maternal hematologic levels and pregnancy outcomes. *Semin Perinatol.* 1981;5:155–162.

Institute of Medicine. Energy requirements, energy intake, and associated weight gain during pregnancy. In *Nutrition During Pregnancy.* Washington, DC: National Academy Press, 1990:137–175.

Institute of Medicine. Iron nutrition during pregnancy. In *Nutrition During Pregnancy.* Washington, DC: National Academy Press, 1990:272–298.

Lewis GJ, Rowe DF. Can a serum ferritin estimation predict which pregnant women need iron? *Br J Clin Pract.* 1986;40:15–16.

Lockitch G (ed.). *Handbook of Diagnostic Biochemistry and Hematology in Normal Pregnancy.* Boca Raton, FL: CRC Press, 1993.

Rucker RB. Nutritional anemias. In *Clinical Nutrition and Dietetics.* 2nd ed. Zeman, FJ (ed.). New York: Macmillan, 1990, pp. 683–700.

Schifman RB, Thomasson JE, Evers JM. Red blood cell zinc protoporphyrin testing for iron deficiency anemia in pregnancy. *Am J Obstet Gynecol.* 1987;157:304–307.

Scholl TO, Hediger ML. Anemia and iron-deficiency anemia: compilation of data on pregnancy outcome. *Am J Clin Nutr.* 1994;59:492S-501S.

U.S. Preventive Services Task Force. Routine iron supplementation during pregnancy. *JAMA,* 1993;270:2846–2854.

# Rheumatoid Arthritis

*Introductory Level*

## Objectives

After completing this case, the student will be able to:

1. Apply working knowledge of pathophysiology to plan nutrition care.
2. Analyze nutrition assessment data to determine baseline nutritional status.
3. Identify potential drug–nutrient interactions and appropriate nutrition interventions for preventing or treating drug–nutrient interactions.
4. Identify and explain the common nutritional risks for the diagnosis of rheumatoid arthritis.
5. Create strategies to maximize calorie and protein intake.
6. Analyze current recommendations for nutritional supplementation.

Mr. Robert Jacobs is admitted to University Hospital to evaluate the current status of his rheumatoid arthritis and to adjust his medical regimen. Mr. Jacobs was diagnosed with rheumatoid arthritis five years ago, and his current medications are not controlling his pain and symptoms. Mr. Jacobs is particularly interested in pursuing alternative medical treatments.

**Name:** R. Jacobs
**DOB:** 5/4 **age** 39
**Physician:** K. Sanders, MD

# ADMISSION DATABASE

| BED# 1 | DATE: 4/18 | TIME: 1300 | TRIAGE STATUS (ER ONLY): ☐ Red ☐ Yellow ☐ Green ☐ White |
|---|---|---|---|

**PRIMARY PERSON TO CONTACT:**
Name: Peter and Myra Jacobs
Home #: 201-345-7890
Work #:

### Initial Vital Signs

| TEMP: 98.8 | RESP: 18 | SAO2: |
|---|---|---|

| HT: 5'10" | WT (lb): 154 (highest wt 165 2 years ago) | B/P: 128/82 | PULSE: 86 |
|---|---|---|---|

**ORIENTATION TO UNIT:** ☒ Call light ☒ Television/telephone ☒ Bathroom ☒ Visiting ☒ Smoking ☒ Meals ☒ Patient rights/responsibilities

| LAST TETANUS unknown | LAST ATE this am | LAST DRANK |
|---|---|---|

## CHIEF COMPLAINT/HX OF PRESENT ILLNESS

"My morning stiffness is worse and lasts almost till noon.

I have problems at night with pain, too."

**PERSONAL ARTICLES:** (Check if retained/describe)
☒ Contacts ☐ R ☐ L          ☐ Dentures ☐ Upper ☐ Lower
☐ Jewelry:
☒ Other: glasses

## ALLERGIES: Meds, Food, IVP Dye, Seafood: Type of Reaction

Eats no pork; eats kosher during religious holidays.

**VALUABLES ENVELOPE:**
☐ Valuables instructions

## PREVIOUS HOSPITALIZATIONS/SURGERIES

**INFORMATION OBTAINED FROM:**
☒ Patient      ☐ Previous record
☐ Family       ☐ Responsible party

Signature  *R. Jacobs*

| Home Medications (including OTC) | Codes: A=Sent home | | B=Sent to pharmacy | | C=Not brought in |
|---|---|---|---|---|---|
| Medication | Dose | Frequency | Time of Last Dose | Code | Patient Understanding of Drug |
| ibuprofen | 500 mg | bid | this am | c | Yes |
| prednisone | 10 mg | daily | this am | c | Yes |
| | | | | | |
| | | | | | |
| | | | | | |
| | | | | | |
| | | | | | |
| | | | | | |
| | | | | | |
| | | | | | |
| | | | | | |

Do you take all medications as prescribed? ☒ Yes ☐ No  If no, why?

## PATIENT/FAMILY HISTORY

| | | |
|---|---|---|
| ☐ Cold in past two weeks | ☐ High blood pressure | ☐ Kidney/urinary problems |
| ☐ Hay fever | ☐ Arthritis | ☐ Gastric/abdominal pain/heartburn |
| ☐ Emphysema/lung problems | ☐ Claustrophobia | ☐ Hearing problems |
| ☐ TB disease/positive TB skin test | ☐ Circulation problems | ☐ Glaucoma/eye problems |
| ☐ Cancer | ☐ Easy bleeding/bruising/anemia | ☐ Back pain |
| ☐ Stroke/past paralysis | ☐ Sickle cell disease | ☐ Seizures |
| ☐ Heart attack | ☐ Liver disease/jaundice | ☒ Other Rheumatoid arthritis |
| ☐ Angina/chest pain | ☐ Thyroid disease | |
| ☐ Heart problems | ☐ Diabetes | |

## RISK SCREENING

Have you had a blood transfusion? ☐ Yes ☒ No
Do you smoke? ☒ Yes ☐ No
If yes, how many pack(s) 1/2/day for 15 years
Does anyone in your household smoke? ☐ Yes ☒ No
Do you drink alcohol? ☒ Yes ☐ No
If yes, how often? weekly      How much? 1-2 drinks
When was your last drink? 4/16/
Do you take any recreational drugs? ☐ Yes ☒ No
If yes, type:_____  Route
Frequency:_____  Date last used:_____/_____/_____

**FOR WOMEN Ages 12–52**

Is there any chance you could be pregnant? ☐ Yes ☐ No
If yes, expected date (EDC):
Gravida/Para:

### ALL WOMEN

Date of last Pap smear:
Do you perform regular breast self-exams? ☐ Yes ☐ No

### ALL MEN

Do you perform regular testicular exams? ☐ Yes ☒ No

Additional comments:

**✗** *Michelle Jenkins, RN*
Signature/Title

**Client name:** Robert Jacobs
**DOB:** 5/4
**Age:** 39
**Sex:** Male
**Education:** Bachelor's degree
**Occupation:** Accountant
**Hours of work:** M–F 9–6 pm—often takes work home on weekends
**Household members:** Lives alone
**Ethnic background:** U.S. born—Caucasian
**Religious affiliation:** Jewish
**Referring physician:** Kevin Sanders, MD (rheumatology)

## Chief complaint:

"My morning stiffness is considerably worse and actually lasts almost to noon. I have noticed problems in the evening as well. I wake up during the night with pain. I have also wondered whether there is anything I can do regarding diet that might help me. I recently began taking an antioxidant combination as well as fish oil capsules. I really don't want to take more medicines."

## Patient history:

*Onset of disease:* Approximately 5 years ago diagnosed with rheumatoid arthritis
*Type of Tx:* Motrin, 1000 mg/day; prednisone, 10 mg/day
*PMH:* No significant illness prior to this diagnosis
*Meds:* See above.
*Smoker:* Yes
*Family Hx: What?* HTN *Who?* Father

## Physical exam:

*General appearance:* Slim 39-year-old white male who moves with some difficulty and pain
*Vitals:* Temp 98.8° F, BP 128/82, HR 86 bpm, RR 22 bpm
*Heart:* Regular rate and rhythm, heart sounds normal
*HEENT:* Slight temporal wasting noted. All other WNL and noncontributory.
*Genitalia:* Normal
*Neurologic:* Alert and oriented, hesitant gait, normal reflexes
*Extremities:* Appearance of slight muscle wasting, joints tender to touch
*Musculoskeletal:* Mild limitation of motion, bony enlargement of the DIP joints of both hands consistent with Heberden's nodes. Shoulders, elbows, wrists, and small joints of the feet show evidence of swelling, warmth, and erythema of these joints.

## Nutrition Hx:

*General:* Appetite fair—hungrier when pain is controlled

*Usual dietary intake:*
| | |
|---|---|
| AM: | Coffee, juice with meds |
| Midmorning: | Doughnut, sweet roll, etc., with coffee |
| Lunch: | Occasionally skips but if eats, it is at fast-food restaurant—burger, sub sandwich, pizza with coffee or soda |

*Dinner:*   Eats out at local restaurants 2–3× per week (meat entree, pasta, salad) or when cooks uses ready-to-eat boxed meals or frozen entrees. Likes to cook but has not had time recently. States that he should follow a kosher diet but admits that he does not except during religious holidays.

*24-hr recall* (in hospital):

*AM:*   1 slice WW toast with 2 pats margarine and 1 container of jam; 3 c black coffee

*Lunch:*   Vegetable soup—1.5 c, 6 saltine crackers, 2 c black coffee

*Dinner:*   Baked fish—6 oz, rice—½ c, chocolate cake—1–2″ square, 2 c decaf coffee

*Food allergies/intolerances/aversions:* "I do not eat pork of any kind. I generally eat kosher during religious holidays."

*Previous MNT?* No

*Food purchase/preparation:* Self

*Vit/min intake:* Centrum multivitamin antioxidant combination; fish oil capsules

**Tx plan:**

Evaluate current status of rheumatoid arthritis. Adjust medical regimen as necessary. Begin treatment with methotrexate in addition to current medications.

# U_H UNIVERSITY HOSPITAL

NAME: R. Jacobs                    DOB: 5/4
AGE: 39                            SEX: M
PHYSICIAN: K. Sanders, MD

\*\*\*\*\*\*\*\*\*\*\*\*\*\*\*\*\*\*\*\*\*\*\*\*\*\*\*\*\*\*\*\*\*\*\*\*\*\*\*\*\*CHEMISTRY\*\*\*\*\*\*\*\*\*\*\*\*\*\*\*\*\*\*\*\*\*\*\*\*\*\*\*\*\*\*\*\*\*\*\*\*\*\*\*\*

| | NORMAL | DAY: 1<br>DATE: 4/18<br>TIME:<br>LOCATION: | UNITS |
|---|---|---|---|
| Albumin | 3.6–5 | 3.8 | g/dL |
| Total protein | 6–8 | 6.0 | g/dL |
| Prealbumin | 19–43 | 32 | mg/dL |
| Transferrin | 200–400 | 220 | mg/dL |
| Sodium | 135–155 | 142 | mmol/L |
| Potassium | 3.5–5.5 | 4.2 | mmol/L |
| Chloride | 98–108 | 104 | mmol/L |
| $PO_4$ | 2.5–4.5 | 4.0 | mmol/L |
| Magnesium | 1.6–2.6 | 1.8 | mmol/L |
| Osmolality | 275–295 | 0 | mmol/kg $H_2O$ |
| Total $CO_2$ | 24–30 | 28 | mmol/L |
| Glucose | 70–120 | 119 | mg/dL |
| BUN | 8–26 | 18 | mg/dL |
| Creatinine | 0.6–1.3 | 0.8 | mg/dL |
| Uric acid | 2.6–6 (women)<br>3.5–7.2 (men) | 6.2 | mg/dL |
| Calcium | 8.7–10.2 | 8.5 | mg/dL |
| Bilirubin | 0.2–1.3 | 0.5 | mg/dL |
| Ammonia ($NH_3$) | 9–33 | | $\mu$mol/L |
| SGPT (ALT) | 10–60 | | U/L |
| SGOT (AST) | 5–40 | | U/L |
| Alk phos | 98–251 | | U/L |
| CPK | 26–140 (women)<br>38–174 (men) | | U/L |
| LDH | 313–618 | | U/L |
| CHOL | 140–199 | 190 | mg/dL |
| HDL-C | 40–85 (women)<br>37–70 (men) | 50 | mg/dL |
| VLDL | | | mg/dL |
| LDL | < 130 | 131 | mg/dL |
| LDL/HDL ratio | < 3.22 (women)<br>< 3.55 (men) | | |
| Apo A | 101–199 (women)<br>94–178 (men) | | mg/dL |
| Apo B | 60–126 (women)<br>63–133 (men) | | mg/dL |
| TG | 35–160 | 155 | mg/dL |
| $T_4$ | 5.4–11.5 | | $\mu$g/dL |
| $T_3$ | 80–200 | | ng/dL |
| $HbA_{1C}$ | 4.8–7.8 | | % |

# U H *UNIVERSITY HOSPITAL*

NAME: R. Jacobs                        DOB: 5/4
AGE: 39                                SEX: M
PHYSICIAN: K. Sanders, MD

\*\*\*\*\*\*\*\*\*\*\*\*\*\*\*\*\*\*\*\*\*\*\*\*\*\*\*\*\*\*\*\*\*\*\*\*\*\*\*\*HEMATOLOGY\*\*\*\*\*\*\*\*\*\*\*\*\*\*\*\*\*\*\*\*\*\*\*\*\*\*\*\*\*\*\*\*\*\*\*\*\*\*\*\*

DAY:                                            1
DATE:                                          4/18
TIME:
LOCATION:

| | NORMAL | | UNITS |
|---|---|---|---|
| WBC | 4.3–10 | 6.0 | $\times 10^3/mm^3$ |
| RBC | 4–5 (women) | 4.8 | $\times 10^6/mm^3$ |
| | 4.5–5.5 (men) | | |
| HGB | 12–16 (women) | 15 | g/dL |
| | 13.5–17.5 (men) | | |
| HCT | 37–47 (women) | 41 | % |
| | 40–54 (men) | | |
| MCV | 84–96 | | fL |
| MCH | 27–34 | | pg |
| MCHC | 31.5–36 | | % |
| RDW | 11.6–16.5 | | % |
| Plt Ct | 140–440 | | $\times 10^3$ |
| Diff TYPE | | | |
| ESR | 0–20 (women) | 33 | mm/hr |
| | 0–15 (men) | | |
| % GRANS | 34.6–79.2 | | % |
| % LYM | 19.6–52.7 | | % |
| SEGS | 50–62 | | % |
| BANDS | 3–6 | | % |
| LYMPHS | 25–40 | | % |
| MONOS | 3–7 | | % |
| EOS | 0–3 | | % |
| TIBC | 65–165 (women) | | $\mu$g/dL |
| | 75–175 (men) | | |
| Ferritin | 18–160 (women) | | $\mu$g/dL |
| | 18–270 (men) | | |
| Vitamin $B_{12}$ | 100–700 | | pg/mL |
| Folate | 2–20 | | ng/mL |
| Total T cells | 812–2318 | | $mm^3$ |
| T-helper cells | 589–1505 | | $mm^3$ |
| T-suppressor cells | 325–997 | | $mm^3$ |
| PT | 11–13 | | sec |

## Case Questions

1.   Describe the inflammatory response that plays a role in the pathophysiology of rheumatoid arthritis. How do corticosteroids and NSAIDs interfere with this inflammatory process?

2.   Calculate %UBW, %IBW, and body mass index (BMI).

3.   Is Mr. Jacobs's weight of concern? Why or why not?

4.   What information in the physician's assessment may lead you to be concerned about muscle stores? What additional anthropometric indices might you evaluate to assess muscle mass or lean body mass?

5.   What may be the possible reasons for any loss of lean body mass?

6.   What laboratory measures correlate with wasting of lean body mass?

7.   Mr. Jacobs states his appetite is fair. What other questions might you ask to further assess his appetite? What are possible causes of his decreased appetite?

8.   What laboratory values will be used to assess nutritional status? Are any significant? Are there others that might be important to assess for patients with rheumatoid arthritis? Explain.

9.   This patient will be started on methotrexate. What are the common drug–nutrient interactions with this medication? Are any other drug–nutrient interactions with his other medications of concern? Explain.

10.   Assess this patient's 24-hour recall using a computerized dietary analysis. Are any nutrients of concern? What additional data are needed for a thorough dietary assessment? Explain your answers.

**11.** What specific dietary interventions would you suggest to maximize his nutritional intake? Be sure to consider micronutrients as well as caloric and protein intake.

**12.** What is the history and rationale for the kosher diet? Does this diet have any nutritional consequences for the patient?

**13.** Should you be concerned about Mr. Jacobs's supplement intake? Is there any rationale for using antioxidants and omega-3 fatty acids in treating rheumatoid arthritis? Explain. What does current research recommend?

# Bibliography

Ariza-Ariza, R, et al. Omega-3 fatty-acids in rheumatoid arthritis: an overview. *Seminar Arthritis Rheum.* 1997;27:366.

American Dietetic Association. Kosher diet. In *Manual of Clinical Dietetics,* 6th ed. Chicago: American Dietetic Association, 2000:785–789.

Anderson LS, Hansen TM. Prospectively measure red cell folate levels in methotrexate treated patients with rheumatoid arthritis: relation to withdrawal and side effects. *J Rheumatol.* 1997;25:830.

Commission on Accreditation for Dietetics Education. Knowledge, skills, and competencies for dietitians. *Accreditation Manual for Dietetics Education Programs,* revised 4th ed. Chicago: American Dietetic Association, 2000.

Escott-Stump S. *Nutrition and Diagnosis Related Care,* 4th ed. Baltimore: Williams & Wilkins, 1998.

ESHA. *Diet Analysis Plus 4.0.* Salem, OR: ESHA Research, 2000.

Franzese TA. Medical nutrition therapy for rheumatic disorders. In Mahan LK, Escott-Stump S (eds.), *Krause's Food, Nutrition, and Diet Therapy,* 10th ed. Philadelphia: Saunders, 2000:970–986.

Future trends in the management of rheumatoid arthritis. Proceedings of the International Rheumatology Round Tables. March 27–28, 1998, and February 19–20, 1999. *Rheumatology.* 1999;38 (Suppl 2):1–53.

Gould BE. *Pathophysiology for the Health-Related Professions.* Philadephia: Saunders, 1997:11–13.

Heliovaara M, et al. Serum antioxidants and risk of rheumatoid arthritis. *Ann Rheum.* 1994;53:51.

Hansen VO, Nielsen L, Kluger E, Thysen M, Emmertsen K, Stengaard-Pedersen K, Hansen EL, Unger B, Andersen PW. Nutritional status of Danish rheumatoid arthritis patients and effects of a diet adjusted in energy intake, fish-meal, and antioxidants. *Scand J Rheumatol.* 1996;25:325–330.

Kremer JM, Bigaouette J. Nutrient intake of patients with rheumatoid arthritis is deficient in pyridoxine, zinc, copper, and magnesium. *J Rheumatol.* 1996;23:990–994.

Lamour A, Le Goff P, Jouquan J, Bendaoud B, Youinou P, Menez J. Some nutritional parameters in patients with rheumatoid arthritis. *Ann Intern Med.* 1995;6:409–412.

Martin RH. The role of nutrition and diet in rheumatoid arthritis. *Proc Nutr Soc.* 1998;57:231.

Pronsky Z. *Food Medication Interactions,* 11th ed. Pottstown, PA: Food Medication Interactions, 1999.

Strano CG, Polito C, Lammarone SS, Di Toro A, Todisco N, Marotta A. Nutritional status in active juvenile chronic arthritis not treated with steroids. *Acta Paediatr.* 1995;84:1010–1013.

Yocum DE, Castro WL, Cornett M. Exercise, education, and behavioral modification as alternative therapy for pain and stress in rheumatic disease. *Rheum Dis Clin North Am.* 2000;26:145–159.

# Polypharmacy of the Elderly: Drug–Nutrient Interactions

*Introductory Level*

## Objectives

After completing this case, the student will be able to:

1. Integrate working knowledge of pharmacology, nutrient–nutrient, and drug–nutrient interaction(s).
2. Interpret pertinent laboratory parameters in the elderly.
3. Identify the unique nutritional needs of the elderly.
4. Assess nutritional risk factors for the elderly patient.
5. Determine appropriate nutrition interventions to correct drug–nutrient interactions and improve nutritional status.

Bob Kaufman, an 85-year-old male, has been brought to the hospital emergency room because of a change in his mental status. Mr. Kaufman suffers from several chronic diseases that are currently treated with multiple medications.

 **UNIVERSITY HOSPITAL**

# ADMISSION DATABASE

Name: Bob Kaufman
DOB: 1/12  age 85
Physician: Curtis Martin, MD

| BED# 1 | DATE: 5/15 | TIME: 1500 | TRIAGE STATUS (ER ONLY): ☐ Red ☐ Yellow ☐ Green ☐ White |
|---|---|---|---|

**Initial Vital Signs**

| TEMP: 97.2 | RESP: 17 | SAO2: |
|---|---|---|

| HT: 5'2" | WT (lb): 196 | B/P: 160/82 | PULSE: 86 |
|---|---|---|---|

| LAST TETANUS unknown | | LAST ATE 12 noon | LAST DRANK 30 minutes ago |
|---|---|---|---|

**PRIMARY PERSON TO CONTACT:**
Name: Megan Smith
Home #: 223-4589
Work #: 222-3421

**ORIENTATION TO UNIT:** ☒ Call light ☒ Television/telephone ☒ Bathroom ☒ Visiting ☒ Smoking ☒ Meals ☒ Patient rights/responsibilities

### CHIEF COMPLAINT/HX OF PRESENT ILLNESS

"We brought my father in because he is more confused. His blood sugar is normal. I thought we should make sure he's OK."

**PERSONAL ARTICLES:** (Check if retained/describe)
☐ Contacts ☐ R ☐ L    ☐ Dentures ☒ Upper ☒ Lower
☐ Jewelry:
☒ Other: eyeglasses

### ALLERGIES: Meds, Food, IVP Dye, Seafood: Type of Reaction

NKA

**VALUABLES ENVELOPE:** no
☐ Valuables instructions

### PREVIOUS HOSPITALIZATIONS/SURGERIES

back surgery-? procedure 15 years ago

TURP 2° prostate CA-10 years ago

lower GI bleed 2° diverticulitis-2 hospitalizations 10 and 12 years ago

**INFORMATION OBTAINED FROM:**
☐ Patient        ☐ Previous record
☒ Family         ☒ Responsible party

Signature *Megan Smith*

### Home Medications (including OTC)    Codes: A=Sent home    B=Sent to pharmacy    C=Not brought in

| Medication | Dose | Frequency | Time of Last Dose | Code | Patient Understanding of Drug |
|---|---|---|---|---|---|
| Diovan | 80 mg | daily | this am | A | no |
| Prilosec | 20 mg | daily | this am | A | no |
| Neurontin | 300 mg | BID | this am | A | no |
| furosemide | 20 mg | 1-2 as needed | this am | A | no |
| Zocor | 20 mg | daily | this am | A | no |
| isosorbide mono | 60 mg | daily | this am | A | no |
| trazodone | 25 mg | at bedtime | last night | A | no |
| aspirin | 325 mg | daily | this am | A | no |
| sodium bicarbonate | 650 mg | 2-TID | this am | A | no |
| NPH insulin/regular insulin | 10 u/3 u | am/before dinner | this am | A | no |
| multivitamins | 1 | daily | this am | A | no |

Do you take all medications as prescribed? ☐ Yes ☒ No   If no, why? Daughter is unclear about meds. Pt is confused.

### PATIENT/FAMILY HISTORY

| | | |
|---|---|---|
| ☐ Cold in past two weeks | ☒ High blood pressure Patient | ☒ Kidney/urinary problems Patient |
| ☐ Hay fever | ☒ Arthritis Patient | ☐ Gastric/abdominal pain/heartburn |
| ☐ Emphysema/lung problems | ☐ Claustrophobia | ☒ Hearing problems Patient |
| ☐ TB disease/positive TB skin test | ☒ Circulation problems Patient | ☐ Glaucoma/eye problems |
| ☒ Cancer Patient | ☒ Easy bleeding/bruising/anemia Patient | ☒ Back pain Patient |
| ☐ Stroke/past paralysis | ☐ Sickle cell disease | ☐ Seizures |
| ☐ Heart attack | ☐ Liver disease/jaundice | ☐ Other |
| ☒ Angina/chest pain Patient | ☐ Thyroid disease | |
| ☐ Heart problems | ☒ Diabetes Patient | |

### RISK SCREENING

Have you had a blood transfusion? ☐ Yes ☒ No
Do you smoke? ☐ Yes ☒ No
If yes, how many pack(s) /day for years
Does anyone in your household smoke? ☐ Yes ☒ No
Do you drink alcohol? ☐ Yes ☒ No
If yes, how often?_____ How much?
When was your last drink? /  /
Do you take any recreational drugs? ☐ Yes ☒ No
If yes, type:_____ Route
Frequency:_____ Date last used:____/____/____

**FOR WOMEN Ages 12–52**

Is there any chance you could be pregnant? ☐ Yes ☐ No
If yes, expected date (EDC):
Gravida/Para:

**ALL WOMEN**

Date of last Pap smear:
Do you perform regular breast self-exams? ☐ Yes ☐ No

**ALL MEN**

Do you perform regular testicular exams? ☐ Yes ☒ No

Additional comments: Information from daughter. Patient is slightly confused on admission.

✗ *Suzanne Miller, RN, BSN*
Signature/Title

**Client name:**  Bob Kaufman
**DOB:**  1/12
**Age:**  85
**Sex:**  Male
**Education:**  High school, one year of college
**Occupation:**  Postal clerk
**Hours of work:**  Retired
**Household members:**  Daughter age 45 and son-in-law age 52, grandsons age 16 and 11—all in good health
**Ethnic background:**  Caucasian
**Religious affiliation:**  Episcopalian
**Referring physician:**  Curtis Martin, MD

## Chief complaint:
"We brought my father to the hospital because he has been dramatically more confused. Sometimes he forgets little things—he is 85, you know, but he generally is not confused. I checked his blood glucose first but that was normal. I thought I had best bring him in to make sure everything was OK."

## Patient history:
*Onset of disease:* Sudden onset of confusion that has been increasing over the past 24 hours. Patient moved to live with daughter and her family almost three years ago. Daughter states that her father is responsible for his own medicine. She is really not even aware of everything that he takes. He does his own insulin injections and his own blood glucose monitoring. Her father still drives almost every day. He keeps his own doctor visits. He does volunteer work at his church and at the local elementary school. Daughter provides most of his meals except for breakfast, which he usually cooks.
*Type of Tx:* Currently treated for CAD, type 2 DM, peripheral neuropathy, and renal insufficiency.
*PMH:* CAD; type 2 DM; renal insufficiency; peripheral neuropathy, osteoarthritis, Hx of prostate CA; diverticulitis/diverticulosis
*Meds:* Diovan; Prilosec; Neurontin; furosemide; isosorbide mononitrate, trazodone; sodium bicarbonate; aspirin; multivitamin; NPH and regular insulin
*Smoker:* No
*Family Hx:* What? CA Who? Mother

## Physical exam:
*General appearance:* Cheerful, obese, elderly gentleman who is obviously confused and appears slightly restless
*Vitals:* Temp 97.2°F, BP 160/82, HR 86 bpm, RR 16 bpm
*Heart:* Regular rate and rhythm; soft systolic murmur
*HEENT:*
　*Eyes:* PERRLA
　*Ears:* Clear
　*Nose:* Clear
　*Throat:* No exudate
*Mouth:* Loose-fitting dentures; membranes dry
*Neurologic:* Inconsistent orientation to time, place, and person
*Extremities:* Significant neuropathy present
*Skin:* Warm to touch; numerous pinpoint hemorrhages; fragile

*Chest/lungs:* Lungs clear to auscultation throughout, bilaterally
*Peripheral vascular:* All pulses present and equal; feet are cool to touch and have slight discoloration
*Abdomen:* Obese; bowel sounds present.

## Nutrition Hx:

*General:* Daughter states that appetite is good—"probably too good!" Daughter states that she prepares most meals. Her father snacks between meals, but she states that she tries to have low-sugar and low-fat choices available. He weighed almost 225# when he came to live with her and her family almost three years ago. His weight has been stable for the past year. Her biggest concern nutritionally is that her father never seems to drink fluids except at mealtime, and she is worried that he doesn't get enough. "I will pour him a glass of water between meals. He will take one sip, and then he just lets it sit there." She states that she tries to keep his calories down and limits simple sugars. That is about as far as they go with diabetic restrictions. She states, "I just don't feel my father will eat anything more restrictive. I figure at 85 we'll just do the best we can."

*Usual dietary intake:*

| | |
|---|---|
| *AM:* | Egg Beaters—1 carton scrambled with 1 T shredded cheese; 2 slices bacon; 1 slice toast; 0.5 c cranberry juice; 3 c coffee with fat-free creamer. About twice a week, he has cornflakes with a banana for breakfast. |
| *Lunch:* | Usually from senior center—diabetic lunch—2–3 oz meat, 1–2 vegetables 0.5 c each, roll, 0.5 c fruit; 6–8 oz iced tea |
| *Dinner:* | 3–4 oz meat, rice, potato, or noodle, 1 slice bread, 0.5 c fresh fruit, 6–8 oz iced tea |
| | *Snacks:* Sugar-free Jello; low-fat yogurt; microwave popcorn usually 2–3 × daily |

*24-hr recall:* Not available

*Food allergies/intolerances/aversions:* NKA
*Previous MNT?* Yes
*Where?* He has attended diabetic classes in the past.
*Food purchase/preparation:* Daughter
*Vit/min intake:* Multivitamin daily
*Anthropometric data:* Ht 5′2″, Wt 196#, UBW 195–225#

## Tx plan:

Admit to Internal Medicine: Dr. Curtis Martin
*Vitals:* Routine; SBGM ac q meal
*Lab:* CBC, SMA
*Head:* CT to R/O CVA
*Diet:* 1800 Kcal ADA diet
*Activity:* Bed rest with supervision
*Meds:* Sliding scale Humulin Regular: < 200 do nothing; 200–300 5u SQ; 300–400 10u SQ; > 400 call MD; Diovan 80 mg; isosorbide mononitrate 60 mg. Continue insulin prescription from home.

## Hospital course:

Head CT normal. *Dx:* Metabolic alkalosis 2° to excessive intake of sodium bicarbonate; mild dehydration. Additional labs consistent with underlying diagnoses of type 2 DM, renal insufficiency. Patient received NS 40 mEq of KCl @ 75 cc/hr for 24 hours. As electrolyte abnormalities resolved, confusion resolved as well. Pt stated prior to discharge that he was confused with medications and there appears to be a misconception on dosage of furosemide and sodium bicarbonate. Discharge medications were adjusted. Pharmacy and nutrition consult ordered prior to discharge.

# UH *UNIVERSITY HOSPITAL*

NAME: Bob Kaufman                 DOB: 1/12
AGE: 85                           SEX: M
PHYSICIAN: Curtis Martin, MD

****************************************CHEMISTRY****************************************

| | NORMAL | Day 1 | Day 2 | UNITS |
|---|---|---|---|---|
| DAY: | | 1 | 2 | |
| DATE: | | | | |
| TIME: | | | | |
| LOCATION: | | | | |
| Albumin | 3.6–5 | 3.5 | | g/dL |
| Total protein | 6–8 | 6.0 | | g/dL |
| Prealbumin | 19–43 | 20 | | mg/dL |
| Transferrin | 200–400 | 210 | | mg/dL |
| Sodium | 135–155 | 145 | 138 | mmol/L |
| Potassium | 3.5–5.5 | 3.4 | 3.8 | mmol/L |
| Chloride | 98–108 | 98 | 99 | mmol/L |
| $PO_4$ | 2.5–4.5 | 4.5 | | mmol/L |
| Magnesium | 1.6–2.6 | 1.7 | | mmol/L |
| Osmolality | 275–295 | 310 | 296 | mmol/kg $H_2O$ |
| Total $CO_2$ | 24–30 | 30 | 27 | mmol/L |
| Glucose | 70–120 | 172 | 155 | mg/dL |
| BUN | 8–26 | 32 | 33 | mg/dL |
| Creatinine | 0.6–1.3 | 1.5 | 1.5 | mg/dL |
| Uric acid | 2.6–6 (women) | 3.7 | | mg/dL |
| | 3.5–7.2 (men) | | | |
| Calcium | 8.7–10.2 | 8.7 | | mg/dL |
| Bilirubin | 0.2–1.3 | 0.3 | | mg/dL |
| Ammonia ($NH_3$) | 9–33 | 10 | | $\mu$mol/L |
| SGPT (ALT) | 10–60 | 22 | | U/L |
| SGOT (AST) | 5–40 | 14 | | U/L |
| Alk phos | 98–251 | 101 | | U/L |
| CPK | 26–140 (women) | 121 | | U/L |
| | 38–174 (men) | | | |
| LDH | 313–618 | 356 | | U/L |
| CHOL | 140–199 | 175 | | mg/dL |
| HDL-C | 40–85 (women) | 41 | | mg/dL |
| | 37–70 (men) | | | |
| VLDL | | | | mg/dL |
| LDL | < 130 | 135 | | mg/dL |
| LDL/HDL ratio | < 3.22 (women) | 3.29 | | |
| | < 3.55 (men) | | | |
| Apo A | 101–199 (women) | | | mg/dL |
| | 94–178 (men) | | | |
| Apo B | 60–126 (women) | | | mg/dL |
| | 63–133 (men) | | | |
| TG | 35–160 | 161 | | mg/dL |
| $T_4$ | 5.4–11.5 | | | $\mu$g/dL |
| $T_3$ | 80–200 | | | ng/dL |
| $HbA_{1C}$ | 4.8–7.8 | 8.2 | | % |

**UH** *UNIVERSITY HOSPITAL*

NAME: Bob Kaufman                          DOB: 1/12
AGE: 85                                    SEX: M
PHYSICIAN: Curtis Martin, MD

\*\*\*\*\*\*\*\*\*\*\*\*\*\*\*\*\*\*\*\*\*\*\*\*\*\*\*\*\*\*\*\*\*\*\*\*\*\*\*\*\*\*\*HEMATOLOGY\*\*\*\*\*\*\*\*\*\*\*\*\*\*\*\*\*\*\*\*\*\*\*\*\*\*\*\*\*\*\*\*\*\*\*\*\*\*\*\*\*\*

DAY:                                              1
DATE:
TIME:
LOCATION:

| | NORMAL | | UNITS |
|---|---|---|---|
| WBC | 4.3–10 | 5.2 | $\times\ 10^3/mm^3$ |
| RBC | 4–5 (women) | 4.5 | $\times\ 10^6/mm^3$ |
| | 4.5–5.5 (men) | | |
| HGB | 12–16 (women) | 13 | g/dL |
| | 13.5–17.5 (men) | | |
| HCT | 37–47 (women) | 40 | % |
| | 40–54 (men) | | |
| MCV | 84–96 | | fL |
| MCH | 27–34 | | pg |
| MCHC | 31.5–36 | | % |
| RDW | 11.6–16.5 | | % |
| Plt Ct | 140–440 | 150 | $\times\ 10^3$ |
| Diff TYPE | | | |
| % GRANS | 34.6–79.2 | 62.8 | % |
| % LYM | 19.6–52.7 | 37.1 | % |
| SEGS | 50–62 | | % |
| BANDS | 3–6 | | % |
| LYMPHS | 25–40 | | % |
| MONOS | 3–7 | | % |
| EOS | 0–3 | | % |
| TIBC | 65–165 (women) | | µg/dL |
| | 75–175 (men) | | |
| Ferritin | 18–160 (women) | | µg/dL |
| | 18–270 (men) | | |
| Vitamin $B_{12}$ | 100–700 | | pg/mL |
| Folate | 2–20 | | ng/mL |
| Total T cells | 812–2318 | | $mm^3$ |
| T-helper cells | 589–1505 | | $mm^3$ |
| T-suppressor cells | 325–997 | | $mm^3$ |
| PT | 11–13 | | sec |

## Case Questions

1. Define *polypharmacy.*

2. Using the following table, list all the medications that Mr. Kaufman was taking at home. Identify the function of each medication.

| Medication | Function |
|---|---|
|  |  |
|  |  |
|  |  |
|  |  |
|  |  |
|  |  |
|  |  |
|  |  |
|  |  |
|  |  |
|  |  |

3. Do you think that Mr. Kaufman's medications represent polypharmacy? What questions do you think would be important to clarify his medications and dosages?

4. What are the drug–nutrient interactions for the medications listed in the table you completed in Question 2?

| Medication | Drug–Nutrient Interaction |
|---|---|
|  |  |
|  |  |
|  |  |
|  |  |
|  |  |
|  |  |
|  |  |
|  |  |
|  |  |
|  |  |
|  |  |

5. What laboratory values are abnormal? Check both the chemistry and arterial blood gases for pertinent labs.

| Laboratory | Normal | Mr. Kaufman's Value |
|---|---|---|
|  |  |  |
|  |  |  |
|  |  |  |
|  |  |  |
|  |  |  |
|  |  |  |
|  |  |  |
|  |  |  |
|  |  |  |
|  |  |  |
|  |  |  |

6. What laboratory values support his medical history of type 2 diabetes mellitus?

7.  What laboratory values support his medical history of renal insufficiency?

8.  Mr. Kaufman has been diagnosed with renal insufficiency; are there also normal changes in renal function that occur with aging?

9.  Mr. Kaufman was diagnosed with mild metabolic alkalosis. What is this?

10.  What laboratory value(s) support this diagnosis of metabolic alkalosis?

11.  Read Mr. Kaufman's history and physical. What symptoms does the patient present with that may be consistent with metabolic alkalosis? Explain.

12.  What medications are the most likely to have contributed to the abnormal lab values and thus this diagnosis? Why?

13.  Mr. Kaufman is 5′2″ tall and weighs 196#. Calculate his body mass index. How would you interpret this value? Should any adjustments be made in the interpretation to account for his age?

14.  Calculate Mr. Kaufman's percent usual body weight. Interpret the significance of this assessment.

15.  When completing a nutritional assessment on an older individual, should specific changes in body composition and energy requirements be considered? If so, which changes?

16.  Estimate Mr. Kaufman's energy and protein needs. What factors should you consider when estimating his requirements?

17.  Mr. Kaufman's daughter expressed concern regarding his fluid intake. Is this a common problem in aging?

**18.**   There are several ways to estimate fluid intake. Calculate Mr. Kaufman's fluid needs by using at least two of these methods. How do they compare? From your evaluation of his usual intake, do you think he is getting enough fluid?

**19.**   Evaluate Mr. Kaufman's usual intake for both caloric and protein intake. How does it compare to the Food Guide Pyramid?

**20.**   Do you think Mr. Kaufman needs to take a multivitamin? In general, do needs for vitamins and minerals change with aging? What reference would you use to determine recommended amounts of the micronutrients?

**21.**   What does the $HbA_{1C}$ measure? What can this value tell you about Mr. Kaufman's overall control over his diabetes?

**22.**   Would you make diabetes teaching a priority in your nutrition counseling of Mr. Kaufman? What methods might you use to help maximize his glucose control? Are there any comments from his daughter that would help you approach nutrition counseling?

# Bibliography

American Dietetic Association. Older Adults. In *Manual of Clinical Dietetics*, 6th ed. Chicago: American Dietetic Association, 2000:141–157.

Bartlett S, Miran M, Taren D, Muramoto M. *Geriatric Nutrition Handbook*. Florence, KY: International Thomson, 1998.

Blumberg J, Couris R. Pharmacology, nutrition, and the elderly: interactions and implications. In Chernoff R (ed.), *Geriatric Nutrition*, 2nd ed. Gaithersberg, MD: Aspen, 1999.

Chernoff R. Thirst and fluid requirements. *Nutr Rev.* 1994; 52(suppl):S3–S5.

Chidester JC, Spangler AA. Fluid intake in the institutionalized elderly. *J Am Diet Assoc.* 1997;97:23–28.

Commission on Accreditation for Dietetics Education. Knowledge, skills, and competencies for dietitians. *Accreditation Manual for Dietetics Education Programs*, 4th revised ed. Chicago: American Dietetic Association, 2000.

Cook MC, Tarren DL. Nutritional implications of medication use and misuse in elderly. *J Fla Med Assoc.* 1990;77: 606.

Gallo RM. *Handbook of Geriatric Assessment*, 3rd ed. Baltimore: Aspen, 2000.

Haken V. Interactions between drugs and nutrients. In Mahan LK, Escott-Stump S (eds.), *Krause's Food, Nutrition, and Diet Therapy*, 10th ed. Philadelphia: Saunders, 2000:399–419.

Holben DH, Hassell JT, Williams JL, Helle B. Fluid intake compared with established standards and symptoms of dehydration among elderly residents of a long-term-care facility. *J Am Diet Assoc.*1999;99:1447–1450.

Johnson R. Energy. In Mahan LK, Escott-Stump S (eds.), *Krause's Food, Nutrition, and Diet Therapy*. 10th ed. Philadelphia: Saunders, 2000:17–30.

Kurpad AV. Protein and amino acid requirements in the elderly. *Eur J Clin Nutr.* 2000;54:S131–S142.

Lindeman RD, Romero LJ, Liang HC, Baumgartner RN, et al. Do elderly persons need to be encouraged to drink more fluids? *J Gerontol.* 55A:M361–M365.

Quandt SA, McDonald J, Arcury TA, Bell RA, Vitolins MZ. Nutritional self-management of elderly widows in rural communities. *Gerontologist.* 2000;40:86–96.

Rolls BJ. Regulation of food and fluid intake in the elderly. *Ann NY Acad Sci.* 1989;561:217–255.

Rolls BJ, Phillips PA. Aging and disturbances of thirst and fluid balance. *Nutr Rev.* 1990;48:137–144.

Standing Committee on the Scientific Evaluation of Dietary Reference Intakes, Food and Nutrition Board, Institute of Medicine. *Dietary Reference Intakes for Calcium, Phosphorus, Magnesium, Vitamin D, Fluoride.* Washington, DC: National Academy Press, 1997.

Standing Committee on the Scientific Evaluation of Dietary Reference Intakes, Food and Nutrition Board, Institute of Medicine. *Dietary Reference Intakes for Thiamin, Riboflavin, Niacin, Vitamin B$_6$, Folate, Vitamin B$_{12}$, Pantothenic Acid, Biotin, and Choline.* Washington, DC: National Academy Press, 1999.

## Case 4

# Athletic Training and Performance: Nutrition Assessment

*Introductory Level*

*Thomas J. Pujol, EdD, FACSM*

## Objectives

After completing this case, the student will be able to:

1. Use nutrition assessment data and physical activity data to determine energy balance.
2. Evaluate laboratory and assessment data to identify typical blood chemistry, hematologic, and anthropometric changes that occur as a result of overtraining syndrome.
3. Identify the signs and symptoms of overtraining syndrome that are not identified by laboratory or assessment data.
4. Explain the importance of nutrition in prevention and treatment of overtraining syndrome.

5. Explain the importance of glycogen replacement and glycemic index on the performance of chronic physical activity.

Jane Reynolds is a 16-year-old high school athlete who has been suffering from an upper respiratory infection. Because of her continued fatigue, weight loss, and general irritability, she is admitted to rule out pneumonia.

# ADMISSION DATABASE

Name: Jane Reynolds
DOB: 3/20  age 16
Physician: R. Henderson

| BED#<br>32 | DATE:<br>2/12 | TIME:<br>1430 | TRIAGE STATUS (ER ONLY):<br>☐ Red  ☐ Yellow  ☐ Green  ☐ White |
|---|---|---|---|

**Initial Vital Signs**

| TEMP:<br>98.6 | RESP:<br>18 | | SAO2:<br>97 |
|---|---|---|---|
| HT:<br>5'5" | WT (lb):<br>105 (UBW:109) | B/P:<br>114/60 | PULSE:<br>82 |
| LAST TETANUS | | LAST ATE<br>this am | LAST DRANK<br>this am |

**PRIMARY PERSON TO CONTACT:**
Name: Mrs. Olivia Reynolds
Home #: 212-555-4322
Work #: same

ORIENTATION TO UNIT: ☒ Call light  ☒ Television/telephone
☒ Bathroom  ☒ Visiting  ☒ Smoking  ☒ Meals
☒ Patient rights/responsibilities

## CHIEF COMPLAINT/HX OF PRESENT ILLNESS

Patient has had a recurring URI for 4 to 6 weeks. Suspected pneumonia.

She has also been irritable and fatigued.

PERSONAL ARTICLES: (Check if retained/describe)
☐ Contacts ☐ R ☐ L          ☐ Dentures ☐ Upper ☐ Lower
☐ Jewelry:
☐ Other:

## ALLERGIES: Meds, Food, IVP Dye, Seafood: Type of Reaction

NKA

VALUABLES ENVELOPE: no
☒ Valuables instructions

## PREVIOUS HOSPITALIZATIONS/SURGERIES

Tonsillectomy-age 6

Appendectomy-age 12

Pneumonia-age 4

INFORMATION OBTAINED FROM:
☒ Patient          ☐ Previous record
☒ Family           ☐ Responsible party

Signature *Olivia Reynolds*

| Home Medications (including OTC) | | Codes: A=Sent home | | B=Sent to pharmacy | | C=Not brought in |
|---|---|---|---|---|---|---|
| Medication | Dose | Frequency | Time of Last Dose | Code | Patient Understanding of Drug |
| Robitussin DM | 1 tsp | 3 to 4 × daily | 7 am today | c | |
| | | | | | |
| | | | | | |
| | | | | | |
| | | | | | |
| | | | | | |
| | | | | | |
| | | | | | |
| | | | | | |
| | | | | | |
| | | | | | |

Do you take all medications as prescribed?  ☒ Yes  ☐ No

## PATIENT/FAMILY HISTORY

| | | |
|---|---|---|
| ☒ Cold in past two weeks Patient | ☒ High blood pressure Father | ☐ Kidney/urinary problems |
| ☐ Hay fever | ☒ Arthritis Maternal grandmother | ☐ Gastric/abdominal pain/heartburn |
| ☐ Emphysema/lung problems | ☒ Claustrophobia Sibling | ☒ Hearing problems Paternal grandmother |
| ☐ TB disease/positive TB skin test | ☒ Circulation problems Paternal grandfather | ☒ Glaucoma/eye problems Paternal grandmother |
| ☒ Cancer Maternal grandmother | ☐ Easy bleeding/bruising/anemia | ☐ Back pain |
| ☐ Stroke/past paralysis | ☐ Sickle cell disease | ☐ Seizures |
| ☐ Heart attack | ☐ Liver disease/jaundice | ☐ Other |
| ☐ Angina/chest pain | ☐ Thyroid disease | |
| ☒ Heart problems Paternal grandfather | ☐ Diabetes | |

## RISK SCREENING

Have you had a blood transfusion?  ☐ Yes  ☒ No
Do you smoke?  ☐ Yes  ☒ No
If yes, how many pack(s)  /day for  years
Does anyone in your household smoke?  ☐ Yes  ☒ No
Do you drink alcohol?  ☐ Yes  ☒ No
If yes, how often?_____  How much?_____
When was your last drink?_____/_____/
Do you take any recreational drugs?  ☐ Yes  ☒ No
If yes, type:_____  Route
Frequency:_____  Date last used:_____/_____/

**FOR WOMEN Ages 12–52**

Is there any chance you could be pregnant?  ☐ Yes  ☒ No
If yes, expected date (EDC):
Gravida/Para:

**ALL WOMEN**

Date of last Pap smear:
Do you perform regular breast self-exams?  ☐ Yes  ☒ No

**ALL MEN**

Do you perform regular testicular exams?  ☐ Yes  ☐ No

Additional comments: Patient is amenorrheic.

✗ *S. Smith, RN, BSN*
Signature/Title

**Client name:** Jane Reynolds
**DOB:** 3/20
**Age:** 16
**Sex:** Female
**Education:** Less than high school *What grade/level?* 10
**Occupation:** Student
**Hours of work:** N/A
**Household members:** Father age 38, mother age 35, brother age 15, sister age 11
**Ethnic background:** Caucasian
**Religious affiliation:** Presbyterian
**Referring physician:** Ralph Henderson, MD (family practice)

## Chief complaint:

Female, age 16 years, complains of URI which is recurring and shin splints that have caused some discomfort for 4 weeks. Jane has lost 4 pounds over the last three weeks. Patient's mother says that Jane has been unusually fatigued and irritable for the last 4 to 6 weeks. Jane's mother has tried to keep her from training for the last week because of the shin pain. Jane has resisted any attempt to withhold her from training for fear that her performance times will suffer. Jane complains that her performance has suffered lately and says that her shin splints are her only real problem.

## Patient history:

*Onset of disease:* URI has been recurring for 4 to 6 weeks, shin splints 4 weeks
*Type of Tx:* URI has been treated with OTC medications, shin splints have been treated with icing and ibuprofen for pain
*PMH:* No history of athletic injury, pneumonia at age 4, bronchitis in infancy
*Meds:* Robitussin DM, Motrin 325 mg
*Smoker:* No
*Family Hx:* What? HTN  Who? Grandfather, father

## Physical exam:

*General appearance:* Patient is a thin adolescent female who looks no different from any other young female distance runner.
*Vitals:* Temp 98.6°F, BP 114/60, HR 82 bpm, RR 18 bpm
*Heart:* Rate is high for an endurance athlete, rhythm is normal, no murmur
*HEENT:*
  *Eyes:* Normal
  *Ears:* Normal
  *Nose:* Normal
  *Throat:* Red
*Genitalia:* Normal
*Neurologic:* Normal
*Extremities:* Normal
*Skin:* Tenting; no flaking; warm and dry
*Chest-lungs:* Congestion from URI, some wheezing
*Peripheral vascular:* Normal
*Abdomen:* Normal

**Nutrition Hx**

*General:* Appetite has not changed greatly. Mother noticed a slight decrease as Jane became more fatigued.

*Usual dietary intake:*

| | |
|---|---|
| *AM:* | Juice, toast with butter |
| *Lunch:* | Apples or oranges (2), sandwich with meat (usually ham), potato chips (about 20) |
| *Afternoon snack:* | Apple or orange, sport nutrition bar (maltodextrin and sucrose; 230 kcal) |
| *Dinner:* | Legumes or green vegetable, meat (chicken or fish; 3 to 5 oz.), pasta or rice Coach has emphasized high carbohydrate consumption; Jane consumes approximately 40 oz. of 8% CHO sports drink per day. Coach suggests all CHO sports drink be consumed in 4 hours after training run. |

*Food allergies/intolerances/aversions:* None

*Previous MNT?* No

*Food purchase/preparation:* Parent(s)

*Vit/min intake:* Iron supplement

*Anthropometric data:* Patient's measured percent fat is 9.8% via skinfold and 16.9% when measured by bioelectrical impedance. Triceps skinfold is 9.5 mm. Measurements are as follows: midarm circumference is 23.0 cm, thigh circumference is 41.2 cm, forearm circumference is 16.4 cm, wrist circumference is 14.0 cm, abdominal circumference is 61 cm.

**Hospital course:** Chest X-ray negative for pneumonia

**Dx:** Amenorrhea; dehydration; URI; overuse injury

**Tx plan:**

Zithromax 300 mg bid; continue ibuprofen and icing for shin pain. Patient should restrict activity to non–weight-bearing exercise for two weeks. Recommend decreasing training frequency to provide rest days. Nutrition consult.

# U<sub>H</sub> *UNIVERSITY HOSPITAL*

NAME: Jane Reynolds                    DOB: 3/20
AGE: 16                                SEX: F
PHYSICIAN: R. Henderson, MD

\*\*\*\*\*\*\*\*\*\*\*\*\*\*\*\*\*\*\*\*\*\*\*\*\*\*\*\*\*\*\*\*\*\*\*\*\*\*\*\*CHEMISTRY\*\*\*\*\*\*\*\*\*\*\*\*\*\*\*\*\*\*\*\*\*\*\*\*\*\*\*\*\*\*\*\*\*\*\*\*\*\*\*\*\*

DAY:                                                      1
DATE:
TIME:
LOCATION:

| | NORMAL | | UNITS |
|---|---|---|---|
| Albumin | 3.6–5 | 3.3 | g/dL |
| Total protein | 6–8 | 6.5 | g/dL |
| Prealbumin | 19–43 | | mg/dL |
| Transferrin | 200–400 | | mg/dL |
| Sodium | 135–155 | 145 | mmol/L |
| Potassium | 3.5–5.5 | 3.8 | mmol/L |
| Chloride | 98–108 | 100 | mmol/L |
| PO$_4$ | 2.5–4.5 | 3.4 | mmol/L |
| Magnesium | 1.6–2.6 | | mmol/L |
| Osmolality | 275–295 | 300 | mmol/kg H$_2$O |
| Total CO$_2$ | 24–30 | | mmol/L |
| Glucose | 70–120 | 65 | mg/dL |
| BUN | 8–26 | 19 | mg/dL |
| Creatinine | 0.6–1.3 | 1.7 | mg/dL |
| Uric acid | 2.6–6 (women) | 5.7 | mg/dL |
| | 3.5–7.2 (men) | | |
| Calcium | 8.7–10.2 | 9.2 | mg/dL |
| Bilirubin | 0.2–1.3 | 0.8 | mg/dL |
| Ammonia (NH$_3$) | 9–33 | 8 | μmol/L |
| SGPT (ALT) | 10–60 | 15 | U/L |
| SGOT (AST) | 5–40 | 17 | U/L |
| Alk phos | 98–251 | 110 | U/L |
| CPK | 26–140 (women) | 165 | U/L |
| | 38–174 (men) | | |
| LDH | 313–618 | | U/L |
| CHOL | 140–199 | 140 | mg/dL |
| HDL-C | 40–85 (women) | 60 | mg/dL |
| | 37–70 (men) | | |
| VLDL | | 13 | mg/dL |
| LDL | < 130 | 67 | mg/dL |
| LDL/HDL ratio | < 3.22 (women) | 1.71 | |
| | < 3.55 (men) | | |
| Apo A | 101–199 (women) | | mg/dL |
| | 94–178 (men) | | |
| Apo B | 60–126 (women) | | mg/dL |
| | 63–133 (men) | | |
| TG | 35–160 | 37 | mg/dL |
| T$_4$ | 5.4–11.5 | | μg/dL |
| T$_3$ | 80–200 | | ng/dL |
| HbA$_{1C}$ | 4.8–7.8 | | % |

# U_H *UNIVERSITY HOSPITAL*

NAME: Jane Reynolds                          DOB: 3/20
AGE: 16                                      SEX: F
PHYSICIAN: R. Henderson, MD

\*\*\*\*\*\*\*\*\*\*\*\*\*\*\*\*\*\*\*\*\*\*\*\*\*\*\*\*\*\*\*\*\*\*\*\*\*\*\*\*HEMATOLOGY\*\*\*\*\*\*\*\*\*\*\*\*\*\*\*\*\*\*\*\*\*\*\*\*\*\*\*\*\*\*\*\*\*\*\*\*\*\*\*

| DAY: | | 1 | |
|------|------|------|------|
| DATE: | | | |
| TIME: | | | |
| LOCATION: | | | |
| | NORMAL | | UNITS |
| WBC | 4.3–10 | 4.1 | $\times\ 10^3/mm^3$ |
| RBC | 4–5 (women) | 4.4 | $\times\ 10^6/mm^3$ |
| | 4.5–5.5 (men) | | |
| HGB | 12–16 (women) | 14 | g/dL |
| | 13.5–17.5 (men) | | |
| HCT | 37–47 (women) | 33 | % |
| | 40–54 (men) | | |
| MCV | 84–96 | 88 | fL |
| MCH | 27–34 | 28 | pg |
| MCHC | 31.5–36 | 34 | % |
| RDW | 11.6–16.5 | | % |
| Plt Ct | 140–440 | | $\times\ 10^3$ |
| Diff TYPE | | | |
| % GRANS | 34.6–79.2 | | % |
| % LYM | 19.6–52.7 | | % |
| SEGS | 50–62 | | % |
| BANDS | 3–6 | | % |
| LYMPHS | 25–40 | | % |
| MONOS | 3–7 | | % |
| EOS | 0–3 | | % |
| TIBC | 65–165 (women) | | µg/dL |
| | 75–175 (men) | | |
| Ferritin | 18–160 (women) | 16 | µg/dL |
| | 18–270 (men) | | |
| Vitamin $B_{12}$ | 100–700 | 245 | pg/mL |
| Folate | 2–20 | 16 | ng/mL |
| Total T cells | 812–2318 | | $mm^3$ |
| T-helper cells | 589–1505 | | $mm^3$ |
| T-suppressor cells | 325–997 | | $mm^3$ |
| PT | 11–13 | | sec |

## Case Questions

1.  After reading the history and physical, what factors support this patient's diagnosis?

2.  After examining the laboratory values, what information, if any, supports the diagnosis of overtraining syndrome?

3.  Why does the coach recommend rest days?

4.  Liver glycogen replenishment after prolonged, vigorous exercise can be difficult. How does the glycemic index of the foods ingested affect glycogen replenishment?

5.  Increased risk of infections is a classic marker of overtraining. In this patient, what factor(s) places this patient at increased risk for infection?

6.  Calculate the patient's body mass index. Using the population as a whole as a reference, how would you evaluate this patient's body mass index?

7.  Calculate the energy needs for the patient using the Harris-Benedict equation for determining basal metabolic rate. Assuming a caloric expenditure of 15 times resting for the patient during exercise plus 100 to 200 kcal in recovery, calculate the daily energy cost for this person given that she runs for 55 to 65 minutes per day.

8.  Based on the description of her normal diet, what conclusions can you make about this patient's nutritional status?

9.  Identify those variables that indicate muscle tissue was broken down to make amino acids available for gluconeogenesis.

10. What do the low serum triglyceride and serum lipid levels suggest?

11.   Describe the anthropometric data (percent fat, upper arm circumference, upper arm muscle area, and so forth) and the implications, given the other information presented.

12.   Why did bioelectrical impedance data indicate a percent fat that is almost 70% higher than the skinfold value? Which measurement are you inclined to accept? Why?

13.   How will changing the patient's training regimen (including more rest days and dietary changes) alter her current condition?

14.   What can the dietitian do to monitor an athlete recovering from overtraining syndrome?

15.   What would be a desirable daily fluid intake for this patient?

16.   How can the dietitian and coach, working together, prevent overtraining syndrome?

## Bibliography

Kuipers H. How much is too much? Performance aspects of overtraining. *Res Qu Exercise and Sport.*1996; 67(S):S65–S69.

Lehmann M, Foster C, Keul J. Overtraining in endurance athletes: a brief review. *Med and Sci in Sports and Exercise.*1993;25:854–862.

Lehmann M, Lormes W, Optiz-Gress A, Steinacker J, Netzer N, Foster C, Gastmann U. Training and overtraining: an overview and experimental results in endurance sports. *J Sports Med and Phys Fitness.* 1997;37:7–17.

Lehmann M, Wieland H, Gastmann U. Influence of an unaccustomed increase in training volume vs. intensity on performance, hematological and blood-chemical parameters in distance runners. *J Sports Med and Phys Fitness.* 1997;37:110–116.

McArdle WD, Katch FI, Katch VL. *Sports and Exercise Nutrition.* Baltimore: Lippincott, Williams & Wilkins, 1999.

Pedersen BK, Rhode T, Zacho M. Immunity in athletes. *J Sports Med and Phys Fitness.* 1996;36:236–245.

# Athletic Training and Performance: The Athlete with Disordered Eating

*Advanced Practice*

*Thomas J. Pujol, EdD, FACSM*
*Jeremy T. Barnes, PhD*
*Mark E. Langenfeld, PhD, FACSM*

## Objectives

After completing this case, the student will be able to:

1.  Use nutrition assessment techniques to determine baseline nutritional status.
2.  Evaluate laboratory and assessment data, and determine significance as it relates to anorexia athletica.
3.  Identify characteristics of anorexia athletica as they differ from anorexia nervosa.
4.  Assess dietary data for nutritional adequacy.
5.  Identify appropriate interventions for an individual affected by this disorder.

Dr. Frank Seymour has elected to admit Debbie Howard, a 16-year-old competitive endurance athlete. Debbie was brought to the emergency room after experiencing dizziness and low blood sugar following a training run. Psychiatric evaluation indicates her to be at high risk for an eating disorder.

 **UNIVERSITY HOSPITAL**

**ADMISSION DATABASE**

Name: D. Howard
DOB: 2/18  age 16
Physician: F. Seymour, MD

| BED# 1 | DATE: 1/8 | TIME: 0300 | TRIAGE STATUS (ER ONLY): ☐ Red ☐ Yellow ☐ Green ☐ White |
|---|---|---|---|

### Initial Vital Signs

| TEMP: 98.5 | RESP: 18 | | SAO2: 98 |
|---|---|---|---|

| HT: 5'4" | WT (lb): 91 (UBW: 103) | | B/P: 108/60 | PULSE: 65 |
|---|---|---|---|---|

| LAST TETANUS | | LAST ATE N/A | LAST DRANK this am |
|---|---|---|---|

**PRIMARY PERSON TO CONTACT:**
Name: Mrs. Patricia Howard
Home #: 535-21-3344
Work #: same

**ORIENTATION TO UNIT:** ☒ Call light ☒ Television/telephone
☒ Bathroom ☒ Visiting ☒ Smoking ☒ Meals
☒ Patient rights/responsibilities

### CHIEF COMPLAINT/HX OF PRESENT ILLNESS

Patient has experienced dizziness at the end of a training run.

**PERSONAL ARTICLES:** (Check if retained/describe)
☐ Contacts ☐ R ☐ L      ☐ Dentures ☐ Upper ☐ Lower
☐ Jewelry:
☐ Other:

### ALLERGIES: Meds, Food, IVP Dye, Seafood: Type of Reaction

NKA

**VALUABLES ENVELOPE:**
☒ Valuables instructions

### PREVIOUS HOSPITALIZATIONS/SURGERIES

Tonsillectomy-age 5

Appendectomy-age 11

**INFORMATION OBTAINED FROM:**
☒ Patient         ☐ Previous record
☒ Family          ☐ Responsible party

Signature  *Patricia Howard*

| Home Medications (including OTC) | | Codes: A=Sent home | | B=Sent to pharmacy | | C=Not brought in |
|---|---|---|---|---|---|---|
| Medication | Dose | Frequency | Time of Last Dose | Code | Patient Understanding of Drug | |
| | | | | | | |
| | | | | | | |
| | | | | | | |
| | | | | | | |
| | | | | | | |
| | | | | | | |
| | | | | | | |
| | | | | | | |
| | | | | | | |
| | | | | | | |

Do you take all medications as prescribed? ☐ Yes ☐ No    If no, why?

### PATIENT/FAMILY HISTORY

| | | |
|---|---|---|
| ☐ Cold in past two weeks | ☒ High blood pressure Paternal grandfather | ☐ Kidney/urinary problems |
| ☐ Hay fever | ☐ Arthritis | ☐ Gastric/abdominal pain/heartburn |
| ☐ Emphysema/lung problems | ☒ Claustrophobia Patient | ☐ Hearing problems |
| ☐ TB disease/positive TB skin test | ☐ Circulation problems | ☐ Glaucoma/eye problems |
| ☐ Cancer | ☒ Easy bleeding/bruising/anemia Sibling | ☐ Back pain |
| ☐ Stroke/past paralysis | ☐ Sickle cell disease | ☐ Seizures |
| ☐ Heart attack | ☐ Liver disease/jaundice | ☐ Other |
| ☐ Angina/chest pain | ☐ Thyroid disease | |
| ☒ Heart problems Paternal grandfather | ☒ Diabetes Sibling | |

### RISK SCREENING

Have you had a blood transfusion?  ☐ Yes  ☒ No
Do you smoke?  ☐ Yes  ☒ No
If yes, how many pack(s)   /day for   years
Does anyone in your household smoke?  ☐ Yes  ☒ No
Do you drink alcohol?  ☐ Yes  ☒ No
If yes, how often?_____  How much?
When was your last drink?_____/_____/
Do you take any recreational drugs?  ☐ Yes  ☒ No
If yes, type:_____  Route
Frequency:_____  Date last used:_____/_____/

**FOR WOMEN Ages 12–52**

Is there any chance you could be pregnant?  ☐ Yes  ☒ No
If yes, expected date (EDC):   /   /
Gravida/Para:

**ALL WOMEN**

Date of last Pap smear:
Do you perform regular breast self-exams?  ☐ Yes  ☒ No

**ALL MEN**

Do you perform regular testicular exams?  ☐ Yes  ☐ No

Additional comments: 16-year-old patient who has not undergone menarche

✗ *Phil Lipe, RN, BSN*
Signature/Title

**Client name:**  Debbie Howard
**DOB:**  2/18
**Age:**  16
**Sex:**  Female
**Education:**  Less than high school  *What grade/level?* 11
**Occupation:**  Student
**Hours of work:**  N/A
**Household members:**  Mother age 35, father age 37, brother age 12
**Ethnic background:**  Caucasian
**Religious affiliation:**  Methodist
**Referring physician:**  Frank Seymour, MD

**Chief complaint:**
Patient was admitted after complaining of dizziness at the end of a training run.

**Patient history:**
*Onset of disease:* Debbie is a premenarchal 16-year-old junior at Paul Keys High School. She is a member of the cross-country and track teams. She is a competitive endurance athlete who has been successful at the regional and state levels in the previous two years. After her training run today, she began to complain of dizziness and could not maintain her balance. Blood glucose at time of admission was 48 mg/dL. Her mother was present at the time of admission to provide some information. Mother says Debbie has lost weight over the last 6 months. Her mother does not know the exact amount of weight lost. Seven months ago Debbie began to train for a marathon (not related to school activities). This extra event required an increase in training volume, thus her loss of weight coincided with this increase in volume. Debbie says she has not lost enough weight and her recent performance has been hampered by her weight. Mother says Debbie trains constantly, even going out on "light runs" after finishing her homework in the evening. When asked directly why she goes on these extra training runs, Debbie responds, "to keep from gaining weight."
*Type of Tx:* None prior to admission
*PMH:* No previous history of dizziness, vertigo, or syncope; no history of cardiovascular or metabolic disease
*Meds:* None
*Smoker:* No
*Family Hx: What?* Other  *Who?* Maternal grandmother, COPD

**Physical exam:**
*General appearance:* Patient is a very thin, pale young woman
*Vitals:* Temp 98.6°F; BP seated 108/68 mm Hg, supine 112/68 mm Hg, standing 98/64 mm Hg: HR 65 bpm; RR 25 bpm
*Heart:* Regular rate and prolonged QT interval on ECG
*HEENT:* Normal
    *Eyes:* Normal
    *Ears:* Normal
    *Nose:* Normal
    *Throat:* Normal
*Teeth:* Normal, no indication of enamel damage
*Genitalia:* Normal

*Neurologic:* Alert and oriented; normal gait, no vertigo
*Extremities:* Normal sensation, no marks, scrapes, or scars from purging
*Skin:* Flaky, no marks from purging; hair is dull and dry
*Chest/lungs:* Respiratory rate high, perhaps acidosis induced
*Peripheral vascular:* Normal bilateral pulse
*Abdomen:* Normal active bowel sounds, soft and nontender

**Nutrition Hx:**
*General:* Appetite decreased from normal over the last several months.

*Usual dietary intake:*

AM:   Skim milk (8 oz) (85 kcal), bagel (200 kcal) w/ peanut butter (1 tbsp) (95 kcals)
Lunch:   (In school cafeteria) Ice milk (1 cup) (184 kcal), banana (105 kcal), pretzels (15 kcal),
   water (16 oz)
PM:   Meat (usually chicken) (140 kcal), 1 vegetable (1 cup) (75 to 100 kcal) plus a salad w/ low
   calorie dressing (90 kcal), skim milk (8 oz.) (85 kcal), water. Debbie provided serving sizes
   and caloric values. Her mother says that she rarely eats at the hours the rest of the family
   does and often will not eat the evening meal until 8 p.m. Breakfast is usually eaten between
   7 and 7:30 a.m., lunch at school is served at 12 p.m. Meals, particularly breakfast and
   lunch, are missed at least twice per week.

*Self-imposed restrictions:* No red meat (because of fat), cheese (taste preference and fat), extra sugar
and salt in prepared food

Subject trains at least 2.5 hours per day, 7 days per week. Training includes approximately 2 hours
of running at 8 miles/hour, which results in an energy expenditure of 0.218 kcal/kg body
weight/min, 145 kcal through stretching and strengthening exercises, and approximately 110 kcal
in recovery.

Eating Disorder Inventory was administered. Results indicate a high score on drive for thinness and
body dissatisfaction. Indicative of preocccupation with weight.

*Anthropometric data:* Skinfold method indicates a percent fat of 10.6%, triceps skinfold of 11.75 mm,
upper arm circumference was measured at 8.75 inches.

*Food allergies/intolerances/aversions:* NKA
*Previous MNT?* No
*Food purchase/preparation:* Parent(s)
*Vit/min intake:* None

**Tx plan:**
*Psychiatric evaluation:* To investigate high scores on subscales of the Eating Disorder Inventory.
*Lab:* Check zinc status and gonadotropin levels. SMA-24 and CBC. Calculation of bone age.
Nutrition consult.
*Exercise recommendations:* Refer to qualified exercise professional for counseling on appropriate and
realistic frequency, duration, and intensity of training for a 16-year-old female; in addition, to estab-
lish reasonable exercise goals as they pertain to caloric expenditure and progression of training.

# U H UNIVERSITY HOSPITAL

```
NAME: D. Howard                          DOB: 2/18
AGE: 16                                  SEX: F
PHYSICIAN: F. Seymour, MD
```

```
**********************************************CHEMISTRY**********************************************
```

```
DAY:                                       1
DATE:                                     4/18
TIME:
LOCATION:
```

| | NORMAL | | UNITS |
|---|---|---|---|
| Albumin | 3.6–5 | 3.8 | g/dL |
| Total protein | 6–8 | 6.0 | g/dL |
| Prealbumin | 19–43 | 22 | mg/dL |
| Transferrin | 200–400 | 220 | mg/dL |
| Sodium | 135–155 | 137 | mmol/L |
| Potassium | 3.5–5.5 | 3.2 | mmol/L |
| Chloride | 98–108 | 104 | mmol/L |
| $PO_4$ | 2.5–4.5 | 4.0 | mmol/L |
| Magnesium | 1.6–2.6 | 1.8 | mmol/L |
| Osmolality | 275–295 | 290 | mmol/kg $H_2O$ |
| Total $CO_2$ | 24–30 | 28 | mmol/L |
| Glucose | 70–120 | 42 | mg/dL |
| BUN | 8–26 | 28 | mg/dL |
| Creatinine | 0.6–1.3 | 1.0 | mg/dL |
| Uric acid | 2.6–6 (women) | 5.9 | mg/dL |
| | 3.5–7.2 (men) | | |
| Calcium | 8.7–10.2 | 8.5 | mg/dL |
| Bilirubin | 0.2–1.3 | 0.5 | mg/dL |
| Ammonia ($NH_3$) | 9–33 | 30 | $\mu$mol/L |
| SGPT (ALT) | 10–60 | 57 | U/L |
| SGOT (AST) | 5–40 | 32 | U/L |
| Alk phos | 98–251 | 95 | U/L |
| CPK | 26–140 (women) | 137 | U/L |
| | 38–174 (men) | | |
| LDH | 313–618 | 310 | U/L |
| CHOL | 140–199 | 185 | mg/dL |
| HDL-C | 40–85 (women) | 40 | mg/dL |
| | 37–70 (men) | | |
| VLDL | | | mg/dL |
| LDL | < 130 | 126 | mg/dL |
| LDL/HDL ratio | < 3.22 (women) | 3.15 | |
| | < 3.55 (men) | | |
| Apo A | 101–199 (women) | 176 | mg/dL |
| | 94–178 (men) | | |
| Apo B | 60–126 (women) | 72 | mg/dL |
| | 63–133 (men) | | |
| TG | 35–160 | 155 | mg/dL |
| $T_4$ | 5.4–11.5 | 6.9 | $\mu$g/dL |
| $T_3$ | 80–200 | 82 | ng/dL |
| $HbA_{1C}$ | 4.8–7.8 | | % |

**U<sub>H</sub>** *UNIVERSITY HOSPITAL*

NAME: D. Howard                          DOB: 2/18
AGE: 16                                   SEX: F
PHYSICIAN: F. Seymour


\*\*\*\*\*\*\*\*\*\*\*\*\*\*\*\*\*\*\*\*\*\*\*\*\*\*\*\*\*\*\*\*\*\*\*\*\*\*HEMATOLOGY\*\*\*\*\*\*\*\*\*\*\*\*\*\*\*\*\*\*\*\*\*\*\*\*\*\*\*\*\*\*\*\*\*\*\*\*\*\*

DAY:                                              1
DATE:                                            4/18
TIME:
LOCATION:

| | NORMAL | | UNITS |
|---|---|---|---|
| WBC | 4.3–10 | 4.4 | $\times\ 10^3/mm^3$ |
| RBC | 4–5 (women) | 4.8 | $\times\ 10^6/mm^3$ |
| | 4.5–5.5 (men) | | |
| HGB | 12–16 (women) | 15 | g/dL |
| | 13.5–17.5 (men) | | |
| HCT | 37–47 (women) | 41 | % |
| | 40–54 (men) | | |
| MCV | 84–96 | | fL |
| MCH | 27–34 | | pg |
| MCHC | 31.5–36 | | g/dL |
| RDW | 11.6–16.5 | | % |
| Plt Ct | 140–440 | | $\times\ 10^3$ |
| Diff TYPE | | | |
| ESR | 0–20 (women) | | mm/hr |
| | 0–15 (men) | | |
| % GRANS | 34.6–79.2 | | % |
| % LYM | 19.6–52.7 | | % |
| SEGS | 50–62 | | % |
| BANDS | 3–6 | | % |
| LYMPHS | 25–40 | | % |
| MONOS | 3–7 | | % |
| EOS | 0–3 | | % |
| TIBC | 65–165 (women) | | μg/dL |
| | 75–175 (men) | | |
| Ferritin | 18–160 (women) | 164 | μg/dL |
| | 18–270 (men) | | |
| Vitamin $B_{12}$ | 100–700 | 685 | pg/mL |
| Folate | 2–20 | 14 | ng/mL |
| Total T cells | 812–2318 | | $mm^3$ |
| T-helper cells | 589–1505 | | $mm^3$ |
| T-suppressor cells | 325–997 | | $mm^3$ |
| PT | 11–13 | | sec |

## Case Questions

1. What essential characteristics of this patient are similar to a typical anorexia nervosa patient? What characteristics differ from an anorexia nervosa patient?

2. Determine Debbie's lean body mass and fat mass.

3. Is there any evidence of a binging–purging cycle in this patient?

4. From the information given in the patient history section, what indicators of anorexia athletica are present? What may have triggered this episode?

5. Calculate Debbie's body mass index. Using the population as a whole as a reference, how would you evaluate her body mass index?

6. What is the percentile rank of the upper arm circumference, upper arm muscle area, and bone-free upper arm muscle area?

7. What is the best method to evaluate Debbie's height and weight? Use the method that you decided on to evaluate height and weight.

8. Using the Harris-Benedict equation and the data provided regarding this patient's caloric expenditure, what is her daily caloric expenditure?

9. Why would an assessment of bone age be appropriate for this patient?

10. What laboratory test data indicate the poor nutritional status of the patient?

11. How would you categorize this patient's menstrual status? What is the relationship between gonadotropin levels and this symptom of anorexia?

12.  Sundgott-Borgen (1993) indicates that 14% of subjects diagnosed with anorexia athletica reported eating two or fewer meals per day. For those who ate breakfast, lunch, and evening meal, a time gap of 3 to 6 hours elapsed between breakfast and lunch. A gap of 7 to 11 hours between lunch and the evening meal was reported. Did this patient fit the profile for eating disordered athletes? Suggest a reason for this wide gap between meals for those who consume three meals, and state a potential danger of this pattern of eating.

13.  What changes would be required for this patient to achieve and maintain a consistent weight?

14.  What case could be made for *not* having this patient keep a food diary?

15.  The diagnosis of anorexia athletica does not mean the patient has a clinical eating disorder; however, the patient with anorexia athletica can lapse into a clinical disorder. What can the dietitian do to prevent this?

16.  Iron and zinc are often deficient in people with eating disorders. What can the dietitian recommend to increase iron and zinc intake?

17.  Considering the low number of calories the patient is consuming, how could the dietitian instruct the parents to improve the nutrient content of her diet?

18.  Determine appropriate behavioral outcomes for this patient.

19.  What type of follow-up would be appropriate for this patient?

20.  Who are the important people to be involved in communication about this client and the prescribed follow-up?

## Bibliography

American Psychiatric Association. *Diagnostic and Statistical Manual of Mental Disorders,* 4th ed. Washington, DC: American Psychiatric Association, 1994.

Clark N, Nelson M, Evans W. Nutrition education for elite female runners. *The Physician and Sports Med.* 1988;16:124–136.

Commission on Accreditation for Dietetics Education. Knowledge, skills, and competencies for dietitians. *Accreditation Manual for Dietetics Education Programs,* rev. 4th ed. Chicago: American Dietetic Association, 2000.

Matejek N, Weimann E, Witzel C, Molenkamp G, Schwidergall S, Bohles H. Hypoleptinaemia in patients with anorexia nervosa and in elite gymnasts with anorexia athletica. *Int J of Sports Med.* 1999;20:451–456.

McArdle WD, Katch FI, Katch VL. *Sports and Exercise Nutrition.* Baltimore: Lippincott, Williams & Wilkins, 1999.

Pugliese MT, Lifshitz F, Grad G, Fort P, Marks-Katz M. Fear of obesity: a cause of short stature and delayed puberty. *N Eng J Med.* 1983;309:513–518.

Sundgott-Borgen J. Eating disorders in female athletes. *Sports Med.* 1994;17:176–188.

Sundgott-Borgen J. Nutrient intake of female elite athletes suffering from eating disorders. *Int J Sports Nutr.* 1993;4:431–442.

Sundgott-Borgen J. Risk and trigger factors for the development of eating disorders in female elite athletes. *Med and Sci in Sports and Exercise.* 1994;26:414–419.

Vaisman N, Rossi MF, Goldberg E, Dibden LJ, Wykes LJ, Pencharz PB. Energy expenditure and body composition in patients with anorexia nervosa. *J Pediatr.* 1988;113:919–924.

# MEDICAL NUTRITION THERAPY FOR CARDIOVASCULAR DISORDERS

## Introduction

Cardiovascular disease is the leading cause of death in the United States. Risk factors for cardiovascular disease include high serum lipid levels, smoking, diabetes mellitus, high blood pressure, obesity, and physical inactivity. Researchers estimate that more than 59 million Americans have one or more forms of cardiovascular disease, so many patients the health care team encounters will have conditions related to cardiovascular disease.

This section includes four of the most common diagnoses: hypertension (HTN), coronary heart disease (CHD), myocardial infarction (MI), and congestive heart failure (CHF). All these diagnoses require a significant medical nutrition therapy component for their care.

Over 50 million people in the United States have hypertension. Hypertension is defined as a systolic blood pressure of 140 mm Hg or higher and diastolic pressure of 90 mm Hg or higher. Essential hypertension, which is the most common form of hypertension, is of unknown etiology. Case 6 focuses on lifestyle modifications as the first step in treatment of hypertension. This case incorporates the pharmacological treatment of hypertension and you will use the most recent information from *Dietary Approaches to Stop Hypertension* (DASH) as the center of the medical nutrition therapy intervention.

Coronary heart disease (CHD) is a complex multifactorial condition. Case 7 provides the opportunity to evaluate these multiple risk factors through all facets of nutrition assessment. We specifically emphasize interpretation of laboratory indices for hyperlipidemia. In this case you will determine the clinical classification and treatment of hyperlipidemia, explore the use of drug therapy to treat dyslipidemias, and develop appropriate nutrition interventions for these diagnoses.

Case 8 focuses on the acute care of an individual suffering a myocardial infarction (MI). One out of every five deaths in the United States can be attributed to MI. Ischemia of the vessels within the heart results in death of the affected component of the heart tissue. This case lets you evaluate pertinent assessment measures for the individual suffering an MI and then develop an appropriate care plan to prevent further cardiac deterioration.

Case 9 addresses the long-term consequences of cardiovascular disease in a patient suffering from congestive heart failure (CHF). In CHF the heart cannot pump effectively and the lack of oxygen and nutrients affects the body's tissues. CHF is a major public health problem in the United States, and its incidence is increasing. Without heart transplant prognosis is poor. This advanced case requires you to integrate understanding of the physiology of several body systems as you address heart failure's metabolic effects. In addition, this case allows you to explore the role of the health care team in palliative care.

# Hypertension

*Introductory Level*

## Objectives

After completing this case, the student will be able to:

1. Describe the physiology of blood pressure and its application to the pathophysiology of hypertension.
2. Collect pertinent information and use nutrition assessment techniques to determine baseline nutritional status.
3. Develop appropriate behavior outcomes for the patient.
4. Identify appropriate MNT goals.
5. Develop appropriate documentation in the medical record.

Mr. Charles Riddle is a 50-year-old high school football coach. He has treated his newly diagnosed hypertension for the past year with lifestyle changes including diet, smoking cessation, and exercise. He is admitted for further evaluation and treatment for his essential hypertension.

# ADMISSION DATABASE

Name: C. Riddle
DOB: 4/14  age 50
Physician: A. Thornton

| BED# 1 | DATE: 10/9 | TIME: 0800 | TRIAGE STATUS (ER ONLY): ☐ Red ☐ Yellow ☐ Green ☐ White |
|---|---|---|---|

**PRIMARY PERSON TO CONTACT:**
Name: Vicki Riddle
Home #: 555-7128
Work #: 555-2157

### Initial Vital Signs

| TEMP: 98.6 | RESP: 15 | SAO2: |
|---|---|---|

| HT: 6'3" | WT (lb): 220 | B/P: 160/100 | PULSE: 80 |
|---|---|---|---|

ORIENTATION TO UNIT: ☒ Call light  ☒ Television/telephone
☒ Bathroom  ☒ Visiting  ☒ Smoking  ☒ Meals
☒ Patient rights/responsibilities

| LAST TETANUS 10 years ago | LAST ATE 0630 | LAST DRANK 0630 |
|---|---|---|

**CHIEF COMPLAINT/HX OF PRESENT ILLNESS**

hypertension

**PERSONAL ARTICLES:** (Check if retained/describe)
☐ Contacts ☐ R ☐ L          ☐ Dentures ☐ Upper ☐ Lower
☐ Jewelry:
☒ Other: glasses

**ALLERGIES: Meds, Food, IVP Dye, Seafood: Type of Reaction**

None

**VALUABLES ENVELOPE:**
☐ Valuables instructions

**PREVIOUS HOSPITALIZATIONS/SURGERIES**

None

**INFORMATION OBTAINED FROM:**
☒ Patient          ☐ Previous record
☐ Family           ☐ Responsible party

Signature  *Charles Riddle*

| Home Medications (including OTC) | | Codes: A=Sent home | B=Sent to pharmacy | | C=Not brought in |
|---|---|---|---|---|---|
| Medication | Dose | Frequency | Time of Last Dose | Code | Patient Understanding of Drug |
|  |  |  |  |  |  |
|  |  |  |  |  |  |
|  |  |  |  |  |  |
|  |  |  |  |  |  |
|  |  |  |  |  |  |
|  |  |  |  |  |  |
|  |  |  |  |  |  |
|  |  |  |  |  |  |
|  |  |  |  |  |  |
|  |  |  |  |  |  |
|  |  |  |  |  |  |

Do you take all medications as prescribed?  ☐ Yes  ☐ No   If no, why?

**PATIENT/FAMILY HISTORY**

| | | |
|---|---|---|
| ☒ Cold in past two weeks Patient | ☒ High blood pressure Mother | ☐ Kidney/urinary problems |
| ☒ Hay fever Patient | ☒ Arthritis Patient | ☐ Gastric/abdominal pain/heartburn |
| ☐ Emphysema/lung problems | ☐ Claustrophobia | ☐ Hearing problems |
| ☐ TB disease/positive TB skin test | ☐ Circulation problems | ☐ Glaucoma/eye problems |
| ☐ Cancer | ☐ Easy bleeding/bruising/anemia | ☐ Back pain |
| ☐ Stroke/past paralysis | ☐ Sickle cell disease | ☐ Seizures |
| ☒ Heart attack Mother | ☐ Liver disease/jaundice | ☐ Other |
| ☐ Angina/chest pain | ☐ Thyroid disease | |
| ☒ Heart problems Mother | ☐ Diabetes | |

**RISK SCREENING**

Have you had a blood transfusion?  ☐ Yes  ☒ No
Do you smoke?  ☐ Yes  ☒ No
If yes, how many pack(s)  /day for  years
Does anyone in your household smoke?  ☐ Yes  ☒ No
Do you drink alcohol?  ☒ Yes  ☐ No
If yes, how often? q pm  How much? 1-2 mixed drinks
When was your last drink?____/____/____
Do you take any recreational drugs?  ☐ Yes  ☒ No
If yes, type:_____  Route
Frequency:_____  Date last used:____/____/____

**FOR WOMEN Ages 12–52**

Is there any chance you could be pregnant?  ☐ Yes  ☐ No
If yes, expected date (EDC):
Gravida/Para:

**ALL WOMEN**

Date of last Pap smear:
Do you perform regular breast self-exams?  ☐ Yes  ☐ No

**ALL MEN**

Do you perform regular testicular exams?  ☐ Yes  ☒ No

Additional comments:

✗ *Connie L. Bussard, RN*
Signature/Title

**Client name:** Charles Riddle
**DOB:** 4/14
**Age:** 50
**Sex:** Male
**Education:** Master's degree
**Occupation:** High school football coach
**Hours of work:** 8:30 AM to 5:30 PM; except during football season, which usually includes 10–12 hour days
**Household members:** Wife age 48, in good health; children are grown and do not live at home.
**Ethnic background:** African American
**Religious affiliation:** Catholic
**Referring physician:** Alan Thornton, MD (cardiology)

## Chief complaint:

"I've tried to cut back on salt, but food just doesn't taste good without it. I want to control this high blood pressure—my mother passed away because her high blood pressure caused her to have a heart attack."

## Patient history:

*Onset of disease:* Mr. Riddle is a 50-yo African American male who works as a football coach at a local high school. He was diagnosed 1 year ago with Stage 2 (essential) HTN. Treatment thus far has been focused on nonpharmacological measures. He began a walking program that has resulted in a 10-pound weight loss that he has been able to maintain during the past year. He walks 30 minutes 4–5 times per week, though he sometimes misses during football season. He was given a diet sheet in the MD's office that outlined a 4-gm Na diet. Mr. Riddle was a 2-pack-a-day smoker but quit ("cold turkey") when he was diagnosed last year. No c/o of any symptoms related to HTN.
*Type of Tx:* Initiation of pharmacologic therapy with thiazide diuretics and reinforcement of lifestyle modifications to decrease fat intake. Rule out Syndrome X.
*PMH:* Not significant before Dx of HTN
*Meds:* Hydrochlorothiazide—25 mg q d
*Smoker:* No—quit 1 year ago
*Family Hx:* What? Mother died of MI related to uncontrolled HTN.

## Physical exam:

*General appearance:* Healthy, middle-aged male who looks his age
*Vitals:* Temp 98.6°F, BP 160/100 mm Hg, HR 80 bpm, RR 15 bpm
*Heart:* Regular rate and rhythm, normal heart sounds—no clicks, murmurs, or gallops
*HEENT:* No carotid bruits
  *Eyes:* No retinopathy, PERRLA
*Genitalia:* Normal uncircumcised male
*Neurologic:* Alert and oriented to person, place, and time
*Extremities:* Noncontributory
*Skin:* Smooth, warm, dry, excellent turgor, no edema
*Chest/lungs:* Lungs clear
*Peripheral vascular:* Pulse 4+ bilaterally, warm, no edema
*Abdomen:* Nontender, no guarding, normal bowel sounds

**Nutrition Hx:**

*General:* Mr. Riddle describes his appetite as "very good." His wife does the majority of grocery shopping and cooking, although Mr. Riddle cooks breakfast for his family on the weekends. He usually eats three meals each day, but during football season he sometimes misses lunch. When he does this, he is really hungry at the evening meal. The family eats out on Friday and Saturday evenings, usually at pizza restaurants or steakhouses (Mr. Riddle usually has 2 regular beers with these meals). He mentions that last year when his HTN was diagnosed, a nurse at the MD's office gave him a sheet of paper with a list of foods to avoid for a 4-gm Na (no added salt) diet. He and his wife tried to comply with the diet guidelines, but they found foods bland and tasteless, and they soon abandoned the effort. Mr. Riddle usually has 1–2 mixed drinks after work "to relax" before the evening meal.

*Usual dietary intake:*

| | |
|---|---|
| AM: | 1 c coffee (black) |
| | Hot (oatmeal with 1 tsp margarine and 2 tsp sugar) or cold (Frosted Mini-Wheats) cereal |
| | ½ c 2% milk |
| | 1 c orange juice |
| Snack: | 2 c coffee (black) |
| | 1 glazed donut |
| Lunch: | At desk when time permits: |
| | 1 can Campbell's tomato bisque soup |
| | 10 saltines |
| | 1 can diet cola |
| After work: | 2 (usually) gin and tonics (3 oz gin with 5 oz tonic) |
| PM: | 6 oz baked chicken (white meat) (seasoned with salt, pepper, garlic) |
| | 1 large baked potato with 2 tsp butter, salt, and pepper |
| | 1 c glazed carrots |
| | Dinner salad with ranch-style dressing (3 tbsp)—lettuce, spinach, croutons, sliced cucumber |
| HS snack: | Butter pecan ice cream (2 c) |

*Food allergies/intolerances/aversions:* None
*Previous MNT?* Yes    *If yes, when:* 1 year ago    *Where?* MD's office
*Food purchase/preparation:* Wife
*Vit/min intake:* Multivitamin/mineral daily
*Current diet order:* 4 gm Na

**Tx plan:**

Urinalysis, hematocrit, blood chemistry to include plasma glucose, potassium, BUN, creatinine, fasting lipid profile, triglycerides, calcium, uric acid
Chest X-ray
EKG
DASH diet
25 g hydrochlorothiazide q d

# U<sub>H</sub> *UNIVERSITY HOSPITAL*

NAME: C. Riddle                           DOB: 4/14
AGE: 50                                    SEX: M
PHYSICIAN: A. Thornton

****************************************CHEMISTRY****************************************

DAY:                                                      Admit
DATE:
TIME:
LOCATION:

| | NORMAL | | UNITS |
|---|---|---|---|
| Albumin | 3.6–5 | 4.6 | g/dL |
| Total protein | 6–8 | 7 | g/dL |
| Prealbumin | 19–43 | 40 | mg/dL |
| Transferrin | 200–400 | 350 | mg/dL |
| Sodium | 135–155 | 142 | mmol/L |
| Potassium | 3.5–5.5 | 5.2 | mmol/L |
| Chloride | 98–108 | 102 | mmol/L |
| $PO_4$ | 2.5–4.5 | 4.1 | mmol/L |
| Magnesium | 1.6–2.6 | 2.1 | mmol/L |
| Osmolality | 275–295 | 275 | mmol/kg $H_2O$ |
| Total $CO_2$ | 24–30 | | mmol/L |
| Glucose | 70–120 | 115 | mg/dL |
| BUN | 8–26 | 20 | mg/dL |
| Creatinine | 0.6–1.3 | 0.9 | mg/dL |
| Uric acid | 2.6–6 (women) | 6.8 | mg/dL |
| | 3.5–7.2 (men) | | |
| Calcium | 8.7–10.2 | 9.2 | mg/dL |
| Bilirubin | 0.2–1.3 | 1.1 | mg/dL |
| Ammonia ($NH_3$) | 9–33 | 29 | $\mu$mol/L |
| SGPT (ALT) | 10–60 | 58 | U/L |
| SGOT (AST) | 5–40 | 39 | U/L |
| Alk phos | 98–251 | 250 | U/L |
| CPK | 26–140 (women) | 100 | U/L |
| | 38–174 (men) | | |
| LDH | 313–618 | 314 | U/L |
| CHOL | 140–199 | 300 | mg/dL |
| HDL-C | 40–85 (women) | 35 | mg/dL |
| | 37–70 (men) | | |
| VLDL | | | mg/dL |
| LDL | < 130 | 135 | mg/dL |
| LDL/HDL ratio | < 3.22 (women) | | |
| | < 3.55 (men) | | |
| Apo A | 101–199 (women) | | mg/dL |
| | 94–178 (men) | | |
| Apo B | 60–126 (women) | | mg/dL |
| | 63–133 (men) | | |
| TG | 35–160 | 250 | mg/dL |
| $T_4$ | 5.4–11.5 | | $\mu$g/dL |
| $T_3$ | 80–200 | | ng/dL |
| $HbA_{1C}$ | 4.8–7.8 | 6.9 | % |

# U<sub>H</sub> *UNIVERSITY HOSPITAL*

NAME: C. Riddle                          DOB: 4/14
AGE: 50                                  SEX: male
PHYSICIAN: A. Thornton

\*\*\*\*\*\*\*\*\*\*\*\*\*\*\*\*\*\*\*\*\*\*\*\*\*\*\*\*\*\*\*\*\*\*\*\*\*\*\*\*\*\*\*\*\*HEMATOLOGY\*\*\*\*\*\*\*\*\*\*\*\*\*\*\*\*\*\*\*\*\*\*\*\*\*\*\*\*\*\*\*\*\*\*\*\*\*\*\*\*\*\*\*\*\*

DAY:                                              Admit
DATE:
TIME:
LOCATION:

|  | NORMAL |  | UNITS |
|---|---|---|---|
| WBC | 4.3–10 | 5.0 | $\times 10^3/mm^3$ |
| RBC | 4–5 (women) | 5.0 | $\times 10^6/mm^3$ |
|  | 4.5–5.5 (men) |  |  |
| HGB | 12–16 (women) | 16.5 | g/dL |
|  | 13.5–17.5 (men) |  |  |
| HCT | 37–47 (women) | 50 | % |
|  | 40–54 (men) |  |  |
| MCV | 84–96 | 90 | fL |
| MCH | 27–34 | 29 | pg |
| MCHC | 31.5–36 | 34 | % |
| RDW | 11.6–16.5 |  | % |
| Plt Ct | 140–440 | 350 | $\times 10^3$ |
| Diff TYPE |  |  |  |
| % GRANS | 34.6–79.2 |  | % |
| % LYM | 19.6–52.7 | 50.1 | % |
| SEGS | 50–62 |  | % |
| BANDS | 3–6 |  | % |
| LYMPHS | 25–40 | 35 | % |
| MONOS | 3–7 |  | % |
| EOS | 0–3 |  | % |
| TIBC | 65–165 (women) |  | $\mu$g/dL |
|  | 75–175 (men) |  |  |
| Ferritin | 18–160 (women) |  | $\mu$g/dL |
|  | 18–270 (men) |  |  |
| Vitamin $B_{12}$ | 100–700 |  | pg/mL |
| Folate | 2–20 |  | ng/mL |
| Total T cells | 812–2318 |  | $mm^3$ |
| T-helper cells | 589–1505 |  | $mm^3$ |
| T-suppressor cells | 325–997 |  | $mm^3$ |
| PT | 11–13 |  | sec |

# U<sub>H</sub> UNIVERSITY HOSPITAL

NAME: C. Riddle                    DOB: 4/14
AGE: 50                            SEX: male
PHYSICIAN: A. Thornton

**********************************************URINALYSIS**********************************************

| | NORMAL | Admit | 2 | d/c | UNITS |
|---|---|---|---|---|---|
| DAY: | | | | | |
| DATE: | | | | | |
| TIME: | | | | | |
| LOCATION: | | | | | |
| Coll meth | | Random specimen | First morning | First morning | |
| Color | | Pale yellow | Pale yellow | Pale yellow | |
| Appear | | Clear | Clear | Clear | |
| Sp grv | 1.003-1.030 | 1.000 | | | |
| pH | 5-7 | | | | |
| Prot | NEG | NEG | | | mg/dL |
| Glu | NEG | NEG | | | mg/dL |
| Ket | NEG | | | | |
| Occ bld | NEG | | | | |
| Ubil | NEG | | | | |
| Nit | NEG | | | | |
| Urobil | < 1.1 | | | | E.U./dL |
| Leu bst | NEG | | | | |
| Prot chk | NEG | | | | |
| WBCs | 0-5 | | | | /HPF |
| RBCs | 0-5 | | | | /HPF |
| EPIs | 0 | | | | /LPF |
| Bact | 0 | | | | |
| Mucus | 0 | | | | |
| Crys | 0 | | | | |
| Casts | 0 | | | | /LPF |
| Yeast | 0 | | | | |

## Case Questions

1. Define blood pressure. How is blood pressure normally regulated?

2. What causes essential hypertension?

3. What are the symptoms of hypertension?

4. How is hypertension diagnosed?

5. List the risk factors for developing hypertension.

6. What risk factors does Mr. Riddle currently have?

7. Hypertension is classified in stages based on the risk of developing CVD. Complete the following table of hypertension classifications.

| | Blood pressure mmHg | |
| --- | --- | --- |
| Category | Systolic BP | Diastolic BP |
| Optimal | and | |
| Normal | and | |
| High–normal | or | |
| Stage 1 (mild) | or | |
| Stage 2 (moderate) | or | |
| Stage 3 (severe) | or | |

8. Given these criteria, which category would Mr. Riddle's admitting blood pressure reading place him in?

**9.**   How is hypertension treated?

**10.**   In the following organ systems, what diseases could possibly manifest as the result of hypertension?

| Organ System | Disease Manifestations |
|---|---|
| Cardiac | |
| Cerebrovascular | |
| Peripheral vascular | |
| Renal | |
| Retinopathy | |

**11.**   Dr. Thornton indicated in his admitting note that he will "rule out Syndrome X." What is Syndrome X?

**12.**   Dr. Thornton ordered the following labs: fasting glucose, cholesterol, triglycerides, creatinine, and uric acid. He also ordered an EKG. In the following table, outline the indication for these tests (why they have been ordered for Mr. Riddle).

| Parameter | Normal Value | Pt's Value | Indication |
|---|---|---|---|
| Glucose | | | |
| BUN | | | |
| Creatinine | | | |
| Total cholesterol | | | |
| HDL cholesterol | | | |
| LDL cholesterol | | | |
| Triglycerides | | | |
| Uric acid | | | |

**13.** Indicate the pharmacological differences among the antihypertensive agents listed in question 7.

| Characteristics | Diuretics | Beta Blockers | Calcium Channel Blockers | ACE Inhibitors | Angiotensin II Receptor Blockers | Alpha-adrenergic Blockers |
|---|---|---|---|---|---|---|
| Mechanism of action | | | | | | |
| Indications or patient characteristics | | | | | | |
| Nutritional side effects and contraindications | | | | | | |

**14.** What are the *relevant* nutritional implications of taking hydrochlorothiazide?

**15.** Assuming medium body frame, how much should Mr. Riddle weigh? Calculate his desirable body weight using ideal body weight (IBW) / Hamwi formula and reference body weight (RBW) / Dietary Guidelines for Americans.

**16.** Calculate Mr. Riddle's body mass index (BMI). What are the health implications of this number?

**17.** Calculate Mr. Riddle's energy needs using the Harris-Benedict equation.

**18.** How would you explain the DASH diet to Mr. Riddle and his wife?

**19.** Using a computer dietary analysis program or food composition table, compare Mr. Riddle's "usual" dietary intake to his prescribed diet (DASH diet).

| Nutrient | Target Level | Patient Intake | % Target Value |
|---|---|---|---|
| Kcal | | | |
| Fat | | | |
| Saturated fat | | | |
| Monounsaturated fat | | | |
| Polyunsaturated fat | | | |
| Carbohydrates | | | |
| Protein | | | |
| Cholesterol | | | |
| Potassium | | | |
| Magnesium | | | |
| Calcium | | | |
| Sodium | | | |
| Fiber | | | |
| Alcohol | | | |

**20.** What nutrients in Mr. Riddle's diet are of major concern to you?

**21.** Compare Mr. Riddle's "usual" food choices with those recommended in the DASH diet.

| Food Groups | DASH Daily Servings | Patient Daily Servings |
|---|---|---|
| Grains, grain products | | |
| Vegetables | | |
| Fruits | | |
| Low fat or nonfat dairy products | | |
| Meats, poultry, fish | | |
| Nuts, seeds, dry beans, peas | | |
| Fats, oils | | |
| Sweets | | |

**22.** After presenting the preceding information to Mr. and Mrs. Riddle, how is it best for the practitioner or the patient (and/or significant other) to evaluate the DASH recommendations and the assessment of his usual intake?

**23.** Determine specific objectives for Mr. Riddle's medical nutrition therapy.

**24.** What are some alternative foods Mr. Riddle can use to make the appropriate modifications in his diet?

| Foods | Alternative(s) | Rationale |
|---|---|---|
| Coffee (3 c per day) | | |
| Oatmeal (w/ margarine & sugar) or Frosted Mini-Wheats | | |
| 2% low-fat milk | | |
| Orange juice | | |
| Glazed donut | | |
| Canned tomato soup | | |
| Saltine crackers | | |
| Diet cola | | |
| Gin and tonic (2–3 per day) | | |
| Baked chicken | | |
| Baked potato w/ butter, salt, and pepper | | |
| Carrots | | |
| Salad w/ ranch-style dressing | | |
| Ice cream | | |

**25.** Mr. Riddle asks you, "A lot of the other faculty have lost weight on that Dr. Atkins diet. Would it be best for me to follow that for awhile to get this weight off?" What can you tell Mr. Riddle about the typical high-protein, low-carbohydrate approach to weight loss?

**26.** Using the SOAP (subjective, objective, assessment, plan) format, write a nutrition note for the patient's medical record.

# Bibliography

American Dietetic Association. *Manual of Clinical Dietetics,* 6th ed. Chicago: American Dietetic Association, 2000.

American Dietetic Association and Morrison Health Care. *Medical Nutrition Therapy across the Continuum of Care.* Chicago: American Dietetic Association, 1998.

American Heart Association. Blood pressure. Available at http://www.americanheart.org. Accessed October 6, 2000.

American Heart Association. For professionals: risk factorsand coronary heart disease. Available at http://www.americanheart.org. Accessed October 10, 2000.

American Heart Association. For professionals: Syndrome X. Available at http://www.americanheart.org. Accessed on October 6, 2000.

Commission on Accreditation for Dietetics Education. Knowledge, skills, and competencies for dietitians. *Accreditation Manual for Dietetics Education Programs,* rev. 4th ed. Chicago: American Dietetic Association, 2000.

*Dietary Approaches to Stop Hypertension* (DASH). Available at http://dash.bwh.harvard.edu. Accessed October 4, 2000.

Dusky L. The Atkins diet. Available at http://my.webmd.com/content/article/3220.136. Accessed May 15, 2001.

Fischbach F. *Manual of Laboratory and Diagnostic Tests,* 6th ed. Philadelphia: Lippincott, 2000.

Henderson SD. Management of hypertension. *Medical Library.* Available at http://www.medical-library.org. Accessed September 22, 2000.

Krummel D. Nutrition in hypertension. In Mahan LK, Escott-Stump S (eds.), *Krause's Food, Nutrition, and Diet Therapy,* 10th ed. Philadelphia: Saunders, 2000: 558–595.

McKenzie CR. Hypertension. In Carey CF, Lee HH, Woeltje KF (eds.), *Manual of Medical Therapeutics,* 29th ed. Philadelphia: Lippincott, Williams & Wilkins, 1998: 61–80.

MedlinePlus Health information. Available at http://www.nlm.nih.gov/medlineplus/druginformation.html. Accessed October 9, 2000.

*The Merck Manual.* Available at http://www.merck.com. Accessed October 9, 2000.

National Heart, Lung, and Blood Institute; National Institutes of Health. *Clinical Guidelines on the Identification, Evaluation, and Treatment of Overweight and Obesity in Adults Executive Summary* (1998). Available at http://www.nhlbi.nih.gov/guidelines/obesity. Accessed October 27, 2000.

Pronsky ZM. *Powers and Moore's Food and Medication Interaction,* 11th ed., Birchrunville, PA: Food–Medication Interactions, 2000.

Sixth Report of the Joint National Committee on Detection, Evaluation, and Treatment of High Blood Pressure (JNC-VI). Available at http://www.nhlbi.nih.gov/guidelines/hypertension. Accessed October 27, 2000.

Stanford EK. Hypertension. *Medical Library.* Available at http://www.medical-library.org. Accessed September 22, 2000.

Virtual Naval Hospital™. *Clinical Section: Cardiovascular Disorders: Hypertension.* Available at http://www.vnh.org/GMO/ClinicalSection/09Hypertension.html. Accessed October 9, 2000.

Windhauser MM, Ernst DB, Karania NM, Crawford SW, Redican SE, Swain JF, Karimbakas JM, Champagne CM, Hoben KP, Evans MA. Translating the Dietary Approach to Stop Hypertension diet from research to practice: dietary and behavior change techniques. *J Am Diet Assoc.* 1999; 99(8, Suppl):S90–S95.

## Case 7

# Cardiovascular Disease with Multiple Risk Factors

*Introductory Level*

*Anne B. Marietta, PhD, RD*

## Objectives

After completing this case, the student will be able to:

1.  Use nutrition assessment techniques to determine baseline nutritional status.
2.  Evaluate laboratory indices for nutritional implications and significance.
3.  Demonstrate understanding of clinical classification and treatment of dyslipidemias.
4.  Identify pharmacologic treatment for dyslipidemias.

5.  Determine appropriate nutritional interventions for respective lipid abnormalities.

Benjamin Gates is a 54-year-old man who has been turned down for life insurance because of an abnormal lipid profile. The company has requested that the patient see a cardiologist for a complete evaluation.

# UNIVERSITY HOSPITAL

## ADMISSION DATABASE

Name: Benjamin Gates
DOB: 9/17  age 54
Physician: J. Hart, MD

| BED# 1 | DATE: 3/15 | TIME: 1500 | TRIAGE STATUS (ER ONLY): ☐ Red ☐ Yellow ☐ Green ☐ White |
|---|---|---|---|

**Initial Vital Signs**

| TEMP: 98.6 | RESP: 25 | | SAO2: |
|---|---|---|---|

| HT: 5'10" | WT (lb): 225 | | B/P: 200/109 | PULSE: 74 |
|---|---|---|---|---|

| LAST TETANUS 3 years ago | | LAST ATE 1200 | LAST DRANK 1400 |
|---|---|---|---|

### CHIEF COMPLAINT/HX OF PRESENT ILLNESS

I am told I have high cholesterol and high blood pressure.

### ALLERGIES: Meds, Food, IVP Dye, Seafood: Type of Reaction

NKA

### PREVIOUS HOSPITALIZATIONS/SURGERIES

Hernia repair 10 years ago

**PRIMARY PERSON TO CONTACT:**
Name: Marilee Gates
Home #: 621-549-0987
Work #: 601-543-8764

ORIENTATION TO UNIT: ☒ Call light ☒ Television/telephone ☒ Bathroom ☒ Visiting ☐ Smoking ☒ Meals ☒ Patient rights/responsibilities

PERSONAL ARTICLES: (Check if retained/describe)
☐ Contacts ☐ R ☐ L   ☐ Dentures ☐ Upper ☐ Lower
☐ Jewelry:
☒ Other: eyeglasses

VALUABLES ENVELOPE:
☐ Valuables instructions

INFORMATION OBTAINED FROM:
☒ Patient   ☐ Previous record
☒ Family   ☐ Responsible party

Signature  *Marilee Gates*

| Home Medications (including OTC) | | Codes: A = Sent home | | B = Sent to pharmacy | | C = Not brought in |
|---|---|---|---|---|---|---|
| Medication | Dose | Frequency | Time of Last Dose | Code | Patient Understanding of Drug | |
| | | | | | | |
| | | | | | | |
| | | | | | | |
| | | | | | | |
| | | | | | | |
| | | | | | | |
| | | | | | | |
| | | | | | | |
| | | | | | | |
| | | | | | | |

Do you take all medications as prescribed?  ☒ Yes  ☐ No   If no, why?

### PATIENT/FAMILY HISTORY

| | | |
|---|---|---|
| ☐ Cold in past two weeks | ☒ High blood pressure Patient | ☐ Kidney/urinary problems |
| ☐ Hay fever | ☐ Arthritis | ☐ Gastric/abdominal pain/heartburn |
| ☐ Emphysema/lung problems | ☐ Claustrophobia | ☐ Hearing problems |
| ☐ TB disease/positive TB skin test | ☐ Circulation problems | ☐ Glaucoma/eye problems |
| ☐ Cancer | ☐ Easy bleeding/bruising/anemia | ☐ Back pain |
| ☐ Stroke/past paralysis | ☐ Sickle cell disease | ☐ Seizures |
| ☐ Heart attack Mother | ☐ Liver disease/jaundice | ☐ Other |
| ☐ Angina/chest pain | ☐ Thyroid disease | |
| ☒ Heart problems Sibling | ☐ Diabetes | |

### RISK SCREENING

Have you had a blood transfusion?  ☐ Yes  ☒ No
Do you smoke?  ☒ Yes  ☐ No
If yes, how many pack(s)  1/day for  30 years
Does anyone in your household smoke?  ☒ Yes  ☐ No
Do you drink alcohol?  ☒ Yes  ☐ No
If yes, how often? occ   How much? 1-2 beers
When was your last drink? 03/01/
Do you take any recreational drugs?  ☐ Yes  ☒ No
If yes, type:_____   Route
Frequency:_____   Date last used:_____/_____/_____

**FOR WOMEN Ages 12–52**

Is there any chance you could be pregnant?  ☐ Yes  ☐ No
If yes, expected date (EDC):
Gravida/Para:

**ALL WOMEN**

Date of last Pap smear:
Do you perform regular breast self-exams?  ☐ Yes  ☐ No

**ALL MEN**

Do you perform regular testicular exams?  ☐ Yes  ☒ No

Additional comments:

**x** *Cindy McIntyre, LPN*
Signature/Title

**Client name:** Benjamin Gates
**DOB:** 9/17
**Age:** 54
**Sex:** Male
**Education:** Associate's degree
**Occupation:** Insurance salesman
**Hours of work:** 50 hrs/week
**Household members:** wife age 53, son age 21
**Ethnic background:** Caucasian
**Religious affiliation:** Southern Baptist
**Referring physician:** Joseph A. Hart, MD (cardiology)

**Chief complaint:**
"I am told I have high cholesterol and high blood pressure. It was so high my doctor wanted me evaluated here."

**Patient history:**
*Onset of disease:* The patient is a 54-year-old male who has been turned down for life insurance because of an abnormal lipid profile. The company has requested that the patient see a cardiologist for a complete evaluation. Patient denies chest pain, SOB, syncope, palpitations, or myocardial infarction.
*Type of Tx:* No prior therapy
*PMH:* Mild obesity for past 10 years. Mild hypertension. No diabetes by history. Thyroid indices are normal. TSH normal.
*Meds:* None at present
*Smoker:* Yes—1 pack/day, 30 years
*Family Hx: What?* CAD    *Who?* 48-year-old brother, s/p–coronary artery surgery

**Physical exam:**
*General appearance:* Obese white male in NAD
*Vitals:* Temp 98.4°F, Ht 70″, Wt 225#, BP 160/100, HR 82, RR 16
*Heart:* PMI sustained in 6th intercostal space in AAL in the left lateral decubitus position, S1 and S2 normal intensity and split. There were no murmurs. A 4th heart sound was present.

*HEENT:*
   *Head:* Normocephalic, no thyromegaly or cysts present
   *Eyes:* EOMI, arteriolar spasm present bilaterally
   *Ears:* Tympanic membranes normal
   *Nose:* WNL
   *Throat:* NNL
*Genitalia:* Normal
*Neurologic:* Oriented to time, place, and person. Cranial nerves intact. DTR present and symmetric
*Extremities:* No clubbing, cyanosis, or edema
*Skin:* No xanthomas or xanthalasma
*Chest/lungs:* Harsh breath sounds present at both bases
*Peripheral vascular:* Peripheral pulses palpable
*Abdomen:* Normal bowel sounds, no masses or organomegaly

**Nutrition Hx:**
*General:* Patient states that he has a good appetite. He eats 3 meals/day and 1 evening snack.

*Usual dietary intake:*

| | |
|---|---|
| *Breakfast:* | (Restaurant) 2 c orange juice, 2 eggs, 2 sl bacon, 2 sl toast with 2 t butter and 1 T jelly or biscuits, black coffee |
| *Lunch:* | (Restaurant) Sandwich, such as hamburger with 4 oz patty, bun, 2 lettuce leaves, 1 T dill pickle slices, 1 oz potato chips, 12 oz diet cola |
| *Dinner:* | (Home) Meat (6 oz steak), potatoes (1 large baked potato) with 2 T sour cream and 2 T butter, vegetable (1 c corn), 2 sl bread and 2 t butter, 1 sl fruit pie or cake, coffee |
| *Evening snack:* | 1 oz pretzels |

*24-hr recall:* N/A

*Food allergies/intolerances/aversions:* NKA
*Previous MNT?* No
*Food purchase/preparation:* Self
*Vit/min intake:* Occasionally takes multivitamin

**Hospital course:**
Patient was evaluated at the request of a life insurance company.

**Dx:**
Physical exam revealed Grade I hypertensive retinopathy, hypertensive heart disease, and early chronic obstructive pulmonary disease. Complete fasting lipid profile was abnormal. EKG was WNL.

**Tx plan:**
Patient was referred for medical nutrition therapy. Therapeutic lifestyle change and drug therapy instituted. Patient will be reassessed in three months.

**U**H *UNIVERSITY HOSPITAL*

NAME: Benjamin Gates                    DOB: 9/17
AGE: 54                                 SEX: M
PHYSICIAN: Joseph A. Hart, MD

\*\*\*\*\*\*\*\*\*\*\*\*\*\*\*\*\*\*\*\*\*\*\*\*\*\*\*\*\*\*\*\*\*\*\*\*\*\*\*\*\*\*\*CHEMISTRY\*\*\*\*\*\*\*\*\*\*\*\*\*\*\*\*\*\*\*\*\*\*\*\*\*\*\*\*\*\*\*\*\*\*\*\*\*\*\*\*\*

DAY:                                    Init      6 Months   9 Months
DATE:
TIME:
LOCATION: Outpatient Lab

| | NORMAL | Init | 6 Months | 9 Months | UNITS |
|---|---|---|---|---|---|
| Albumin | 3.6–5 | 4.5 | 4.3 | 4.4 | g/dL |
| Total protein | 6–8 | 7 | 7 | 6.8 | g/dL |
| Prealbumin | 19–43 | 41 | 39 | 40 | mg/dL |
| Transferrin | 200–400 | 355 | 366 | 345 | mg/dL |
| Sodium | 135–155 | 138 | 137 | 140 | mmol/L |
| Potassium | 3.5–5.5 | 4.5 | 3.6 | 3.9 | mmol/L |
| Chloride | 98–108 | 101 | 100 | 98 | mmol/L |
| $PO_4$ | 2.5–4.5 | 3.5 | 3.5 | 3.5 | mmol/L |
| Magnesium | 1.6–2.6 | 2.3 | 2.3 | 2.3 | mmol/L |
| Osmolality | 275–295 | 288 | 290 | 289 | mmol/kg $H_2O$ |
| Total $CO_2$ | 24–30 | 30 | 29 | 29 | mmol/L |
| Glucose | 70–120 | 90 | 88 | 96 | mg/dL |
| BUN | 8–26 | 20 | 15 | 22 | mg/dL |
| Creatinine | 0.6–1.3 | 1.1 | 1.1 | 1.2 | mg/dL |
| Uric acid | 2.6–6 (women) | 7.0 | 7.2 | 7.2 | mg/dL |
| | 3.5–7.2 (men) | | | | |
| Calcium | 8.7–10.2 | 9.0 | 9.1 | 9.0 | mg/dL |
| Bilirubin | 0.2–1.3 | .8 | 1.1 | .9 | mg/dL |
| Ammonia ($NH_3$) | 9–33 | 18 | 22 | 19 | μmol/L |
| SGPT (ALT) | 10–60 | 45 | 50 | 48 | U/L |
| SGOT (AST) | 5–40 | 34 | 35 | 38 | U/L |
| Alk phos | 98–251 | 115 | 111 | 112 | U/L |
| CPK | 26–140 (women) | 125 | 134 | 130 | U/L |
| | 38–174 (men) | | | | |
| LDH | 313–618 | 323 | 350 | 310 | U/L |
| CHOL | 140–199 | 270 | 230 | 210 | mg/dL |
| HDL–C | 40–85 (women) | 30 | 35 | 38 | mg/dL |
| | 37–70 (men) | | | | |
| VLDL | | | | | mg/dL |
| LDL | < 130 | 210 | 169 | 147 | mg/dL |
| LDL/HDL ratio | < 3.22 (women) | 7.0 | 4.8 | 3.9 | |
| | < 3.55 (men) | | | | |
| Apo A | 101–199 (women) | 75 | 100 | 110 | mg/dL |
| | 94–178 (men) | | | | |
| Apo B | 60–126 (women) | 140 | 120 | 115 | mg/dL |
| | 63–133 (men) | | | | |
| TG | 35–160 | 150 | 130 | 125 | mg/dL |
| $T_4$ | 5.4–11.5 | 8.6 | 8.5 | 7.8 | μg/dL |
| $T_3$ | 80–200 | 150 | 155 | 145 | ng/dL |
| $HbA_{1C}$ | 4.8–7.8 | 6.2 | 6.3 | 7.2 | % |

# U·H UNIVERSITY HOSPITAL

NAME: Benjamin Gates                    DOB: 9/17
AGE: 54                                 SEX: M
PHYSICIAN: Joseph A. Hart, MD

\*\*\*\*\*\*\*\*\*\*\*\*\*\*\*\*\*\*\*\*\*\*\*\*\*\*\*\*\*\*\*\*\*\*\*\*\*\*\*\*HEMATOLOGY\*\*\*\*\*\*\*\*\*\*\*\*\*\*\*\*\*\*\*\*\*\*\*\*\*\*\*\*\*\*\*\*\*\*\*\*\*\*

DAY:                                    Init      6 Months    9 Months
DATE:
TIME:
LOCATION:
                   NORMAL                                                 UNITS

| | NORMAL | Init | 6 Months | 9 Months | UNITS |
|---|---|---|---|---|---|
| WBC | 4.3–10 | 6.1 | 5.7 | 5.5 | $\times 10^3/mm^3$ |
| RBC | 4–5 (women) | 5.3 | 5.0 | 5.25 | $\times 10^6/mm^3$ |
| | 4.5–5.5 (men) | | | | |
| HGB | 12–16 (women) | 15.2 | 14.8 | 14.5 | g/dL |
| | 13.5–17.5 (men) | | | | |
| HCT | 37–47 (women) | 45 | 44 | 44 | % |
| | 40–54 (men) | | | | |
| MCV | 84–96 | 88 | 90 | 85 | fL |
| MCH | 27–34 | 29 | 30 | 29 | pg |
| MCHC | 31.5–36 | 35 | 32 | 33 | % |
| RDW | 11.6–16.5 | 13.5 | 13.4 | 13.6 | % |
| Plt Ct | 140–440 | 430 | 350 | 366 | $\times 10^3$ |
| Diff TYPE | | | | | |
| % GRANS | 34.6–79.2 | 66.2 | 60.1 | 55.2 | % |
| % LYM | 19.6–52.7 | 33.8 | 39.9 | 42.1 | % |
| SEGS | 50–62 | 60 | 52.1 | 51 | % |
| BANDS | 3–6 | 1 | 2 | 2 | % |
| LYMPHS | 25–40 | 33 | 40 | 32 | % |
| MONOS | 3–7 | 5 | 3 | 4 | % |
| EOS | 0–3 | 1 | 3 | 2 | % |
| TIBC | 65–165 (women) | 172 | 155 | 157 | μg/dL |
| | 75–175 (men) | | | | |
| Ferritin | 18–160 (women) | 263 | 255 | 241 | μg/dL |
| | 18–270 (men) | | | | |
| Vitamin B$_{12}$ | 100–700 | 600 | 705 | 633 | pg/mL |
| Folate | 2–20 | 3 | 5 | 8 | ng/mL |
| PT | 11–13 | 12.5 | 12.7 | 12.2 | $mm^3$ |

# U<sub>H</sub> *UNIVERSITY HOSPITAL*

NAME: Benjamin Gates                    DOB: 9/17
AGE: 54                                 SEX: M
PHYSICIAN: Joseph A. Hart, MD

\*\*\*\*\*\*\*\*\*\*\*\*\*\*\*\*\*\*\*\*\*\*\*\*\*\*\*\*\*\*\*\*\*\*\*\*\*\*\*\*\*\*\*URINALYSIS\*\*\*\*\*\*\*\*\*\*\*\*\*\*\*\*\*\*\*\*\*\*\*\*\*\*\*\*\*\*\*\*\*\*\*\*\*\*\*\*\*\*

DAY:                          Init            6 Months        9 Months
DATE:
TIME: 8 AM
LOCATION: Office
            NORMAL                                                    UNITS
---------------------------------------------------------------------------

| | NORMAL | Init | 6 Months | 9 Months | UNITS |
|---|---|---|---|---|---|
| Coll meth | | Random specimen | Random specimen | First morning | |
| Color | | Pale yellow | Pale yellow | Pale yellow | |
| Appear | | Clear | Clear | Clear | |
| Sp grv | 1.003-1.030 | 1.025 | 1.021 | 1.024 | |
| pH | 5-7 | 7.0 | 5.0 | 6.0 | |
| Prot | NEG | Negative | Negative | Negative | mg/dL |
| Glu | NEG | Negative | Negative | Negative | mg/dL |
| Ket | NEG | Trace | 1+ | 2+ | |
| Occ bld | NEG | Negative | Negative | Negative | |
| Ubil | NEG | Negative | Negative | Negative | |
| Nit | NEG | Negative | Negative | Negative | |
| Urobil | < 1.1 | 0.02 | .01 | Negative | EU/dL |
| Leu bst | NEG | Negative | Negative | Negative | |
| Prot chk | NEG | Negative | Negative | Negative | |
| WBCs | 0-5 | 0 | 0 | 0 | /HPF |
| RBCs | 0-5 | 0 | 0 | 0 | /HPF |
| EPIs | 0 | Rare | 0 | 0 | /LPF |
| Bact | 0 | 0 | 0 | 0 | |
| Mucus | 0 | 0 | 0 | 0 | |
| Crys | 0 | 0 | 0 | 0 | |
| Casts | 0 | 0 | 0 | 0 | /LPF |
| Yeast | 0 | 0 | 0 | 0 | |

## Case Questions

1. What medical and social history is pertinent for determining Mr. Gates's CHD risk category?

2. Are there any additional questions you would like to ask Mr. Gates to help you assess his nutritional needs?

3. What is the patient's usual intake of calories and dietary fat, based on the nutrition information obtained from him?

4. What is the patient's BMI, and how would you interpret its significance?

5. Compare the following blood levels with the values obtained in Mr. Gates's lipid profile. Which ones are abnormal? What is their significance in determining Mr. Gates's risk factors?

| Parameter | Normal Value | Patient's Value | Significance |
|---|---|---|---|
| Cholesterol | | | |
| HDL-C | | | |
| VLDL | | | |
| LDL-C | | | |
| LDL/HDL ratio | | | |
| Apo A | | | |
| Apo B | | | |
| TG | | | |

6. What is the significance of apolipoprotein A and apolipoprotein B?

7. List your biochemical goals for Mr. Gates's profile.

8.  Calculate Mr. Gates's ideal body weight and % IBW.


9.  What is Mr. Gates's resting and total energy expenditure?


10. What overall goals for Mr. Gates's medical/nutritional needs would you establish?


11. How many calories would you recommend for Mr. Gates based on the goals you set for him? What energy distribution would you suggest?


12. Assuming that the foods in his 24-hour recall are typical of his eating pattern, what would be the basis of your medical nutrition therapy?


13. What tools would be useful to you in your educational session with Mr. Gates?


14. What would you want to reevaluate in three to four weeks at a follow-up appointment?


15. What are some possible barriers to compliance?


16. Identify the major sources of saturated fat and cholesterol in Mr. Gates's diet. What suggestions would you make for substitutions and/or other changes that would help Mr. Gates reach his medical nutrition therapy goals?


17. The most recent recommendations suggest the therapeutic use of stanol esters. What are these, and what is the rationale for their use?


18. When you ask Mr. Gates how much weight he would like to lose, he tells you he would like to weigh 175, which is what he weighed most of his adult life. Is this reasonable? What would you suggest as a goal for weight loss for Mr. Gates?

**19.**   How quickly should Mr. Gates lose this weight?

**20.**   Mr. Gates's physician has decided to prescribe a diuretic to be taken daily (hydrochlorothiazide) and an HMGCoA reductase inhibitor (Zocor). What changes can be expected in his lipid profile as a result of taking these medications? What are the pertinent drug–nutrient interactions and medical side effects for each?

**21.**   How does a HMGCoA reductase inhibitor work to lower serum lipid? What other classes of medications can be used to treat hypercholesterolemia? Why was Zocor most probably chosen for Mr. Gates?

**22.**   Evaluate Mr. Gates's labs at 6 months and then at 9 months. Have the biochemical goals been met with the current regimen?

# Bibliography

Cardiovascular disease. Available at http:// www.americanheart.org/Heart_and_Stroke_A_Z_ Guide/cvds.html. Accessed May 1, 2001.

Commission on Accreditation for Dietetics Education. Knowledge, skills, and competencies for dietitians. *Accreditation Manual for Dietetics Education Programs,* revised 4th ed. Chicago: American Dietetic Association, 2000: 29–45.

*Guidelines Treating Overweight and Obesity.* Bethesda, MD: National Institutes of Health, 2001. Available at http:// www.nhlbi.nih.gov/guidelines/obesity/e_txtbk/txgd/ 4311.htm. Accessed May 20, 2001.

Kris-Etherton P, Burns J (eds.). *Cardiovascular Nutrition: Strategies and Tools for Disease Management and Prevention.* Chicago: American Dietetic Association, 1998.

Krummel D. Nutrition in cardiovascular disease. In Mahan LK, Escott-Stump S. (eds.), *Krause's Food, Nutrition, and Diet Therapy,* 10th ed. Philadelphia: Saunders, 2000: 566–595.

*Medical Nutrition Therapy Across the Continuum of Care: Client Protocols.* Chicago: American Dietetic Association and Morrison Health Care, 1998.

*Third Report of the Expert Panel on Detection, Evaluation, and Treatment of High Blood Cholesterol in Adults (Adult Treatment Panel III).* Bethesda, MD: National Institutes of Health, 2001. Available at http:// www.nhlbi.nih.gov/guidelines/cholesterol/ index.htm. Accessed May 20, 2001.

## Case 8

# Myocardial Infarction

*Introductory Level*

*Anne B. Marietta, PhD, RD*

## Objectives

After completing this case, the student will be able to:

1. Use nutrition assessment techniques to determine baseline nutritional status.
2. Evaluate laboratory indices for nutritional implications and significance.
3. Demonstrate understanding of clinical classification and treatment of hyperlipidemia.
4. Identify pharmacologic side effects of drug therapy for low cardiac output.
5. Determine appropriate nutritional interventions for post-MI care.

Mr. Klosterman, a 61-year-old man, is admitted through the emergency room of University Hospital after experiencing a sudden onset of severe precordial pain on the way home from work. Mr. Klosterman is found to have suffered a myocardial infarction and is treated with an emergency angioplasty of the infarct-related artery.

 **UNIVERSITY HOSPITAL**

# ADMISSION DATABASE

Name: James Klosterman
DOB: 12/1  age 61
Physician: Regina H. Smith, MD

| BED# 1 | DATE: 3/25 | TIME: 1000 | TRIAGE STATUS (ER ONLY): ☐ Red ☐ Yellow ☐ Green ☐ White |
|---|---|---|---|

**Initial Vital Signs**

| TEMP: 98.4 | RESP: 20 | SAO2: 80 |
|---|---|---|

| HT: 5'10" | WT (lb): 185 | B/P: 118/78 | PULSE: 92 |
|---|---|---|---|

| LAST TETANUS 2000 | LAST ATE 800 | LAST DRANK 800 |
|---|---|---|

**PRIMARY PERSON TO CONTACT:**
Name: Sally Klosterman
Home #: 404-321-9214
Work #: 404-322-1822

ORIENTATION TO UNIT: ☒ Call light ☒ Television/telephone ☒ Bathroom ☒ Visiting ☒ Smoking ☒ Meals ☒ Patient rights/responsibilities

### CHIEF COMPLAINT/HX OF PRESENT ILLNESS

Severe unrelenting chest pain for past 1.5 hrs

PERSONAL ARTICLES: (Check if retained/describe)
☐ Contacts ☐ R ☐ L ☐ Dentures ☐ Upper ☐ Lower
☐ Jewelry:
☐ Other:

### ALLERGIES: Meds, Food, IVP Dye, Seafood: Type of Reaction

Sulfa-hives

VALUABLES ENVELOPE:
☐ Valuables instructions

### PREVIOUS HOSPITALIZATIONS/SURGERIES

INFORMATION OBTAINED FROM:
☒ Patient   ☐ Previous record
☒ Family    ☐ Responsible party

Signature  *Sally Klosterman*

| Home Medications (including OTC) | Codes: A=Sent home | | B=Sent to pharmacy | | C=Not brought in |
|---|---|---|---|---|---|
| Medication | Dose | Frequency | Time of Last Dose | Code | Patient Understanding of Drug |
| none | | | | | |
| | | | | | |
| | | | | | |
| | | | | | |
| | | | | | |
| | | | | | |
| | | | | | |
| | | | | | |
| | | | | | |
| | | | | | |
| | | | | | |

Do you take all medications as prescribed?   ☐ Yes   ☐ No

### PATIENT/FAMILY HISTORY

| | | |
|---|---|---|
| ☐ Cold in past two weeks | ☐ High blood pressure | ☐ Kidney/urinary problems |
| ☐ Hay fever | ☐ Arthritis | ☐ Gastric/abdominal pain/heartburn |
| ☒ Emphysema/lung problems Patient | ☐ Claustrophobia | ☐ Hearing problems |
| ☐ TB disease/positive TB skin test | ☐ Circulation problems | ☐ Glaucoma/eye problems |
| ☐ Cancer Maternal grandmother | ☐ Easy bleeding/bruising/anemia | ☐ Back pain |
| ☐ Stroke/past paralysis | ☐ Sickle cell disease | ☐ Seizures |
| ☐ Heart attack | ☐ Liver disease/jaundice | ☐ Other |
| ☒ Angina/chest pain Patient | ☐ Thyroid disease | |
| ☐ Heart problems | ☐ Diabetes | |

### RISK SCREENING

Have you had a blood transfusion?   ☐ Yes   ☐ No
Do you smoke?   ☒ Yes   ☐ No
If yes, how many pack(s) 1 /day for 40 years
Does anyone in your household smoke?   ☒ Yes   ☐ No
Do you drink alcohol?   ☒ Yes   ☐ No
If yes, how often? 1 drink/day   How much?
When was your last drink? last night/_____/
Do you take any recreational drugs?   ☐ Yes   ☒ No
If yes, type:_____   Route
Frequency:_____   Date last used:_____/_____/

**FOR WOMEN Ages 12–52**
Is there any chance you could be pregnant?   ☐ Yes   ☐ No
If yes, expected date (EDC):
Gravida/Para:

**ALL WOMEN**
Date of last Pap smear:
Do you perform regular breast self-exams?   ☐ Yes   ☐ No

**ALL MEN**
Do you perform regular testicular exams?   ☒ Yes   ☐ No

Additional comments:

✗ *Mark Settle, RN*
Signature/Title

**Client name:** James Klosterman
**DOB:** 12/1
**Age:** 61
**Sex:** Male
**Education:** BS degree
**Occupation:** Computer programmer
**Hours of work:** 40/wk
**Household members:** Wife age 61
**Ethnic background:** German
**Religious affiliation:** Lutheran
**Referring physician:** Regina H. Smith, MD (internal medicine)

**Chief complaint:**
Severe, unrelenting precordial chest pain for the past 1.5 hours.

**Patient history:**
*Onset of disease:* 61-yo male who noted the sudden onset of severe precordial pain on the way home from work. The pain is described as pressure-like pain radiating to the jaw and left arm. The patient has noted an episode of emesis and nausea. He denies palpitations or syncope. He denies prior history of pain. He admits to smoking cigarettes 1 pack/day for 40 years. He denies hypertension, diabetes, or high cholesterol. He denies SOB.
*Type of Tx:* Hospitalization, emergency coronary angiography with angioplasty of infarct-related artery, coronary care unit, rhythm monitoring, bed rest, sequential cardiograms, and cardiac enzymes
*PMH:* Surgery cholecystectomy 10 years ago, appendectomy 30 years ago
*Meds:* None. *Allergies:* Sulfa drugs
*Smoker:* Yes—40 years, 1 pack per day
*Family Hx: What?* CAD  *Who?* Father—MI age 59

**Physical exam:**
*General appearance:* Mildly obese male in acute distress from chest pain
*Vitals:* Temp 98.4°F, BP 118/78, HR 92, RR 20
*Heart:* PMI 5 ICS MCL focal. $S_1$ normal intensity. $S_2$ normal intensity and split. $S_4$ gallop at the apex No murmurs, clicks, or rubs

*HEENT:*
  *Head:* Normocephalic
  *Eyes:* EOMI, fundoscopic exam WNL. No evidence of atherosclerosis, diabetic retinopathy, or early hypertensive changes
  *Ears:* TM normal bilaterally
  *Nose:* WNL
  *Throat:* Tonsils not infected, uvula midline, gag normal
*Genitalia:* Grossly physiologic
*Neurologic:* No focal localizing abnormalities. DTR symmetric bilaterally
*Extremities:* No C, C, E
*Skin:* Diaphoretic and pale
*Chest/lungs:* Lungs clear to auscultation and percussion
*Peripheral vascular:* PPP

*Abdomen:* RLQ scar and midline suprapubic scar. BS WNL. No hepatomegaly, splenomegaly, masses, inguinal lymph nodes, or abdominal bruits.
*Height/Weight:* 70″, 185#

## Nutrition Hx:

*General:* Appetite good. Has been trying to change some things in his diet. Wife indicates that she has been using "corn oil" instead of butter and has tried not to fry foods as often.

*Usual dietary intake:*

| | |
|---|---|
| *Midmorning snack:* | 1 large cinnamon raisin bagel with 1 T fat-free cream cheese, 8 oz orange juice, coffee |
| *Lunch:* | 1 c canned vegetable beef soup, sandwich with 4 oz roast beef, lettuce, tomato, dill pickles, 2 t mayonnaise, 1 small apple, 8 oz 2% milk |
| *Dinner:* | 2 lean pork chops (3 oz each), 1 large baked potato, 2 t margarine, ½ c green beans, ½ c coleslaw (cabbage with 1 T salad dressing), 1 sl apple pie |
| *Snack:* | 8 oz 2% milk, 1 oz pretzels |

*Food allergies/intolerances/aversions:* None
*Previous MNT?* Yes    *If yes, when:* last year    *Where?* Community dietitian
*Food purchase/preparation:* Spouse
*Vit/min intake:* None

## Tx plan:

Intravenous heparin—5000 units bolus followed by 1000 unit/hour continuous infusion with a PTT at 2 × control
Chewable aspirin 160 mg PO and continued as qd
Lopressor 50 mg bid
Lidocaine prn
NPO until procedure completed
Type and cross for 6 units of packed cells

## Hospital course:

The patient's chest pain resolved after two sublingual NTG at three-minute intervals and 2 mgm of IV morphine. In the cath lab the patient was found to have a totally occluded distal right coronary artery and a 70% in the left circumflex coronary artery. The left anterior descending was patent. Angioplasty of the distal right coronary artery resulted in a patent infarct-related artery with near normal flow. A stent was left in place to stabilize the patient and limit infarct size. Left ventricular ejection fraction was normal at 42%, and a posterobasilar scar was present with hypokinesis. A consult was made for nutrition counseling and for referral to cardiac rehabilitation. Pt was discharged on the following medications: Lopressor 50 mg qd; lisinopril 10 mg qd; Nitro-bid 9 mg bid; NTG, 4 mg sublingually prn for chest pain; ASA 81 mg qd.

# U<sub>H</sub> *UNIVERSITY HOSPITAL*

NAME: James Klosterman          DOB: 12/1
AGE: 61                         SEX: M
PHYSICIAN: Regina H. Smith, MD

**\*\*\*\*\*\*\*\*\*\*\*\*\*\*\*\*\*\*\*\*\*\*\*\*\*\*\*\*\*\*\*\*\*\*\*\*\*\*\*\*\*\*\*\*\*\*CHEMISTRY\*\*\*\*\*\*\*\*\*\*\*\*\*\*\*\*\*\*\*\*\*\*\*\*\*\*\*\*\*\*\*\*\*\*\*\*\*\*\*\*\*\*\*\*\***

DAY:                            Day 1     Day 2     Day 4
DATE:
TIME:
LOCATION: Outpatient Lab

| | NORMAL | Day 1 | Day 2 | Day 4 | UNITS |
|---|---|---|---|---|---|
| Albumin | 3.6–5 | 4.2 | 4.3 | 4.2 | g/dL |
| Total protein | 6–8 | 6.0 | 5.9 | 6.1 | g/dL |
| Prealbumin | 19–43 | 35 | 36 | 34 | mg/dL |
| Transferrin | 200–400 | 300 | 310 | 320 | mg/dL |
| Sodium | 135–155 | 141 | 142 | 138 | mmol/L |
| Potassium | 3.5–5.5 | 4.2 | 4.1 | 3.9 | mmol/L |
| Chloride | 98–108 | 104 | 102 | 100 | mmol/L |
| $PO_4$ | 2.5–4.5 | 3.1 | 3.2 | 3.0 | mmol/L |
| Magnesium | 1.6–2.6 | 2.3 | 2.3 | 2.2 | mmol/L |
| Osmolality | 275–295 | 292 | 290 | 291 | mmol/kg $H_2O$ |
| Total $CO_2$ | 24–30 | 20 | 24 | 26 | mmol/L |
| Glucose | 70–120 | 136 | 106 | 104 | mg/dL |
| BUN | 8–26 | 14 | 16 | 16 | mg/dL |
| Creatinine | 0.6–1.3 | 1.1 | 1.1 | 1.1 | mg/dL |
| Uric acid | 2.6–6 (women) | 7.0 | 6.8 | 6.6 | mg/dL |
| | 3.5–7.2 (men) | | | | |
| Calcium | 8.7–10.2 | 9.4 | 9.4 | 9.4 | mg/dL |
| Bilirubin | 0.2–1.3 | .6 | .8 | .7 | mg/dL |
| Ammonia ($NH_3$) | 9–33 | 26 | 22 | 25 | μmol/L |
| SGPT (ALT) | 10–60 | 30 | 215 | 185 | U/L |
| SGOT (AST) | 5–40 | 25 | 245 | 175 | U/L |
| Alk phos | 98–251 | 150 | 145 | 140 | U/L |
| CPK | 26–140 (women) | 75 | 500 | 335 | U/L |
| | 38–174 (men) | | | | |
| CPK-MB | 0 | 0 | 75 | 55 | % |
| LDH | 313–618 | 325 | 635 | 365 | U/L |
| CHOL | 140–199 | 220 | 210 | 200 | mg/dL |
| HDL-C | 40–85 (women) | 30 | 32 | 33 | mg/dL |
| | 37–70 (men) | | | | |
| VLDL | | 45 | 44 | 40 | mg/dL |
| LDL | < 130 | 160 | 150 | 141 | mg/dL |
| LDL/HDL ratio | < 3.22 (women) | 5.3 | 4.7 | 4.3 | |
| | < 3.55 (men) | | | | |
| Apo A | 101–199 (women) | 72 | 80 | 98 | mg/dL |
| | 94–178 (men) | | | | |
| Apo B | 60–126 (women) | 115 | 110 | 105 | mg/dL |
| | 63–133 (men) | | | | |
| TG | 35–160 | 150 | 140 | 130 | mg/dL |
| $T_4$ | 5.4–11.5 | 7.6 | 7.8 | 7.4 | μg/dL |
| $T_3$ | 80–200 | 150 | 165 | 156 | ng/dL |
| $HbA_{1C}$ | 4.8–7.8 | 7.2 | 7.1 | 7.2 | % |
| Troponin I | < 0.5 | 2.4 | 2.8 | | ng/dL |
| Troponin T | < 0.5 | 2.1 | 2.7 | | ng/dL |

# U H *UNIVERSITY HOSPITAL*

NAME: James Klosterman                         DOB: 12/1
AGE: 61                                        SEX: M
PHYSICIAN: Regina H. Smith, MD

\*\*\*\*\*\*\*\*\*\*\*\*\*\*\*\*\*\*\*\*\*\*\*\*\*\*\*\*\*\*\*\*\*\*\*\*\*\*\*\*\*\*\*\*HEMATOLOGY\*\*\*\*\*\*\*\*\*\*\*\*\*\*\*\*\*\*\*\*\*\*\*\*\*\*\*\*\*\*\*\*\*\*\*\*\*\*\*\*\*\*\*\*

DAY:                                    1          3          7
DATE:
TIME:
LOCATION:

| | NORMAL | | | | UNITS |
|---|---|---|---|---|---|
| WBC | 4.3–10 | 11,000 | 9320 | 8800 | $\times\ 10^3/mm^3$ |
| RBC | 4–5 (women) | 4.7 | 4.75 | 4.68 | $\times\ 10^6/mm^3$ |
| | 4.5–5.5 (men) | | | | |
| HGB | 12–16 (women) | 15 | 14.8 | 14.4 | g/dL |
| | 13.5–17.5 (men) | | | | |
| HCT | 37–47 (women) | 45 | 45 | 44 | % |
| | 40–54 (men) | | | | |
| MCV | 84–96 | 91 | 92 | 90 | fL |
| MCH | 27–34 | 30 | 31 | 30 | pg |
| MCHC | 31.5–36 | 33 | 32 | 33 | % |
| RDW | 11.6–16.5 | 13.2 | 12.8 | 13.0 | % |
| Plt Ct | 140–440 | 320 | 295 | 280 | $\times\ 10^3$ |
| Diff TYPE | | | | | |
| % GRANS | 34.6–79.2 | 86 | 80 | 78 | % |
| % LYM | 19.6–52.7 | 14 | 20 | 22 | % |
| SEGS | 50–62 | 84 | 80 | 78 | % |
| BANDS | 3–6 | 2 | 0 | 0 | % |
| LYMPHS | 25–40 | 14 | 20 | 22 | % |
| MONOS | 3–7 | 0 | 0 | 0 | % |
| EOS | 0–3 | 0 | 0 | 0 | % |
| TIBC | 65–165 (women) | 138 | 146 | 132 | µg/dL |
| | 75–175 (men) | | | | |
| Ferritin | 18–160 (women) | 190 | 208 | 196 | µg/dL |
| | 18–270 (men) | | | | |
| Vitamin $B_{12}$ | 100–700 | 300 | 280 | 270 | pg/mL |
| Folate | 2–20 | 8 | 9 | 11 | ng/mL |
| Total T cells | 812–2318 | 1600 | 1600 | 1600 | $mm^3$ |
| T-helper cells | 589–1505 | 800 | 800 | 800 | $mm^3$ |
| T-suppressor cells | 325–997 | 400 | 400 | 400 | $mm^3$ |
| PT | 11–13 | 12.6 | 12.6 | 12.4 | sec |

# U<sub>H</sub> UNIVERSITY HOSPITAL

NAME: James Klosterman                DOB: 12/1
AGE: 61                               SEX: M
PHYSICIAN: Regina H. Smith, MD

\*\*\*\*\*\*\*\*\*\*\*\*\*\*\*\*\*\*\*\*\*\*\*\*\*\*\*\*\*\*\*\*\*\*\*\*\*\*\*\*\*\*\*\*URINALYSIS\*\*\*\*\*\*\*\*\*\*\*\*\*\*\*\*\*\*\*\*\*\*\*\*\*\*\*\*\*\*\*\*\*\*\*\*\*\*\*\*\*\*

| | NORMAL | 1 | 2 | 3 | UNITS |
|---|---|---|---|---|---|
| DAY: | | 1 | 2 | 3 | |
| DATE: | | | | | |
| TIME: | | | | | |
| LOCATION: | | | | | |
| Coll meth | | First morning | First morning | First morning | |
| Color | | Pale yellow | Pale yellow | Pale yellow | |
| Appear | | Clear | Clear | Clear | |
| Sp grv | 1.003-1.030 | 1.020 | 1.015 | 1.018 | |
| pH | 5-7 | 5.8 | 5 | 6 | |
| Prot | NEG | Negative | Negative | Negative | mg/dL |
| Glu | NEG | Negative | Negative | Negative | mg/dL |
| Ket | NEG | Trace | Negative | Negative | |
| Occ bld | NEG | Negative | Negative | Negative | |
| Ubil | NEG | Negative | Negative | Negative | |
| Nit | NEG | Negative | Negative | Negative | |
| Urobil | < 1.1 | Negative | Negative | Trace | EU/dL |
| Leu bst | NEG | Negative | Negative | Negative | |
| Prot chk | NEG | Negative | Negative | Negative | |
| WBCs | 0-5 | 0 | 0 | 0 | /HPF |
| RBCs | 0-5 | 0 | 0 | 0 | /HPF |
| EPIs | 0 | 0 | 0 | 0 | /LPF |
| Bact | 0 | 0 | 0 | 0 | |
| Mucus | 0 | 0 | 0 | 0 | |
| Crys | 0 | 0 | 0 | 0 | |
| Casts | 0 | 0 | 0 | 0 | /LPF |
| Yeast | 0 | 0 | 0 | 0 | |

**UH** *UNIVERSITY HOSPITAL*

# MEDICATION ADMINISTRATION RECORD

Name: James Klosterman

Physician: Regina H. Smith, MD

Allergies: Sulfa

Admit Date: 3/25

Diagnosis: Acute myocardial infarction

| MEDICATIONS | 2331/0730 | 0731/1530 | 1531/2330 | ✓ |
|---|---|---|---|---|
| Heparin<br>5000 U bolus IV<br>1000 U/hour continuous IV infusion<br><br>Start: 3/25          Stop: 4/2 | 1000 | 1000 | 1000 | |
| Aspirin 81 mg po q daily<br><br><br>Start: 3/25          Stop: 4/2 | 1000 | 1000 | 1000 | |
| Lisinopril 10 mg q daily<br><br><br>Start: 3/25          Stop: 4/2 | 1000 | 1000 | 1000 | |
| Nitro-Bid 9 mg po bid<br><br><br>Start: 3/25          Stop: 4/2 | 1000 | 1000 | 1000 | |
| NTG .4 mg sl prn<br><br><br>Start: 3/25          Stop: 4/2 | 1000 | 1000 | 1000 | |

| MEDICATIONS NOT GIVEN AND REASON | | | | |
|---|---|---|---|---|
| MEDICATION | TIME | INITIALS | | |
| | | | | |
| | | | | |
| | | | | |
| | | | | |
| | | | | |

## Case Questions

1. Mr. Klosterman had a myocardial infarction. Explain what happened to his heart.

2. What medical procedure was done to the heart area with the infarction, and what was its purpose?

3. Examine the chemistry results for Mr. Klosterman. Which labs are consistent with the MI diagnosis? Explain. Why were the levels higher on day 2?

4. What do you think should be the ultimate goal for this individual's heart health?

5. After his infarction and procedure on admission, Mr. Klosterman was hospitalized. What diet should be prescribed initially?

6. What risk factors indicated in his medical record can be addressed through medical nutrition therapy?

7. What is abnormal about his lipid profile? Indicate the abnormal levels.

8. What goals do you have for changing his lipid profile?

9. You talk with Mr. Klosterman and his wife, a math teacher at the local high school. They are friendly and seem cooperative. They are both anxious to learn what they can do to prevent another heart attack. What questions will you ask them to assess how to best help them?

10. This patient is a computer programmer. He does get some exercise daily. He walks his dog outside for about 15 minutes at a leisurely pace. Calculate his energy need. How many grams of protein should he have daily?

**11.**   What is his ideal body weight?

**12.**   From Mr. Klosterman and his wife, you obtain a record of the foods he ate the day before his heart attack. It was a typical day. He usually does not eat breakfast but does eat a mid-morning snack, lunch, dinner, and an evening snack.

*Mid-morning snack*
1 large cinnamon raisin bagel with 1 T fat-free cream cheese
8 oz orange juice
Coffee

*Lunch*
1 cup canned vegetable beef soup
Sandwich with 4 oz roast beef
Lettuce, tomato, dill pickles, 2 t mayonnaise
1 small apple
8 oz 2% milk

*Dinner*
2 lean pork chops (3 oz each)
1 large baked potato, 2 t margarine
½ cup green beans
½ c coleslaw (cabbage with 1 T salad dressing)
1 sl apple pie

*Snack*
8 oz 2% milk
1 oz pretzels

Calculate the total number of calories he consumed as well as the energy distribution of calories for protein, carbohydrate, and fat. Use the exchange system.

**13.**   What medical nutrition therapy do you recommend for this patient? How does this diet compare with the analysis of his 24-hour recall?

**14.**   What other issues might you consider to support the success of his lifestyle change?

**15.**   What tools could you use that would be helpful in instructing him?

**16.**   What barriers to compliance do you see with Mr. Klosterman?

**17.**   Mr. Klosterman and his wife ask about specific vitamin supplements. "My roommate here in the hospital told me I should be taking vitamin E and—I think it was folate." What do you think about Vitamin E and folate supplementation for this patient? What is your rationale?

**18.**   What would you want to assess in three to four weeks when he and his wife return for additional counseling?

**19.**   Mr. Klosterman was prescribed the following medications on discharge. What are the food–medication interactions for this list of medications?

Lopressor 50 mg q daily
Lisinopril 10 mg q daily
Nitro-Bid 9.0 mg bid
NTG .4 mg sl prn chest pain
ASA 81 mg q daily

## Bibliography

Cardiovascular disease. Available at http://www.americanheart.org/Heart_and_Stroke_A_Z_Guide/cvds.html. Accessed May 1, 2001.

Commission on Accreditation for Dietetics Education. Knowledge, skills, and competencies for dietitians. *Accreditation Manual for Dietetics Education Programs,* rev. 4th ed. Chicago: American Dietetic Association, 2000: 29–45.

Kris-Etherton P, Burns J (eds.). *Cardiovascular Nutrition: Strategies and Tools for Disease Management and Prevention.* Chicago: American Dietetic Association, 1998.

Krummel D. Nutrition in cardiovascular disease. In Mahan LK, Escott-Stump S, eds. *Krause's Food, Nutrition, and Diet Therapy,* 10th ed. Philadelphia: Saunders, 2000: 558–595.

*Medical Nutrition Therapy Across the Continuum of Care: Client Protocols.* Chicago: American Dietetic Association and Morrison Health Care, 1998.

Pronsky, Z. *Food Medication Interactions,* 11th ed. Pottstown, PA: Food Medication Interactions, 2000.

*Third Report of the Expert Panel on Detection, Evaluation, and Treatment of High Blood Cholesterol in Adults (Adult Treatment Panel III).* Bethesda, MD: National Institutes of Health, 2001. Available at http://www.nhlbi.nih.gov/guidelines/cholesterol/index.htm. Accessed May 20, 2001.

# Congestive Heart Failure with Resulting Cardiac Cachexia

*Advanced Practice*

*Anne B. Marietta, PhD, RD*

## Objectives

After completing this case, the student will be able to:

1. Use nutrition assessment information to determine baseline nutritional status.
2. Correlate patient's signs and symptoms with pathophysiology of congestive heart failure.
3. Evaluate laboratory indices for nutritional implications and significance.
4. Demonstrate understanding of nutrition support options for congestive heart failure.
5. Identify role of pharmacologic intervention and drug–nutrient interactions.
6. Determine appropriate nutritional interventions for the patient with congestive heart failure and cardiac cachexia.

Dr. Charles Peterman, an 85-year-old retired physician, is admitted with acute symptoms related to his congestive heart failure. Dr. Peterman has a long history of cardiac disease, including a previous myocardial infarction and mitral valve disease.

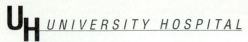 **UNIVERSITY HOSPITAL**

# ADMISSION DATABASE

Name: Charles Peterman
DOB: 4/15  age 85
Physician: D. Schmidt, MD

| BED# | DATE: | TIME: | TRIAGE STATUS (ER ONLY): |
|---|---|---|---|
| 2 | 3/31 | 1600 | ☐ Red  ☐ Yellow  ☐ Green  ☐ White |

### Initial Vital Signs

| TEMP: | RESP: | | SAO2: | |
|---|---|---|---|---|
| 101 | 25 | | 80 mmHg | |

| HT: | WT (lb): | B/P: | PULSE: |
|---|---|---|---|
| 5'10" | 165 | 90/70 | 101 |

| LAST TETANUS | LAST ATE | LAST DRANK |
|---|---|---|
| 35 years ago | lunch | 1400 |

### CHIEF COMPLAINT/HX OF PRESENT ILLNESS

Passed out

85-yo male with chronic class IV CHF who has history of remote
myocardial infarction

### ALLERGIES: Meds, Food, IVP Dye, Seafood: Type of Reaction

Shellfish, aspirin, ibuprofen–hives

### PREVIOUS HOSPITALIZATIONS/SURGERIES

1970 Acute diverticulitis

---

**PRIMARY PERSON TO CONTACT:**
Name: Jean Peterman
Home #: 412-561-8556
Work #: NA

ORIENTATION TO UNIT: ☒ Call light  ☒ Television/telephone
☒ Bathroom  ☒ Visiting  ☒ Smoking  ☒ Meals
☒ Patient rights/responsibilities

PERSONAL ARTICLES: (Check if retained/describe)
☐ Contacts ☐ R ☐ L          ☐ Dentures ☐ Upper ☐ Lower
☒ Jewelry: wedding band
☐ Other:

VALUABLES ENVELOPE:
☐ Valuables instructions

INFORMATION OBTAINED FROM:
☐ Patient          ☐ Previous record
☒ Family           ☐ Responsible party

Signature  *Jean Peterman*

---

| Home Medications (including OTC) | Codes: A=Sent home | | B=Sent to pharmacy | | C=Not brought in |
|---|---|---|---|---|---|
| Medication | Dose | Frequency | Time of Last Dose | Code | Patient Understanding of Drug |
| Lanoxin | .125 mg | q daily | 8 am | c | yes |
| Lasix | 80 mg | bid | 5 pm | c | yes |
| lisinopril | 30 mg | q daily | 8 am | c | yes |
| Centrum Silver | 2 | q daily | 8 am | c | yes |
| Lopressor | 25 mg | q daily | 8 am | c | yes |
| Zocor | 20 mg | q hs | 9 pm | c | yes |
| calcium carbonate | 500 mg | bid | 5 pm | c | yes |
| Metamucil | Tbs | bid | 6 pm | c | yes |
| Aldactone | 25 mg | q daily | 8 am | c | yes |
| | | | | | |
| | | | | | |

Do you take all medications as prescribed?   ☒ Yes   ☐ No   If no, why?

### PATIENT/FAMILY HISTORY

| | | |
|---|---|---|
| ☐ Cold in past two weeks | ☒ High blood pressure Patient | ☒ Kidney/urinary problems Patient |
| ☒ Hay fever Patient | ☒ Arthritis Patient | ☒ Gastric/abdominal pain/heartburn Patient |
| ☐ Emphysema/lung problems | ☐ Claustrophobia | ☒ Hearing problems Patient |
| ☐ TB disease/positive TB skin test | ☐ Circulation problems | ☐ Glaucoma/eye problems |
| ☐ Cancer | ☐ Easy bleeding/bruising/anemia | ☒ Back pain Patient |
| ☐ Stroke/past paralysis | ☐ Sickle cell disease | ☐ Seizures |
| ☒ Heart attack Patient | ☐ Liver disease/jaundice | ☐ Other |
| ☐ Angina/chest pain | ☐ Thyroid disease | |
| ☒ Heart problems Patient | ☐ Diabetes | |

### RISK SCREENING

Have you had a blood transfusion?   ☐ Yes   ☒ No
Do you smoke?   ☐ Yes   ☒ No
If yes, how many pack(s)   /day for   years
Does anyone in your household smoke?   ☐ Yes   ☒ No
Do you drink alcohol?   ☐ Yes   ☒ No
If yes, how often?   How much?
When was your last drink? ____/____/____
Do you take any recreational drugs?   ☐ Yes   ☒ No
If yes, type:_____   Route
Frequency:_____   Date last used:____/____/____

**FOR WOMEN Ages 12–52**

Is there any chance you could be pregnant?   ☐ Yes   ☐ No
If yes, expected date (EDC):
Gravida/Para:

**ALL WOMEN**

Date of last Pap smear:
Do you perform regular breast self-exams?   ☐ Yes   ☐ No

**ALL MEN**

Do you perform regular testicular exams?   ☒ Yes   ☐ No

---

Additional comments:

✗ *Samuel Layton, RN*
Signature/Title

**Client name:** Charles Peterman
**DOB:** 4/15
**Age:** 85
**Sex:** Male
**Education:** Postgraduate
**Occupation:** Physician
**Hours of work:** Retired
**Household members:** wife age 82, in good health
**Ethnic background:** Caucasian
**Religious affiliation:** Presbyterian
**Referring physician:** Douglas A. Schmidt, MD (cardiology)

**Chief complaint:**
Patient collapsed at home and was brought to the emergency room by ambulance.

**Patient history:**
*Onset of disease:* CHF × 2 yrs
*Type of Tx:* Medical Tx of CAD, HTN, and CHF
*PMH:* Long-standing history of CAD, HTN, mitral valve insufficiency, previous anterior MI
*Meds:* Lanoxin .125 mg q daily, Lasix 80 mg bid, Aldactone 25 mg q daily, lisinopril 30 mg po q daily, Lopressor 25 mg q daily, Zocor 20 mg q daily, Metamucil 1 Tbsp bid, calcium carbonate 500 mg bid, Centrum 2 tablets q daily
*Smoker:* No
*Family Hx: What?* HTN, CAD    *Who?* Parents

**Physical exam:**
*General appearance:* Elderly male in acute distress
*Vitals:* Temp 98°F, pulse 110, RR 24, BP 90/70
*Heart:* Diffuse PMI in AAL in LLD, Grade II holosystolic murmur at the apex radiating to the left sternal border; first heart sound diminished and second heart sound preserved, third heart sound present
*Skin:* Gray, moist

*HEENT:*
   *Eyes:* Ophthalmoscopic exam reveals AV crossing changes and arteriolar spasm
   *Ears:* WNL
   *Nose:* WNL
   *Throat:* Jugular venous distension in sitting position with a positive hepatojugular reflux
*Chest/lungs:* Rales in both bases posteriorly
*Abdomen:* Ascites, no masses, liver tender to A&P
*Genitalia:* WNL
*Extremities:* 4 + pedal edema
*Peripheral vascular:* WNL
*Neurologic:* WNL
*Height/Weight:* 70″, 165#
*Usual dietary intake:* Generally likes all foods but has recently been eating only soft foods, esp. ice cream, Ensure 2 cans/day

*24-hr recall:* NPO
*Food allergies/intolerances/aversions:* Shellfish
*Previous MNT?* No
*Food purchase/preparation:* Wife

**Nutrition Hx:**
*General:* Wife reports that Dr. Peterman's appetite has been poor for last 6 months, with no real weight loss that she can determine. He has difficulty eating due to SOB and nausea.
*Usual dietary intake:* Generally likes all foods but has recently been eating only soft foods, esp. ice cream. Tries to drink 2 cans Ensure each day.
*24-hour recall:* NPO
*Food allergies/intolerances/aversions:* Shellfish
*Previous MNT:* No
*Food purchase/preparation:* Spouse
*Vit/min intake:* Centrum Silver 2/day, calcium supplement 1000 mg/day

**Tx plan:**
Admit to CCU.
Parenteral dopamine and IV diuretics
Telemetry, vitals q 1 hr × 8, q 2 hrs × 8 for 24 hours
Daily ECG and chest X-rays
Echocardiogram
Chem 24, urinalysis, strict I&O's

**Hospital course:**
Swan-Ganz catheter inserted. Echocardiogram indicated severe cardiomegaly secondary to end-stage congestive heart failure. Enteral feeding initiated but discontinued due to severe diarrhea. Patient had a living will that stated he wanted no other extraordinary measures taken to prolong his life. The patient was able to express his wishes verbally: He requested oral feedings and palliative care only. Patient expired after two-week hospitalization.

# U_H UNIVERSITY HOSPITAL

NAME: Charles Peterman        DOB: 4/15
AGE: 85                       SEX: M
PHYSICIAN: Douglas Schmidt, MD

\*\*\*\*\*\*\*\*\*\*\*\*\*\*\*\*\*\*\*\*\*\*\*\*\*\*\*\*\*\*\*\*\*\*\*\*\*\*\*\*\*\*\*\*\*\*\*CHEMISTRY\*\*\*\*\*\*\*\*\*\*\*\*\*\*\*\*\*\*\*\*\*\*\*\*\*\*\*\*\*\*\*\*\*\*\*\*\*\*\*\*\*\*\*\*\*

DAY:                              1          3          7
DATE:
TIME:
LOCATION:

| | NORMAL | 1 | 3 | 7 | UNITS |
|---|---|---|---|---|---|
| Albumin | 3.6–5 | 2.8 | 2.7 | 2.6 | g/dL |
| Total protein | 6–8 | 5.8 | 5.6 | 5.5 | g/dL |
| Prealbumin | 19–43 | 15 | 11 | 10 | mg/dL |
| Transferrin | 200–400 | 350 | 355 | 352 | mg/dL |
| Sodium | 135–155 | 132 | 133 | 133 | mmol/L |
| Potassium | 3.5–5.5 | 3.7 | 3.6 | 3.8 | mmol/L |
| Chloride | 98–108 | 98 | 100 | 99 | mmol/L |
| $PO_4$ | 2.5–4.5 | 4.0 | 3.8 | 3.6 | mmol/L |
| Magnesium | 1.6–2.6 | 2.0 | 1.9 | 1.8 | mmol/L |
| Osmolality | 275–295 | 292 | 288 | 290 | mmol/kg $H_2O$ |
| Total $CO_2$ | 24–30 | 26 | 24 | 25 | mmol/L |
| Glucose | 70–120 | 110 | 106 | 102 | mg/dL |
| BUN | 8–26 | 32 | 34 | 30 | mg/dL |
| Creatinine | 0.6–1.3 | 1.6 | 1.7 | 1.5 | mg/dL |
| Uric acid | 2.6–6 (women) | 6.0 | 6.4 | 6.7 | mg/dL |
| | 3.5–7.2 (men) | | | | |
| Calcium | 8.7–10.2 | 9.0 | 8.8 | 8.9 | mg/dL |
| Bilirubin | 0.2–1.3 | 1.0 | 1.1 | .9 | mg/dL |
| Ammonia ($NH_3$) | 9–33 | 32 | 30 | 34 | $\mu$mol/L |
| SGPT (ALT) | 10–60 | 100 | 120 | 115 | U/L |
| SGOT (AST) | 5–40 | 70 | 80 | 85 | U/L |
| Alk phos | 98–251 | 200 | 190 | 200 | U/L |
| CPK | 26–140 (women) | 150 | 175 | 200 | U/L |
| | 38–174 (men) | | | | |
| LDH | 313–618 | 350 | 450 | 556 | U/L |
| CHOL | 140–199 | 240 | 220 | 210 | mg/dL |
| HDL-C | 40–85 (women) | 30 | 31 | 30 | mg/dL |
| | 37–70 (men) | | | | |
| VLDL | | 40 | 42 | 39 | mg/dL |
| LDL | < 130 | 180 | 160 | 152 | mg/dL |
| LDL/HDL ratio | < 3.22 (women) | 6 | 5.2 | 5.1 | |
| | < 3.55 (men) | | | | |
| Apo A | 101–199 (women) | 60 | 65 | 70 | mg/dL |
| | 94–178 (men) | | | | |
| Apo B | 60–126 (women) | 140 | 138 | 136 | mg/dL |
| | 63–133 (men) | | | | |
| TG | 35–160 | 150 | 145 | 140 | mg/dL |
| $T_4$ | 5.4–11.5 | 8.0 | 7.8 | 7.6 | $\mu$g/dL |
| $T_3$ | 80–200 | 160 | 156 | 150 | ng/dL |
| $HbA_{1C}$ | 4.8–7.8 | 6.8 | 7.0 | 7.0 | % |

# U H UNIVERSITY HOSPITAL

NAME: Charles Peterman                DOB: 4/15
AGE: 85                               SEX: M
PHYSICIAN: Douglas Schmidt, MD

\*\*\*\*\*\*\*\*\*\*\*\*\*\*\*\*\*\*\*\*\*\*\*\*\*\*\*\*\*\*\*\*\*\*\*\*\*\*HEMATOLOGY\*\*\*\*\*\*\*\*\*\*\*\*\*\*\*\*\*\*\*\*\*\*\*\*\*\*\*\*\*\*\*\*\*\*\*\*\*\*

DAY:                                  1        3        7
DATE:
TIME:
LOCATION:

| | NORMAL | | | | UNITS |
|---|---|---|---|---|---|
| WBC | 4.3–10 | 11 | 10.5 | 9.8 | $\times\ 10^3/mm^3$ |
| RBC | 4–5 (women) | 5 | 5.1 | 5.2 | $\times\ 10^6/mm^3$ |
| | 4.5–5.5 (men) | | | | |
| HGB | 12–16 (women) | 14 | 14.3 | 14.5 | g/dL |
| | 13.5–17.5 (men) | | | | |
| HCT | 37–47 (women) | 41 | 42 | 42 | % |
| | 40–54 (men) | | | | |
| MCV | 84–96 | 90 | 89 | 91 | fL |
| MCH | 27–34 | 31 | 31 | 30 | pg |
| MCHC | 31.5–36 | 33 | 34 | 32 | % |
| RDW | 11.6–16.5 | 12 | 13 | 12 | % |
| Plt Ct | 140–440 | 300 | 290 | 310 | $\times\ 10^3$ |
| Diff TYPE | | | | | |
| % GRANS | 34.6–79.2 | 76 | 82 | 72 | % |
| % LYM | 19.6–52.7 | 24 | 18 | 28 | % |
| SEGS | 50–62 | 65 | 73 | 66 | % |
| BANDS | 3–6 | 11 | 9 | 6 | % |
| LYMPHS | 25–40 | 20 | 17 | 26 | % |
| MONOS | 3–7 | 4 | 1 | 2 | % |
| EOS | 0–3 | | | | % |
| TIBC | 65–165 (women) | 160 | 155 | 162 | µg/dL |
| | 75–175 (men) | | | | |
| Ferritin | 18–160 (women) | 100 | 96 | 98 | µg/dL |
| | 18–270 (men) | | | | |
| Vitamin $B_{12}$ | 100–700 | 110 | 120 | 116 | pg/mL |
| Folate | 2–20 | 10 | 8 | 12 | ng/mL |
| Total T cells | 812–2318 | 1000 | 1100 | 1200 | $mm^3$ |
| T-helper cells | 589–1505 | 800 | 860 | 840 | $mm^3$ |
| T-suppressor cells | 325–997 | 460 | 440 | 500 | $mm^3$ |
| PT | 11–13 | 12.2 | 12.3 | 12.3 | sec |

## Case Questions

1. Dr. Peterman has a two-year history of congestive heart failure. Describe the pathophysiology of congestive heart failure.

2. Find the signs and symptoms in the patient's physical examination that are consistent with congestive heart failure.

3. Dr. Peterman's admitting diagnosis was cardiac cachexia. What is cardiac cachexia? What are the characteristic symptoms?

4. What role does the underlying heart disease play in the development of cardiac cachexia?

5. After reading the history and physical, do you find any indication of Dr. Peterman's fluid status? If so, what?

6. How did Dr. Peterman's fluid status change over the first week of his hospital stay? What indices would you examine to determine this evaluation?

7. Which biochemical markers are significant for Dr. Peterman's case in determining fluid status?

8. The following chart lists the drugs that were prescribed for Dr. Peterman. Give the rationale for the use of each drug for this patient. In addition, list the food–medication interactions.

| Medication and Dose | Rationale for Use | Food–Medication Interaction |
|---|---|---|
| Lanoxin .125 mg IV qd | | |
| Lasix 80 mg IV push | | |
| dopamine 30 mcg/kg/min | | |

9.  What general recommendations would you make regarding the nutritional care of Dr. Peterman?

10. Calculate his energy needs. His weight on admission was 165 lbs. His weight three days later was 145 lbs. Height is 5'10"; age is 85.

11. What are Dr. Peterman's fluid needs?

12. Two days after admission, Dr. Peterman was put on enteral feeding but was not able to tolerate it because of diarrhea. What recommendations could be made to improve tolerance to the tube feeding?

13. The tube feeding was discontinued because of continued problems. Parenteral nutrition cannot be considered at this time because of need to severely restrict fluid. What recommendations could you make to optimize Dr. Peterman's oral intake?

14. This patient had a living will that expressed his wishes regarding life support measures and requested palliative care only. What is a living will? What is palliative care?

15. Dr. Peterman is not receiving parenteral or enteral nutritional support. What is the role of the registered dietitian during palliative care?

## Bibliography

American Dietetic Association. Gastroesophageal reflux disease. In *Manual of Clinical Dietetics,* 6th ed. Chicago: American Dietetic Association, 2000:401–405.

Gallagher-Allred C. *Nutritional Care of the Terminally Ill.* Rockville, MD: Aspen, 1989.

Gomberg-Maitland M, Baran DA, and Fuster V. Treatment of congestive heart failure: guidelines for the primary care physician and the health failure specialist. *Arch Intern Med* 2001;161:342–352.

Gould B. Cardiovascular and lymphatic disorders. In *Pathophysiology for the Health Related Professions.* Philadelphia: Saunders, 1997:159–212.

Katz AM. *Heart Failure: Pathophysiology, Molecular Biology, Clinical Management.* Philadelphia: Lippincott/ Williams and Wilkins, 2000.

Krummel D. Medical nutrition therapy for heart failure and transplant. In Mahan LK, Escott-Stump S, eds., *Krause's Food, Nutrition, and Diet Therapy,* 10th ed. Philadelphia: Saunders, 2000:801–814.

# Unit Three

# MEDICAL NUTRITION THERAPY FOR GASTROINTESTINAL DISORDERS

## Introduction

The eight cases presented in this section cover a wide array of diagnoses that ultimately affect normal digestion and absorption. These conditions use medical nutrition therapy as a cornerstone for their treatment. In some disorders, such as celiac disease, medical nutrition therapy is the *only* treatment. It is also important to understand that because of the symptoms the patient experiences, nutritional status is often in jeopardy. Nausea, vomiting, diarrhea, constipation, and malabsorption are common with these disorders. Interventions in these cases are focused on treating such symptoms to restore nutritional health.

Case 10 targets gastroesophageal reflux disease (GERD). More than 20 million Americans suffer from symptoms of gastroesophageal reflux daily, and more than 100 million suffer occasional symptoms. Gastroesophageal reflux disease most frequently results from lower esophageal sphincter (LES) incompetence. Factors that influence LES competence include both physical and lifestyle factors. This case identifies the common symptoms of GERD and challenges you to develop and analyze both nutritional and medical care for this patient.

Cases 11 and 12 focus on peptic ulcer disease treated pharmacologically and surgically. Peptic ulcer disease involves ulcerations that penetrate the submucosa, usually in the antrum of the stomach or in the duodenum. Erosion may proceed to other levels of tissue and can eventually perforate. The breakdown in tissue allows continued insult by the highly acidic environment of the stomach. *Helicobacter pylori* has been established as a major cause of chronic gastritis and peptic ulcer disease.

Medical nutrition therapy for peptic ulcer disease is highly individualized. Treatment plans should avoid foods that increase gastric secretions and restrict any particular food or beverage that the patient does not tolerate. In this case, you will design appropriate nutrition interventions for peptic ulcer disease.

Refractory peptic ulcer disease, addressed in Case 12, results in hemorrhage and perforation that require surgical intervention. Nutritional complications often accompany gastric surgery, such as dumping syndrome and malabsorption. This case introduces the transition from enteral nutrition support to appropriate oral diet for postoperative use.

The next five cases target conditions affecting the large intestine. These conditions, whose etiologies are all different, involve the symptoms of diarrhea, constipation, and in several, malabsorption. In all the cases, medical nutrition therapy is one of the major modes of treatment. Case 13 addresses the metabolic complications of diarrhea and dehydration. This pediatric case allows you to interpret nutrition assessment for children and to plan appropriate reintroduction of solid food to help the patient recover from acute diarrhea.

Celiac disease, explored in Case 14, is an autoimmune disease that destroys the mucosa of the small intestine. This reaction is caused by exposure to gliadin, which is found in the gluten portion of grain. Treatment for this disease is total avoidance of wheat, rye, and barley.

This case explores new diagnostic procedures for celiac disease, secondary malabsorption syndromes, and the use of medical nutrition therapy.

Case 15 examines diverticulosis, a condition associated with both age and low fiber intake. A diet low in fiber increases colonic intraluminal pressure as the body strives to move the small amount of stool through the colon. This increased pressure results in herniations of the colon wall, which are diverticula. Long-term treatment includes a transition to a high-fiber diet. This case involves the care of acute diverticulitis and the transition to preventive care.

Irritable bowel syndrome, discussed in Case 16, is another condition of the large intestine that often presents symptoms of both diarrhea and constipation. In the past several years, the medical profession has made significant progress in understanding and treating this syndrome. This case lets you explore the latest treatments and applications of medical nutrition therapy.

The final two cases in this section, 17 and 18, target inflammatory bowel disease. Crohn's disease and ulcerative colitis are two conditions that fall under the diagnosis of inflammatory bowel disease. Both these conditions dramatically affect nutritional status and often require nutritional support during periods of exacerbation. These cases involve effects on digestion and absorption, diagnosis of malnutrition, and enteral and parenteral nutrition support.

# Gastroesophageal Reflux Disease

*Introductory Level*

## Objectives

After completing this case, the student will be able to:

1. Integrate the principles of pathophysiology to support principles of medical nutrition therapy in gastroesophageal reflux disease (GERD).
2. Demonstrate a working knowledge of pathophysiology related to nutrition care.
3. Interpret pertinent laboratory parameters.
4. Delineate basic principles of drug action required for medical treatment of gastroesophageal reflux disease.

5. Establish a care plan to support successful medical nutrition therapy for the individual patient.

Jack Nelson is admitted to University Hospital for evaluation of his increasing complaints of severe indigestion. Intraesophageal pH monitoring and barium esophagram support a diagnosis of gastroesophageal reflux disease.

**UNIVERSITY HOSPITAL**

# ADMISSION DATABASE

Name: Jack Nelson
DOB: 7/22    age 48
Physician: J. Phelps, MD

| BED#<br>1 | DATE:<br>9/22 | TIME:<br>0900 | TRIAGE STATUS (ER ONLY):<br>☐ Red  ☐ Yellow  ☐ Green  ☐ White |
|---|---|---|---|

**Initial Vital Signs**

| TEMP:<br>98.6 | RESP:<br>16 | SAO2: |
|---|---|---|

| HT:<br>5′9″ | WT (lb):<br>215 | B/P:<br>119/75 | PULSE:<br>90 |
|---|---|---|---|

| LAST TETANUS<br>1 year ago | LAST ATE<br>11 pm | LAST DRANK<br>this am |
|---|---|---|

**PRIMARY PERSON TO CONTACT:**
Name: Mary Nelson
Home #: 612-444-5689
Work #: 612-453-5689

**ORIENTATION TO UNIT:** ☒ Call light  ☒ Television/telephone  ☒ Bathroom  ☒ Visiting  ☒ Smoking  ☒ Meals  ☒ Patient rights/responsibilities

### CHIEF COMPLAINT/HX OF PRESENT ILLNESS

"My wife insisted that I come see someone. The pain was so bad that I was afraid I was having a heart attack."

**PERSONAL ARTICLES:** (Check if retained/describe)
☐ Contacts  ☐ R  ☐ L          ☐ Dentures  ☐ Upper  ☐ Lower
☐ Jewelry:
☒ Other: eyeglasses

### ALLERGIES: Meds, Food, IVP Dye, Seafood: Type of Reaction

NKA

**VALUABLES ENVELOPE:** no
☐ Valuables instructions

### PREVIOUS HOSPITALIZATIONS/SURGERIES

s/y R knee arthroplasty 5 years ago

**INFORMATION OBTAINED FROM:**
☒ Patient          ☐ Previous record
☒ Family          ☐ Responsible party

Signature  *Jack Nelson*

### Home Medications (including OTC)          Codes: A=Sent home          B=Sent to pharmacy          C=Not brought in

| Medication | Dose | Frequency | Time of Last Dose | Code | Patient Understanding of Drug |
|---|---|---|---|---|---|
| atenolol | 50 mg | q am | this am | c | yes |
| aspirin | 325 mg | q am | this am | c | yes |
| ibuprofen | 500 mg | bid | this am | c | yes |
| | | | | | |
| | | | | | |
| | | | | | |
| | | | | | |
| | | | | | |
| | | | | | |
| | | | | | |

Do you take all medications as prescribed?    ☒ Yes    ☐ No    If no, why?

### PATIENT/FAMILY HISTORY

| | | |
|---|---|---|
| ☐ Cold in past two weeks | ☒ High blood pressure Patient | ☐ Kidney/urinary problems |
| ☐ Hay fever | ☐ Arthritis | ☒ Gastric/abdominal pain/heartburn Patient |
| ☐ Emphysema/lung problems | ☐ Claustrophobia | ☐ Hearing problems |
| ☐ TB disease/positive TB skin test | ☐ Circulation problems | ☐ Glaucoma/eye problems |
| ☐ Cancer | ☐ Easy bleeding/bruising/anemia | ☐ Back pain |
| ☐ Stroke/past paralysis | ☐ Sickle cell disease | ☐ Seizures |
| ☒ Heart attack Father | ☐ Liver disease/jaundice | ☐ Other |
| ☒ Angina/chest pain Father | ☐ Thyroid disease | |
| ☒ Heart problems Mother | ☐ Diabetes | |

### RISK SCREENING

Have you had a blood transfusion?  ☐ Yes  ☒ No
Do you smoke?  ☐ Yes  ☒ No
If yes, how many pack(s) ___ /day for ___ years
Does anyone in your household smoke?  ☐ Yes  ☒ No
Do you drink alcohol?  ☒ Yes  ☐ No
If yes, how often? 3-4 × week  How much? 1-2 beers
When was your last drink? last pm / ___ / ___
Do you take any recreational drugs?  ☐ Yes  ☒ No
If yes, type:_____  Route _____
Frequency:_____  Date last used:_____/_____/_____

**FOR WOMEN Ages 12–52**

Is there any chance you could be pregnant?  ☐ Yes  ☐ No
If yes, expected date (EDC):
Gravida/Para:

**ALL WOMEN**

Date of last Pap smear:
Do you perform regular breast self-exams?  ☐ Yes  ☐ No

**ALL MEN**

Do you perform regular testicular exams?  ☒ Yes  ☐ No

Additional comments:

x *Cathy Mosely, RN*
Signature/Title

**Client name:** Jack Nelson
**DOB:** 7/22
**Age:** 48
**Sex:** Male
**Education:** PhD
**Occupation:** University professor
**Hours of work:** M–F  Works consistently in evenings and weekends as well
**Household members:** Wife age 42, 2 sons ages 10 and 16—all in good health
**Ethnic background:** Caucasian
**Religious affiliation:** Protestant
**Referring physician:** Patricia Phelps, MD (family practice)

## Chief complaint:

"My wife insisted that I come see someone. I am taking Tums constantly and am really uncomfortable from this constant indigestion! It was so bad yesterday that I was afraid I was having a heart attack. I also recently hurt my shoulder when I was coaching my son's baseball team, but as long as I take Advil I am able to cope with that pain."

## Patient history:

*Onset of disease:* Has been experiencing increased indigestion over last year. Previously only at night but now almost constantly
*Type of Tx:* Taking OTC antacids
*PMH:* Essential HTN—Dx 1 year ago; s/p R knee arthroplasty 5 years ago
*Meds:* Atenolol 50 mg qd; 325 mg aspirin qd; multivitamin qd; 500 mg ibuprofen bid for last month
*Smoker:* Yes
*Family Hx: What?* CAD  *Who?* Father

## Physical exam:

*General appearance:* Mildly obese 48-year-old white male in mild distress
*Vitals:* Temp 98.6°F, BP 119/75, HR 90 BPM/normal, RR 16 BPM
*Heart:* Noncontributory
*HEENT:* Noncontributory
*Rectal:* No hemorrhoids seen or felt; prostate not enlarged or soft; stool—slight Heme $\oplus$
*Neurologic:* Oriented ×4
*Extremities:* No edema; normal strength, sensations, and DTR
*Skin:* Warm, dry
*Chest/lungs:* Lungs clear to auscultation and percussion
*Peripheral vascular:* Pulses full—no bruits
*Abdomen:* No distention. BS present in all regions. Liver percusses approx 8 cm at the midclavicular line, one fingerbreadth below the right costal margin. Epigastric tenderness without rebound or guarding

## Nutrition Hx:

*General:* Patient relates that he has gained almost 35 lbs since his knee surgery. He attributes this to a decrease in ability to run and has not found a consistent replacement for exercise. Patient states that he plays with his children on the weekends but that is the most exercise that he receives. He recently has been chair of his department at the university, and this position has been very stressful. He states that he probably has been eating and drinking more over the last year, which he attributes to stress. He is worried about his family history of heart disease, which is why he takes an aspirin each day. He has not really followed any diet restrictions.

*Usual dietary intake:*

| | |
|---|---|
| AM: | 1½–2 c dry cereal (Cheerios, bran flakes, Crispix); ½–¾ c skim milk; 16–32 oz orange juice |
| Lunch: | 1.5 oz ham on ww bagel, 1 apple or other fruit; 1 c chips, diet soda |
| Snack when he comes home: | Handful of crackers, cookies, or chips; 1–2 16 oz beers |
| PM: | 6–9 oz of meat (grilled, baked usually), pasta, rice, or potatoes—1–2 c; fresh fruit, salad, or other vegetable, bread, iced tea |
| Late PM: | Ice cream, popcorn, or crackers. Drinks 5–6 12-oz diet sodas daily as well as Iced tea. Relates that his family's schedule has been increasingly busy so that they order pizza or stop for fast food 1–2 times per week instead of cooking. |

*24-hr recall*

| | |
|---|---|
| (at home PTA): | Crispix—2 c, 1 c skim milk, 16 oz orange juice |
| At work: | 3 12-oz Diet Pepsis |
| Lunch: | Grilled chicken sandwich from McDonalds, small french fries, 32 oz iced tea |
| Late afternoon: | 2 c chips, 1 beer |
| Dinner: | 1 breast, fried, from Kentucky Fried Chicken; 1.5 c potato salad, ¼ c green bean casserole, ½ c fruit salad, 1 c baked beans, iced tea |
| Bedtime: | 2 c ice cream mixed with 1 c skim milk for milkshake |

*Food allergies/intolerances/aversions* (specify): Fried foods seem to make the indigestion worse.
*Previous MNT?* No
*Food purchase/preparation:* Wife or eats out
*Vit/min intake:* One-A-Day for Men multivitamin daily

**Tx plan:**
Ambulatory 24-hour pH monitoring with intraesophageal pH electrode and recorder
Barium esophagram—request radiologist to attempt to demonstrate reflux using abdominal pressure and positional changes
Endoscopy with biopsy to r/o *H. pylori* infection

**Hospital course:**
pH monitoring and barium esophagram support diagnosis of gastroesophageal reflux disease with negative biopsy for *H. pylori*. Endoscopy indicates no ulcerations or lesions but generalized gastritis present. Begin lansoprazole 30 mg q am. Decrease aspirin to 80 mg qd. Consult to orthopedics for shoulder injury. D/C self-medication of ibuprofen qd. Nutrition consult.

# U<sub>H</sub> *UNIVERSITY HOSPITAL*

NAME: Jack Nelson                    DOB: 7/22
AGE: 48                              SEX: M
PHYSICIAN: P. Phelps, MD

\*\*\*\*\*\*\*\*\*\*\*\*\*\*\*\*\*\*\*\*\*\*\*\*\*\*\*\*\*\*\*\*\*\*\*\*\*\*\*\*\*\*CHEMISTRY\*\*\*\*\*\*\*\*\*\*\*\*\*\*\*\*\*\*\*\*\*\*\*\*\*\*\*\*\*\*\*\*\*\*\*\*\*\*\*\*\*

| DAY: | | Admit | |
|------|--------|-------|-------|
| DATE: | | 9/22 | |
| TIME: | | | |
| LOCATION: | | | |
| | NORMAL | | UNITS |
| Albumin | 3.6–5 | 4.9 | g/dL |
| Total protein | 6–8 | 7.2 | g/dL |
| Prealbumin | 19–43 | 33 | mg/dL |
| Transferrin | 200–400 | 350 | mg/dL |
| Sodium | 135–155 | 144 | mmol/L |
| Potassium | 3.5–5.5 | 4.5 | mmol/L |
| Chloride | 98–108 | 102 | mmol/L |
| $PO_4$ | 2.5–4.5 | 3.8 | mmol/L |
| Magnesium | 1.6–2.6 | 2.0 | mmol/L |
| Osmolality | 275–295 | 282 | mmol/kg $H_2O$ |
| Total $CO_2$ | 24–30 | 28 | mmol/L |
| Glucose | 70–120 | 110 | mg/dL |
| BUN | 8–26 | 9 | mg/dL |
| Creatinine | 0.6–1.3 | 0.7 | mg/dL |
| Uric acid | 2.6–6 (women) | | mg/dL |
| | 3.5–7.2 (men) | | |
| Calcium | 8.7–10.2 | 9.1 | mg/dL |
| Bilirubin | 0.2–1.3 | 0.8 | mg/dL |
| Ammonia ($NH_3$) | 9–33 | | $\mu$mol/L |
| SGPT (ALT) | 10–60 | 30 | U/L |
| SGOT (AST) | 5–40 | 22 | U/L |
| Alk phos | 98–251 | 156 | U/L |
| CPK | 26–140 (women) | 100 | U/L |
| | 38–174 (men) | | |
| LDH | 313–618 | 400 | U/L |
| CHOL | 140–199 | 220 | mg/dL |
| HDL–C | 40–85 (women) | 20 | mg/dL |
| | 37–70 (men) | | |
| VLDL | | | mg/dL |
| LDL | < 130 | 165 | mg/dL |
| LDL/HDL ratio | < 3.22 (women) | | |
| | < 3.55 (men) | | |
| Apo A | 101–199 (women) | | mg/dL |
| | 94–178 (men) | | |
| Apo B | 60–126 (women) | | mg/dL |
| | 63–133 (men) | | |
| TG | 35–160 | 178 | mg/dL |
| $T_4$ | 5.4–11.5 | | $\mu$g/dL |
| $T_3$ | 80–200 | | ng/dL |
| HbA$_{1C}$ | 4.8–7.8 | | % |

# UH UNIVERSITY HOSPITAL

NAME: Jack Nelson                      DOB: 7/22
AGE: 48                                SEX: M
PHYSICIAN: P. Phelps, MD

\*\*\*\*\*\*\*\*\*\*\*\*\*\*\*\*\*\*\*\*\*\*\*\*\*\*\*\*\*\*\*\*\*\*\*\*\*\*\*\*\*\*HEMATOLOGY\*\*\*\*\*\*\*\*\*\*\*\*\*\*\*\*\*\*\*\*\*\*\*\*\*\*\*\*\*\*\*\*\*\*\*\*\*\*\*\*\*

| | NORMAL | Admit 9/22 | UNITS |
|---|---|---|---|
| DAY: | | | |
| DATE: | | | |
| TIME: | | | |
| LOCATION: | | | |
| WBC | 4.3–10 | 5.6 | $\times\ 10^3/mm^3$ |
| RBC | 4–5 (women) | 5.2 | $\times\ 10^6/mm^3$ |
| | 4.5–5.5 (men) | | |
| HGB | 12–16 (women) | 14.0 | g/dL |
| | 13.5–17.5 (men) | | |
| HCT | 37–47 (women) | 40 | % |
| | 40–54 (men) | | |
| MCV | 84–96 | 85 | fL |
| MCH | 27–34 | 28 | pg |
| MCHC | 31.5–36 | 32 | % |
| RDW | 11.6–16.5 | | % |
| Plt Ct | 140–440 | | $\times\ 10^3$ |
| Diff TYPE | | | |
| % GRANS | 34.6–79.2 | | % |
| % LYM | 19.6–52.7 | | % |
| SEGS | 50–62 | | % |
| BANDS | 3–6 | | % |
| LYMPHS | 25–40 | | % |
| MONOS | 3–7 | | % |
| EOS | 0–3 | | % |
| TIBC | 65–165 (women) | | μg/dL |
| | 75–175 (men) | | |
| Ferritin | 18–160 (women) | | μg/dL |
| | 18–270 (men) | | |
| Vitamin $B_{12}$ | 100–700 | | pg/mL |
| Folate | 2–20 | | ng/mL |

## Case Questions

1.  How is acid produced and controlled within the gastrointestinal tract?

2.  What role does lower esophageal sphincter (LES) pressure play in the etiology of gastro-esophageal reflux disease?

3.  What are the complications of gastroesophageal reflux disease?

4.  What is *H. pylori,* and why did the physician want to biopsy the patient for *H. pylori?*

5.  Identify the patient's signs and symptoms that could suggest the diagnosis of gastroesophageal reflux disease.

6.  Describe the diagnostic tests performed for this patient.

7.  What risk factors does the patient present with that might contribute to his diagnosis? (Be sure to consider lifestyle, medical, and nutritional factors.)

8.  Calculate this patient's IBW, %IBW, %UBW, and BMI. What does this assessment of weight tell you? Could this contribute to his diagnosis?

9.  The MD has decreased this patient's dose of daily aspirin and recommended discontinuing his ibuprofen. Why? How do aspirin and NSAIDs affect gastroesophageal disease?

10. The MD has prescribed lansoprazole. What class of medication is this? What is the basic mechanism of the drug? What other drugs are available in this class? What other groups of medications are used to treat GERD?

11. Complete a computerized nutrient analysis for this patient's usual intake and 24-hour recall. Calculate this patient's energy and protein needs. How does his caloric intake compare to your calculated requirements?

**12.** What nutritional goals would you set for this patient within his care plan?

**13.** Would you recommend a special diet for him? Outline necessary modifications for him within his 24-hour recall, which you can use as a teaching tool.

| Food Item | Modification | Rationale |
|---|---|---|
| Crispix | | |
| Skim milk | | |
| Orange juice | | |
| Diet Pepsi | | |
| Grilled chicken sandwich | | |
| French fries | | |
| Iced tea | | |
| Chips | | |
| Beer | | |
| Fried chicken | | |
| Potato salad | | |
| Green bean casserole | | |
| Fruit salad | | |
| Baked beans | | |
| Milkshake | | |

**14.** What other components of lifestyle modification would you address to help in treating his disorder?

# Bibliography

American Dietetic Association. Gastroesophageal reflux disease. In *Manual of Clinical Dietetics,* 6th ed. Chicago: American Dietetic Association, 2000:401–405.

American Gastroenterological Association. (1996). *American Gastroenterological Association Medical Position Statement: Guidelines on the Use of Esophageal pH Recording.* Available at http://www.wbsaunders.com/gastro/policy/v110n6p1981.html. Accessed September 1, 2000.

Beyer PL. Medical nutrition therapy for upper gastrointestinal tract disorders. In Mahan LK, Escott-Stump S (eds.), *Krause's Food, Nutrition, and Diet Therapy,* 10th ed. Philadelphia: Saunders, 2000:649–650.

Bjorkman DJ. Current status of nonsteroidal anti-inflammatory drug (NSAID) use in the United States: risk factors and frequency of complications. *Am J Med.* 1999;107(6A):3S-10S.

Commission on Accreditation for Dietetics Education. Knowledge, skills, and competencies for dietitians. *Accreditation Manual for Dietetics Education Programs,* revised 4th ed. Chicago: American Dietetic Association, 2000:29–45.

Dixon MF. Pathophysiology of *Helicobacter pylori* infection. *Scand J Gastroenterol.* 1994;201(Suppl):7–10.

Escott-Stump S. *Nutrition and Diagnosis Related Care,* 4th ed. Baltimore: Williams & Wilkins, 1998.

McQuaid K. Dyspepsia. In Feldman M, Scharschmidt BF, Sleisenger MH (eds.), *Gastrointestinal and Liver Disease,* 6th ed. Philadelphia: Saunders, 1998.

Meyers BM, et al. Effect of red pepper and black pepper on the stomach. *Am J Gastroenterol.* 1987;82:211.

O'Connor HJ. Review article: *Helicobacter pylori* and gastro-oesophageal reflux disease: clinical implications and management. *Aliment-Pharmacol-Ther.* 1999;12(2):117–127.

Pagana KD, Pagana TJ. *Mosby's Diagnostic and Laboratory Test Reference.* St. Louis: Mosby Year Book, 1992.

Pronsky ZM. *Food-Medication Interactions,* 11th ed. Pottstown, PA: Food-Medication Interactions, 1999.

Rodriguez S, et al. Meal type affects heartburn severity. *Dig Dis Sci.* 1998;43:485.

Soll AH. Peptic ulcer and dyspepsia. *Clin Cornerstone.* 1999;1(5):29–41.

Talley NJ, et al. AGA technical review: evaluation of dyspepsia. *Gastroenterology.* 1998;114:582.

# Ulcer Disease, Part One: Medical Treatment*

*Introductory Level*

## Objectives

After completing this case, the student will be able to:

1. Delineate etiology and risk factors for development of ulcer disease.
2. Identify classes of medications used to treat ulcer disease and determine the possible drug–nutrient interactions.
3. Assess nutritional status through interpretation of laboratory and anthropometric data.
4. Use nutritional assessment data to diagnose malnutrition.
5. Apply appropriate medical nutrition therapy recommendations for ulcer disease.

Maria Rodriguez has been treated as an outpatient for her gastroesophageal reflux disease. Her increasing symptoms lead her to be admitted for further gastrointestinal workup. Her endoscopy reveals a 2-cm duodenal ulcer with generalized gastritis. Medical treatment for her ulcer disease is initiated.

---

*References for this case are the same as those for Case 12. See the bibliography on page 132.

# ADMISSION DATABASE

Name: Maria Rodriguez
DOB: 12/19  age 38
Physician: A. Gustaf, MD

| BED# 2 | DATE: 8/25 | TIME: 1400 | TRIAGE STATUS (ER ONLY): ☐ Red ☐ Yellow ☐ Green ☐ White |
|---|---|---|---|

**Initial Vital Signs**

| TEMP: 101.2 | RESP: 29 | | SAO2: |
|---|---|---|---|

**PRIMARY PERSON TO CONTACT:**
Name: Emilio Santiago (brother)
Home #: 504-212-7890
Work #: 504-213-4563

| HT: 5'2" | WT (lb): 110 usual wt 145 | B/P: 92/65 | PULSE: 72 |
|---|---|---|---|

**ORIENTATION TO UNIT:** ☒ Call light ☒ Television/telephone ☒ Bathroom ☒ Visiting ☒ Smoking ☒ Meals ☒ Patient rights/responsibilities

| LAST TETANUS 1 year ago | LAST ATE last pm | LAST DRANK last pm |
|---|---|---|

### CHIEF COMPLAINT/HX OF PRESENT ILLNESS

"The cimetidine I used helped my pain but now it is sharper. I can't eat and I've lost weight."

**PERSONAL ARTICLES:** (Check if retained/describe)
☐ Contacts ☐ R ☐ L   ☐ Dentures ☐ Upper ☐ Lower
☒ Jewelry: wedding band
☐ Other:

### ALLERGIES: Meds, Food, IVP Dye, Seafood: Type of Reaction

Codeine causes nausea and vomiting.

**VALUABLES ENVELOPE:** none
☐ Valuables instructions

### PREVIOUS HOSPITALIZATIONS/SURGERIES

for delivery of her two daughters only

**INFORMATION OBTAINED FROM:**
☒ Patient   ☐ Previous record
☐ Family   ☐ Responsible party

Signature  *Maria S. Rodriguez*

| Home Medications (including OTC) | | Codes: A = Sent home | B = Sent to pharmacy | | C = Not brought in |
|---|---|---|---|---|---|
| Medication | Dose | Frequency | Time of Last Dose | Code | Patient Understanding of Drug |
| cimetidine | 800 mg | daily | yesterday am | c | yes |
| | | | | | |
| | | | | | |
| | | | | | |
| | | | | | |
| | | | | | |
| | | | | | |
| | | | | | |
| | | | | | |
| | | | | | |

Do you take all medications as prescribed?   ☒ Yes   ☐ No   If no, why?

### PATIENT/FAMILY HISTORY

| | | |
|---|---|---|
| ☐ Cold in past two weeks | ☐ High blood pressure | ☐ Kidney/urinary problems |
| ☐ Hay fever | ☐ Arthritis | ☒ Gastric/abdominal pain/heartburn Patient |
| ☐ Emphysema/lung problems | ☐ Claustrophobia | ☐ Hearing problems |
| ☐ TB disease/positive TB skin test | ☐ Circulation problems | ☐ Glaucoma/eye problems |
| ☐ Cancer | ☐ Easy bleeding/bruising/anemia | ☐ Back pain |
| ☐ Stroke/past paralysis | ☐ Sickle cell disease | ☐ Seizures |
| ☐ Heart attack | ☐ Liver disease/jaundice | ☒ Other Father and grandfather had ulcer disease. |
| ☐ Angina/chest pain | ☐ Thyroid disease | |
| ☐ Heart problems | ☒ Diabetes Maternal grandmother | |

### RISK SCREENING

Have you had a blood transfusion?   ☐ Yes   ☒ No
Do you smoke?   ☒ Yes   ☐ No
If yes, how many pack(s)  1.5/day for   15  years
Does anyone in your household smoke?   ☒ Yes   ☐ No
Do you drink alcohol?   ☐ Yes   ☒ No
If yes, how often?_____   How much?_____
When was your last drink?____/____/____
Do you take any recreational drugs?   ☐ Yes   ☒ No
If yes, type:_____   Route_____
Frequency:_____   Date last used:____/____/____

**FOR WOMEN Ages 12–52**

Is there any chance you could be pregnant?   ☐ Yes   ☒ No
If yes, expected date (EDC):
Gravida/Para: 2/2

**ALL WOMEN**

Date of last Pap smear:  /02 /this year
Do you perform regular breast self-exams?   ☒ Yes   ☐ No

**ALL MEN**

Do you perform regular testicular exams?   ☐ Yes   ☐ No

Additional comments: Daughters are with patient's brother. Patient has been a widow for the past 3 years.

x *Abigail Kidd, RN*
Signature/Title

**Client name:** Maria Rodriguez
**DOB:** 12/19
**Age:** 38
**Sex:** Female
**Education:** Associate's degree
**Occupation:** Works in computer programming for local firm
**Hours of work:** M–F 9–5
**Household members:** 2 daughters ages 12 and 14, in good health
**Ethnic background:** Hispanic
**Religious affiliation:** Catholic
**Referring physician:** Anna Gustaf, MD (gastroenterologist)

**Chief complaint:**
"I have had problems with chronic indigestion for the last year. The cimetidine helped for a while but now the pain is much worse than it ever has been. It is much sharper, and none of my medications seem to be helping. I can't seem to eat anything. I have lost a lot of weight. I also seem short of breath."

**Patient history:**
*Onset of disease:* Diagnosed with GERD approx 11 months ago
*Type of Tx:* Treated with histamine 2-receptor antagonist: cimetidine—800 mg/day
*PMH:* Gravida 2 para 2. No other significant history
*Meds:* Cimetidine 800 mg qd
*Smoker:* Yes
*Family Hx: What?* DM, PUD  *Who?* DM: maternal grandmother, PUD: father and grandfather

**Physical exam:**
*General appearance:* 38-year-old Hispanic female—thin, pale, and in obvious distress
*Vitals:* Temp 100.3°F, BP 95/60, HR 70 bpm, RR 28 bpm
*Heart:* Regular rate and rhythm, heart sounds normal
*HEENT:* Noncontributory
*Genitalia:* Normal. Rectal: Hard stool in vault; Heme $\oplus$
*Neurologic:* Alert and oriented
*Extremities:* Noncontributory
*Skin:* Warm and dry to touch
*Chest/lungs:* Rapid breath sounds; lungs clear
*Abdomen:* Tender with guarding, decreased bowel sounds

**Nutrition Hx:**
*General:* Patient describes appetite as poor. States that she is afraid to eat because it makes the pain worse. Specific food intolerances include anything fried or "spicy," coffee, and chocolate. Patient relates her usual weight to be about 145 lb. The last time she weighed herself was 6 weeks ago. Her admission weight is 110 lb.

*Usual dietary intake:* (prior to current illness)

*AM:*           Coffee, dry toast. On weekends, cooked large breakfasts for family, which included omelets, rice or grits, or pancakes, waffles, fruit.

*Lunch:*        Sandwich from home, fruit, cookies

*Dinner:*       Rice, some type of meat, fresh vegetables, coffee. Has previously drunk 8–10 c coffee daily. 1–2 sodas each day.

*24-hr recall:* Has been NPO since admission. Diet to be advanced to clear liquids today.

*Food allergies/intolerances/aversions:* See Nutrition Hx.

*Previous MNT?* No

*Food purchase/preparation:* Self and daughters

*Vit/min intake:* None

**Tx plan:**

S/p endoscopy that revealed 2-cm duodenal ulcer with generalized gastritis. Biopsy positive for *Helicobacter pylori*. *Rx:* 14-day course of bismuth subsalicylate 525 mg qid, metronidazole 250 mg qid, tetracycline 500 mg qid. Omeprazole 20 mg bid × 28 days. Nutrition consult.

## Case Questions

1.  Identify this patient's risk factors for ulcer disease.

2.  How has smoking been related to ulcer disease?

3.  How is *H. pylori* related to ulcer disease?

4.  This patient was prescribed four different medications for treatment of her infection. What are the current recommendations for treatment of *H. pylori* infection?

| Drug | Action | Nutrition/GI Side Effect |
|---|---|---|
| Metronidazole | | |
| Tetracycline | | |
| Bismuth subsalicylate | | |
| Omeprazole | | |

5.  What are the possible drug–nutrient side effects from Mrs. Rodriguez's prescribed regimen? Which drug–nutrient side effects are most pertinent to her current nutritional status?

6.  Assess this patient's available anthropometric data. Calculate IBW, %IBW, UBW, %UBW, and BMI. Which of these is the most pertinent in identifying the patient's nutrition risk? Why?

7.  What other anthropometric measures could be used to further confirm her nutritional status?

8. Using her admission chemistry and hematology values, which biochemical measures are abnormal? Explain. (Refer to the laboratory forms in Case 12.)

   a. Which values can be used to further assess her nutritional status? Explain.

   b. Which laboratory measures are related to her diagnosis of duodenal ulcer? Why would they be abnormal?

9. Do you think this patient is malnourished? If so, why? What criteria can be used to diagnose malnutrition? Within what category does this patient fit?

10. Estimate this patient's energy and protein requirements.

11. This patient has been NPO, and her diet is to be advanced to clear liquids today. What might be the RD's recommendations for medical nutrition therapy as the diet is advanced? What is your rationale?

12. This patient has had a very poor appetite. What interventions could be attempted in order to increase her caloric and protein intake?

13. What nutrition education should this patient receive prior to discharge?

14. Do any lifestyle issues need to be addressed with this patient? Explain.

# Case 12

# Ulcer Disease, Part Two: Surgical Treatment

*Introductory Level*

## Objectives

After completing this case, the student will be able to:

1. Identify surgical procedures used to treat refractory ulcer disease and implications for maintenance of nutritional status.
2. Identify appropriate medical nutrition therapy recommendations for prevention of dumping syndrome.
3. Assess nutritional status through interpretation of laboratory and anthropometric data.
4. Demonstrate ability to calculate enteral nutrition formulations.
5. Demonstrate ability to evaluate standard enteral nutritional regimen.

Mrs. Rodriguez is readmitted through the emergency room with hematemesis, vomiting, and diarrhea two days after her previous discharge. She undergoes a gastrojejunostomy to treat her perforated duodenal ulcer.

# UNIVERSITY HOSPITAL

## ADMISSION DATABASE

Name: Maria Rodriguez
DOB: 12/19  age 38
Physician: A. Gustaf, MD

| BED# 1 | DATE: 8/30 | TIME: 1700 | TRIAGE STATUS (ER ONLY): ☐ Red ☒ Yellow ☐ Green ☐ White |
|---|---|---|---|

**Initial Vital Signs**

| TEMP: 102 | RESP: 32 | SAO2: | |
|---|---|---|---|

| HT: 5'2" | WT (lb): 110 usual wt 145 | B/P: 78/60 | PULSE: 68 |
|---|---|---|---|

| LAST TETANUS 1 year ago | LAST ATE yesterday | LAST DRANK water 1 hour ago |
|---|---|---|

PRIMARY PERSON TO CONTACT:
Name: Emilio Santiago (Brother)
Home #: 504-212-7090
Work #: 504-213-4563

ORIENTATION TO UNIT: ☒ Call light ☒ Television/telephone ☒ Bathroom ☒ Visiting ☒ Smoking ☒ Meals ☒ Patient rights/responsibilities

### CHIEF COMPLAINT/HX OF PRESENT ILLNESS

"I was just here. I have been vomiting and I have diarrhea.

My pain is terrible. There is blood in my vomit and in my diarrhea."

PERSONAL ARTICLES: (Check if retained/describe)
☐ Contacts ☐ R ☐ L          ☐ Dentures ☐ Upper ☐ Lower
☒ Jewelry: wedding band
☐ Other:

### ALLERGIES: Meds, Food, IVP Dye, Seafood: Type of Reaction

codeine causes nausea and vomiting.

VALUABLES ENVELOPE:
☐ Valuables instructions

### PREVIOUS HOSPITALIZATIONS/SURGERIES

for delivery of her two daughters only

d/c 2 days previous after diagnosis of duodenal ulcer

INFORMATION OBTAINED FROM:
☒ Patient          ☒ Previous record
☒ Family           ☐ Responsible party

Signature  *Maria Rodriguez*

### Home Medications (including OTC)

Codes: A=Sent home     B=Sent to pharmacy     C=Not brought in

| Medication | Dose | Frequency | Time of Last Dose | Code | Patient Understanding of Drug |
|---|---|---|---|---|---|
| bismuth subsalicylate | 525 mg | qid | this am | | yes |
| metronidazole | 250 mg | qid | this am | c | yes |
| tetracycline | 500 mg | qid | this am | | yes |
| omeprazole | 20 mg | bid | this am | c | yes |
| | | | | | |
| | | | | | |
| | | | | | |
| | | | | | |
| | | | | | |
| | | | | | |
| | | | | | |

Do you take all medications as prescribed?  ☒ Yes  ☐ No  If no, why?

### PATIENT/FAMILY HISTORY

| | | |
|---|---|---|
| ☐ Cold in past two weeks | ☐ High blood pressure | ☐ Kidney/urinary problems |
| ☐ Hay fever | ☐ Arthritis | ☒ Gastric/abdominal pain/heartburn Patient |
| ☐ Emphysema/lung problems | ☐ Claustrophobia | ☐ Hearing problems |
| ☐ TB disease/positive TB skin test | ☐ Circulation problems | ☐ Glaucoma/eye problems |
| ☐ Cancer | ☐ Easy bleeding/bruising/anemia | ☐ Back pain |
| ☐ Stroke/past paralysis | ☐ Sickle cell disease | ☐ Seizures |
| ☐ Heart attack | ☐ Liver disease/jaundice | ☐ Other |
| ☐ Angina/chest pain | ☐ Thyroid disease | |
| ☐ Heart problems | ☒ Diabetes Maternal grandmother | |

### RISK SCREENING

Have you had a blood transfusion?  ☐ Yes  ☒ No
Do you smoke?  ☒ Yes  ☐ No
If yes, how many pack(s)  1.5/day for  15 years
Does anyone in your household smoke?  ☒ Yes  ☐ No
Do you drink alcohol?  ☐ Yes  ☒ No
If yes, how often?_____  How much?
When was your last drink?_____/_____/_____
Do you take any recreational drugs?  ☐ Yes  ☒ No
If yes, type:_____  Route
Frequency:_____  Date last used:_____/_____/_____

**FOR WOMEN Ages 12–52**

Is there any chance you could be pregnant?  ☐ Yes  ☒ No
If yes, expected date (EDC):
Gravida/Para: 2/2

**ALL WOMEN**

Date of last Pap smear:  /02 /this year
Do you perform regular breast self-exams?  ☐ Yes  ☒ No

**ALL MEN**

Do you perform regular testicular exams?  ☐ Yes  ☐ No

Additional comments:

✗ *Sophia McMillan, RN*
Signature/Title

**Client name:**  Maria Rodriguez
**DOB:**  12/19
**Age:**  38
**Sex:**  Female
**Education:**  Associate's degree
**Occupation:**  Works in computer programming for local firm
**Hours of work:**  M–F 9–5
**Household members:**  2 daughters ages 12 and 14, in good health
**Ethnic background:**  Hispanic
**Religious affiliation:**  Catholic
**Referring physician:**  Anna Gustaf, MD (gastroenterologist)

## Chief complaint:

"I was just here. I have been vomiting and I have diarrhea. My pain is terrible. There is blood in my vomit and in my diarrhea."

## Patient history:

*Onset of disease:* Diagnosed with GERD approx. 11 months ago; diagnosed with duodenal ulcer 3 days ago
*Type of Tx:* 14-day course of bismuth subsalicylate 525 mg qid; metronidazole 250 mg qid, tetracycline 500 mg qid. Omeprazole 20 mg bid × 28 days
*PMH:* Gravida 2 para 2. No other significant history
*Meds:* See above.
*Smoker:* Yes
*Family Hx:* What? DM, PUD  Who? DM: maternal grandmother, PUD: father and grandfather

## Physical exam:

*General appearance:* 38-year-old Hispanic female—thin, pale and in acute distress
*Vitals:* Temp: 101.3°F, BP 78/60, HR 68 bpm, RR 32 bpm
*Heart:* Regular rate and rhythm, heart sounds normal
*HEENT:* Noncontributory
*Genitalia:* Normal
*Rectal:* Not performed
*Neurologic:* Alert and oriented
*Extremities:* Noncontributory
*Skin:* Warm and dry to touch
*Chest/lungs:* Rapid breath sounds, lungs clear
*Abdomen:* Tender with guarding, absent bowel sounds

## Tx plan:

Surgical consult for possible perforated duodenal ulcer. On 8/31, a gastrojejunostomy was completed.

**Tx:**

Patient is now s/p gastrojejunostomy secondary to perforated duodenal ulcer. Feeding jejunostomy was placed during surgery, and patient is receiving Vital HN @ 25 cc/hr via continuous drip. Orders have been left to advance the enteral feeding to 50 cc/hr. She is receiving only ice chips by mouth.

**Nutrition Hx:**

Patient relates that she understands about the feeding she is receiving through her tube. She explained that she was unable to eat at all after her last admission and wonders how long it will be before she can eat again. Her physicians have told her they might like her to try something by mouth in the next few days.

# U<sub>H</sub> *UNIVERSITY HOSPITAL*

NAME: Maria Rodriguez                    DOB: 12/19
AGE: 38                                  SEX: F
PHYSICIAN: M. Gustaf, MD

\*\*\*\*\*\*\*\*\*\*\*\*\*\*\*\*\*\*\*\*\*\*\*\*\*\*\*\*\*\*\*\*\*\*\*\*\*\*\*\*\*\*\*\*\*\*CHEMISTRY\*\*\*\*\*\*\*\*\*\*\*\*\*\*\*\*\*\*\*\*\*\*\*\*\*\*\*\*\*\*\*\*\*\*\*\*\*\*\*\*\*\*\*\*\*

| DAY: | | Admit 1 | Admit 2 | Postop Day 3 | |
|------|--------|---------|---------|--------------|------|
| DATE: | | 8/25 | 8/30 | 9/3 | |
| TIME: | | 1500 | 0800 | 0600 | |
| LOCATION: | | | | | |
| | NORMAL | | | | UNITS |
| Albumin | 3.6–5 | 3.4 | 3.0 | 3.3 | g/dL |
| Total protein | 6–8 | 5.9 | 5.5 | 6.0 | g/dL |
| Prealbumin | 19–43 | 18 | 15 | 14 | mg/dL |
| Transferrin | 200–400 | 380 | 425 | 419 | mg/dL |
| Sodium | 135–155 | 138 | 141 | 140 | mmol/L |
| Potassium | 3.5–5.5 | 3.7 | 4.5 | 4.2 | mmol/L |
| Chloride | 98–108 | 99 | 103 | 101 | mmol/L |
| $PO_4$ | 2.5–4.5 | 3.2 | 3.7 | 3.5 | mmol/L |
| Magnesium | 1.6–2.6 | 1.7 | 1.9 | 1.7 | mmol/L |
| Osmolality | 275–295 | 286 | 295 | 292 | mmol/kg $H_2O$ |
| Total $CO_2$ | 24–30 | 25 | 26 | 24 | mmol/L |
| Glucose | 70–120 | 92 | 80 | 128 | mg/dL |
| BUN | 8–26 | 11 | 24 | 15 | mg/dL |
| Creatinine | 0.6–1.3 | 0.8 | 1.1 | 0.9 | mg/dL |
| Uric acid | 2.6–6 (women) | | | | mg/dL |
| | 3.5–7.2 (men) | | | | |
| Calcium | 8.7–10.2 | 8.8 | 9.0 | 8.7 | mg/dL |
| Bilirubin | 0.2–1.3 | 0.8 | 1.3 | 0.6 | mg/dL |
| Ammonia ($NH_3$) | 9–33 | 10 | 11 | 10 | μmol/L |
| SGPT (ALT) | 10–60 | 22 | 30 | 24 | U/L |
| SGOT (AST) | 5–40 | 15 | 31 | 17 | U/L |
| Alk phos | 98–251 | 131 | 145 | 133 | U/L |
| CPK | 26–140 (women) | | | | U/L |
| | 38–174 (men) | | | | |
| LDH | 313–618 | | | | U/L |
| CHOL | 140–199 | | | | mg/dL |
| HDL-C | 40–85 (women) | | | | mg/dL |
| | 37–70 (men) | | | | |
| VLDL | | | | | mg/dL |
| LDL | < 130 | | | | mg/dL |
| LDL/HDL ratio | < 3.22 (women) | | | | |
| | < 3.55 (men) | | | | |
| Apo A | 101–199 (women) | | | | mg/dL |
| | 94–178 (men) | | | | |
| Apo B | 60–126 (women) | | | | mg/dL |
| | 63–133 (men) | | | | |
| TG | 35–160 | | | | mg/dL |
| $T_4$ | 5.4–11.5 | | | | μg/dL |
| $T_3$ | 80–200 | | | | ng/dL |
| $HbA_{1C}$ | 4.8–7.8 | | | | % |

# U$_H$ UNIVERSITY HOSPITAL

NAME: Maria Rodriguez               DOB: 12/19
AGE: 38                             SEX: F
PHYSICIAN: M. Gustaf, MD

**********************************************HEMATOLOGY**********************************************

| | NORMAL | Admit 1 | Admit 2 | Postop Day 3 | UNITS |
|---|---|---|---|---|---|
| DAY: | | Admit 1 | Admit 2 | Postop Day 3 | |
| DATE: | | 8/25 | 8/30 | 9/3 | |
| TIME: | | | | | |
| LOCATION: | | | | | |
| WBC | 4.3–10 | 15 | 16.3 | 12.5 | $\times\ 10^3/mm^3$ |
| RBC | 4–5 (women) | | | | $\times\ 10^6/mm^3$ |
| | 4.5–5.5 (men) | | | | |
| HGB | 12–16 (women) | 10.5 | 11.2 | 10.2 | g/dL |
| | 13.5–17.5 (men) | | | | |
| HCT | 37–47 (women) | 30 | 33 | 31 | % |
| | 40–54 (men) | | | | |
| MCV | 84–96 | 85 | 91 | 86 | fL |
| MCH | 27–34 | | | | pg |
| MCHC | 31.5–36 | 30 | 31 | 28.5 | % |
| RDW | 11.6–16.5 | 20.1 | 19.5 | 22 | % |
| Plt Ct | 140–440 | 330 | 345 | 356 | $\times\ 10^3$ |
| Diff TYPE | | | | | |
| % GRANS | 34.6–79.2 | | | | % |
| % LYM | 19.6–52.7 | | | | % |
| SEGS | 50–62 | 83 | 87 | 78 | % |
| BANDS | 3–6 | 5 | 6 | 4 | % |
| LYMPHS | 25–40 | 10 | 12 | 22 | % |
| MONOS | 3–7 | 3 | 5 | 4 | % |
| EOS | 0–3 | 1 | 2 | 3 | % |
| TIBC | 65–165 (women) | 221 | 241 | 232 | µg/dL |
| | 75–175 (men) | | | | |
| Ferritin | 18–160 (women) | | | | µg/dL |
| | 18–270 (men) | | | | |
| Vitamin B$_{12}$ | 100–700 | | | | pg/mL |
| Folate | 2–20 | | | | ng/mL |
| Total T cells | 812–2318 | | | | mm$^3$ |
| T-helper cells | 589–1505 | | | | mm$^3$ |
| T-suppressor cells | 325–997 | | | | mm$^3$ |
| PT | 11–13 | | | | sec |

# UNIVERSITY HOSPITAL

Name: Maria Rodriguez
Physician: A. Gustaf, MD

## PATIENT CARE SUMMARY SHEET

Date: 9/3    Room: 1145    Wt Yesterday: 110 lb    Today: 111 lb    Post/dialysis ___ lb

| Temp °F | NIGHTS | | | | | | | | DAYS | | | | | | | | EVENINGS | | | | | | | |
|---|---|---|---|---|---|---|---|---|---|---|---|---|---|---|---|---|---|---|---|---|---|---|---|---|
| | 00 | 01 | 02 | 03 | 04 | 05 | 06 | 07 | 08 | 09 | 10 | 11 | 12 | 13 | 14 | 15 | 16 | 17 | 18 | 19 | 20 | 21 | 22 | 23 |
| 105 | | | | | | | | | | | | | | | | | | | | | | | | |
| 104 | | | | | | | | | | | | | | | | | | | | | | | | |
| 103 | | | | | | | | | | | | | | | | | | | | | | | | |
| 102 | | | | | | | | | | | | | | | | | | | | | | | | |
| 101 | | | | | | | | | | | | | | | | | | | | | | | | |
| 100 | | | | | | | × | | | | | | | | | | | | | | | | | |
| 99 | | | | | | | | | | | | | | | | × | | | | | | | | × |
| 98 | | | | | | | | | | | | | | | | | | | | | | | | |
| 97 | | | | | | | | | | | | | | | | | | | | | | | | |
| 96 | | | | | | | | | | | | | | | | | | | | | | | | |

| | NIGHTS | DAYS | EVENINGS |
|---|---|---|---|
| Pulse | 82 | 78 | 80 |
| Respiration | 28 | 27 | 28 |
| BP | 109/78 | 110/82 | 115/72 |
| Blood Glucose | 122 | 115 | 82 |
| Appetite/Assist | NPO | NPO | NPO |
| INTAKE | 175 cc | 200 cc | 190 cc |
| Oral | | | |
| IV | 380 | 400 | 400 |

| TF Formula/Flush | 00 | 01 | 02 | 03 | 04 | 05 | 06 | 07 | 08 | 09 | 10 | 11 | 12 | 13 | 14 | 15 | 16 | 17 | 18 | 19 | 20 | 21 | 22 | 23 |
|---|---|---|---|---|---|---|---|---|---|---|---|---|---|---|---|---|---|---|---|---|---|---|---|---|
| | 25 | 25 | 25 | 25 | 25 | 25 | 25 | 25/50 | 25 | | 25 | | 25 | 25 | 25 | 25/50 | | 25 | | | | 25 | 25 | 25/50 |

| | NIGHTS | DAYS | EVENINGS |
|---|---|---|---|
| Shift Total | 805 | 800 | 740 |
| OUTPUT | | | |
| Void | | | |
| Cath. | 480 | 550 | |
| Emesis | | | |
| BM | 128 | 200 | |
| Drains | 220 | 275 | 320 |
| Shift Total | 828 | 975 | 920 |
| Gain | | | +180 |
| Loss | −18 | −175 | |
| Signatures | L. Smith, RN | M. Taylor, RN | N. Parrish, RN |

## Case Questions

1. Explain the surgical procedure that the patient received.

2. What are the potential nutritional complications after this surgical procedure?

3. The most common physical side effects from this surgery are the development of early or late dumping syndrome. Describe each of these syndromes, including symptoms the patient might experience, etiology, and treatment.

4. This patient was started on an enteral feeding postoperatively. Why do you think this decision was made?

5. What type of enteral formula is Vital HN? Is it an appropriate choice for this patient?

6. Why was the enteral formula started at 25 cc/hr?

7. Is the current enteral prescription meeting this patient's nutritional needs? Compare her energy and protein requirements to what is provided by the formula. If her needs are not met, what should be the goal for her enteral support?

8. What would the RD assess to monitor tolerance to the enteral feeding?

9. Go to the patient care summary sheet. For postoperative day 2, how much enteral nutrition did the patient receive? How does this compare to what was prescribed?

10. When evaluating the patient care summary sheet, you notice the patient has gained 1 pound in 24 hours. Should you address this in your nutrition note as an improvement in nutritional status?

**11.**   As this patient is advanced to solid food, what modifications in diet would the RD address? Why? What would be a typical first meal for this patient?

**12.**   What other considerations would you give to Mrs. Rodriguez to maximize her tolerance to solid food?

**13.**   Mrs. Rodriguez asks for you to come to her room because she is concerned that she may have to follow this diet forever. What might you tell her?

**14.**   Should Mrs. Rodriguez be on any type of vitamin/mineral supplementation at home when she is discharged? Would you make any recommendations for specific types?

**15.**   Why might Mrs. Rodriguez be at risk for iron deficiency anemia, pernicious anemia, and/or megaloblastic anemia secondary to folate deficiency and/or poor vitamin $B_{12}$ absorption?

**16.**   Will the oral vitamin/mineral supplement be adequate to prevent the anemias discussed in Question 15? Explain.

# Bibliography

American Dietetic Association. Nutrition management of gastroesophageal reflux. In *Manual of Clinical Dietetics.* Chicago: American Dietetic Association, 2000:401–405.

Baron JH. Peptic ulcer. *Mount Sinai J Med.* 2000;67:58–62.

Beyer PL. Medical nutrition therapy for upper gastrointestinal tract disorders. In Mahan LK, Escott-Stump S (eds.), *Krause's Food, Nutrition, and Diet Therapy,* 10th ed. Philadelphia: Saunders, 2000:649–650.

Brown LF, Wilson DE. Gastroduodenal ulcers: causes, diagnosis, prevention and treatment. *Comprehensive Therapy.* 1999;25:30–38.

Commission on Accreditation for Dietetics Education. Knowledge, skills, and competencies for dietitians. *Accreditation Manual for Dietetics Education Programs,* rev. 4th ed. Chicago: American Dietetic Association, 2000.

Escott-Stump S. *Nutrition and Diagnosis Related Care,* 4th ed. Baltimore: Williams & Wilkins, 1998.

Graham DY, et al. Recognizing peptic ulcer disease: keys to clinical and laboratory diagnosis. *Postgrad Med.* 1999;113–116,121–123,127–128.

*Helicobacter pylori* in peptic ulcer disease. *NIH Consensus Statement* 1994;12(1,January 7–9):1–23.

Jamieson GG. Current status of indications for surgery in peptic ulcer disease. *World J Surg.* 2000;24:256–258.

Kiyota K, Habu Y, Sugano Y, Inokuchi H, Mizuno S, Kimoto K, Kawai K. Comparison of 1-week and 2-week triple therapy with omeprazole amoxicillin, and clarithromycin in peptic ulcer patients with *Helicobacter pylori* infection: results of a randomized trial. *J Gastroenterol.* 1999;34 (Suppl 11):76–79.

Navuluri R, Yue S. Case report. Understanding peptic ulcer disease pharmacotherapeutics. *Nurse Pract.* 1999;24:128,130–132.

Qureshi WA, Graham DY. Diagnosis and management of *Helicobacter pylori* infection. *Clin Cornerstone.* 1999;1:18–28.

Soll AH. Medical treatment of peptic ulcer disease: practice guidelines. *JAMA.* 1996;275(8,February 28): 622–629.

Spechler SJ. Epidemiology of natural history of gastro-oesophageal reflux disease. *Digestion.* 1992;51 (1 suppl):24–29.

# Case 13

## Infectious Diarrhea with Resulting Dehydration

*Introductory Level*

### Objectives

After completing this case, the student will be able to:

1. Identify effects of infection and dehydration on normal physiology.
2. Interpret laboratory parameters for nutritional significance.
3. Demonstrate ability to determine nutrient, fluid, and electrolyte requirements for children.
4. Prescribe appropriate medical nutrition therapy for dehydration and malabsorption resulting from diarrhea.
5. Develop and implement transitional feeding plans.

Sam Jones is admitted to the pediatric unit of University Hospital with severe dehydration secondary to diarrhea. His medical evaluation reveals a diagnosis of *E. coli* 0157:H7.

# UNIVERSITY HOSPITAL

## ADMISSION DATABASE

Name: Sam Jones
DOB: 4/13  age 8
Physician: C. Fraser, MD

| BED#<br>2 | DATE:<br>7/22 | TIME:<br>1500 | TRIAGE STATUS (ER ONLY):<br>☐ Red  ☐ Yellow  ☐ Green  ☐ White |
|---|---|---|---|

**Initial Vital Signs**

| TEMP:<br>102.3 | RESP:<br>17 | SAO2: |
|---|---|---|

| HT:<br>4′1″ | WT (lb): 50<br>UBW 54 | B/P:<br>90/70 | PULSE:<br>72 |
|---|---|---|---|

| LAST TETANUS<br>3 yrs ago | LAST ATE<br>last am | LAST DRANK<br>today at noon |
|---|---|---|

**PRIMARY PERSON TO CONTACT:**
Name: Violet and Philip Jones
Home #: 312-256-7892
Work #: 312-257-7721

ORIENTATION TO UNIT: ☒ Call light  ☒ Television/telephone
☒ Bathroom  ☒ Visiting  ☒ Smoking  ☒ Meals
☒ Patient rights/responsibilities

### CHIEF COMPLAINT/HX OF PRESENT ILLNESS

"We thought he had the flu or some kind of virus. He has had diarrhea for over 4 days now. He just has not gotten any better."

PERSONAL ARTICLES: (Check if retained/describe)
☐ Contacts ☐ R ☐ L      ☐ Dentures ☐ Upper ☐ Lower
☐ Jewelry:
☒ Other: glasses

### ALLERGIES: Meds, Food, IVP Dye, Seafood: Type of Reaction

bee stings–respiratory difficulty and large amounts of swelling

VALUABLES ENVELOPE: no
☐ Valuables instructions

### PREVIOUS HOSPITALIZATIONS/SURGERIES

INFORMATION OBTAINED FROM:
☒ Patient          ☐ Previous record
☒ Family           ☐ Responsible party

Signature  *Violet Jones (for Sam Jones)*

### Home Medications (including OTC)    Codes: A=Sent home    B=Sent to pharmacy    C=Not brought in

| Medication | Dose | Frequency | Time of Last Dose | Code | Patient Understanding of Drug |
|---|---|---|---|---|---|
|  |  |  |  |  |  |
|  |  |  |  |  |  |
|  |  |  |  |  |  |
|  |  |  |  |  |  |
|  |  |  |  |  |  |
|  |  |  |  |  |  |
|  |  |  |  |  |  |
|  |  |  |  |  |  |
|  |  |  |  |  |  |
|  |  |  |  |  |  |
|  |  |  |  |  |  |

Do you take all medications as prescribed?  ☐ Yes  ☐ No    If no, why?

### PATIENT/FAMILY HISTORY

| | | |
|---|---|---|
| ☐ Cold in past two weeks | ☒ High blood pressure Father | ☐ Kidney/urinary problems |
| ☐ Hay fever | ☐ Arthritis | ☐ Gastric/abdominal pain/heartburn |
| ☐ Emphysema/lung problems | ☐ Claustrophobia | ☐ Hearing problems |
| ☐ TB disease/positive TB skin test | ☐ Circulation problems | ☐ Glaucoma/eye problems |
| ☐ Cancer | ☐ Easy bleeding/bruising/anemia | ☐ Back pain |
| ☐ Stroke/past paralysis | ☐ Sickle cell disease | ☐ Seizures |
| ☐ Heart attack | ☐ Liver disease/jaundice | ☐ Other |
| ☐ Angina/chest pain | ☐ Thyroid disease | |
| ☐ Heart problems | ☐ Diabetes | |

### RISK SCREENING

Have you had a blood transfusion?  ☐ Yes  ☒ No
Do you smoke?  ☐ Yes  ☐ No
If yes, how many pack(s) /day for years
Does anyone in your household smoke?  ☐ Yes  ☒ No
Do you drink alcohol?  ☐ Yes  ☐ No
If yes, how often?_____  How much?
When was your last drink?_____/_____/_____
Do you take any recreational drugs?  ☐ Yes  ☐ No
If yes, type:_____  Route
Frequency:_____  Date last used:_____/_____/_____

**FOR WOMEN Ages 12–52**

Is there any chance you could be pregnant?  ☐ Yes  ☐ No
If yes, expected date (EDC):
Gravida/Para:

**ALL WOMEN**

Date of last Pap smear:
Do you perform regular breast self-exams?  ☐ Yes  ☐ No

**ALL MEN**

Do you perform regular testicular exams?  ☐ Yes  ☐ No

Additional comments:

✗ *Gina Miller, RN*
Signature/Title

**Client name:** Sam Jones
**DOB:** 4/13
**Age:** 8
**Sex:** Male
**Education:** Less than high school   *What grade/level?* 3rd grade
**Occupation:** Student
**Hours of work:** N/A
**Household members:** Father age 48, mother age 39, brother age 11, sister age 10—all well
**Ethnic background:** African American
**Religious affiliation:** African Methodist Episcopal church
**Referring physician:** C. Fraser, MD

**Chief complaint:**
"We thought he had the flu or some kind of virus. He has had diarrhea for over 4 days now. He just has not gotten any better. We are really worried—he just seems so weak and listless."

**Patient history:**
Parents describe that the family spent last weekend at an amusement water park. Sam, their 8-year-old son, began having diarrhea and running a fever Sunday morning. They decided to cut their weekend trip short, thinking that he had gotten the flu or some type of viral illness. Now, four days later, he is still running a fever and the diarrhea has gotten worse instead of better. They have been giving him soft foods, soups, and liquids since he got sick. He has had very little to eat in the last 24 hours, and parents state that it has been difficult for him to even drink anything. They also note that there seems to be blood in the diarrhea now. His parents estimate that Sam has had anywhere from 8 to 15 diarrhea episodes in the past 24 hours. The other two children also have had diarrhea but have since improved. They have talked with their pediatrician several times, but Sam has not been seen since by his MD. They have given Sam over-the-counter meds for diarrhea, including Pepto-Bismol and Kaopectate.
*Onset of disease:* Five days previous
*Type of Tx:* None at present
*Meds:* Pepto-Bismol and Kaopectate
*Smoker:* No
*Family Hx: What?* HTN   *Who?* Father

**Physical exam:**
*General appearance:* Lethargic 8-year-old African-American male
*Vitals:* Temp 102.3°F, BP 90/70 (orthostatic ↓ 75/62), HR 92 bpm, RR 17 bpm
*Heart:* Moderately elevated pulse
*HEENT:*
   *Eyes:* Sunken; sclera clear without evidence of tears
   *Ears:* Clear
   *Nose:* Dry mucous membranes
   *Throat:* Dry mucous membranes, no inflammation
*Genitalia:* Unremarkable
*Neurologic:* Alert, oriented × 3; irritable
*Extremities:* No joint deformity or muscle tenderness. No edema

*Skin:* Warm, dry; reduced capillary refill (approximately 2 seconds)
*Chest/lungs:* Clear to auscultation and percussion
*Abdomen:* Tender, nondistended, minimal bowel sounds

## Nutrition Hx:

*General:* Prior to admission, good appetite with consumption of a wide variety of foods except for vegetables

*Usual dietary intake:*

| | |
|---|---|
| *AM:* | Cereal, toast or bagel, juice |
| *Lunch:* | Sandwich, chips, fruit, cookies, milk |
| *Dinner:* | All meats, pasta or rice, fruit, milk |
| *Snacks:* | Juice, fruit, cookies, crackers |
| *24-hour recall:* | Parents estimate that child has had less than 6 oz of Gatorade in past 24 hours and that has had to be strongly encouraged through sips. |

*Food allergies/intolerances/aversions:* NKA
*Previous MNT?* No
*Food purchase/preparation:* Parent(s)
*Vit/min intake:* Flintstones vitamin daily

## Dx:

Moderate dehydration R/O bacterial vs. viral gastroenteritis

## Tx plan:

D5 ¼ Normal saline with 40 mEq KCl/L 20 mL per kg/hr for 3 hours. Increase to 100 mL/kg over next 7 hours, then decrease to 100 mL/hr. Begin Pedialyte 30 cc q hr as tolerated. Fecal smear for RBC and leukocytes. Stool culture.

## Hospital course:

Fecal smear with gross blood and leukocytes. Diagnosis of *E. coli* 0157:H7 secondary from contamination at water park. Local health department follow-up indicates additional 15 cases from visitors at park during same time period. Most likely source of contamination is playground fountain. Investigation of filtration and chlorination system is underway.

U<sub>H</sub> *UNIVERSITY HOSPITAL*

NAME: Sam Jones                                 DOB: 4/13
AGE: 8                                          SEX: M
PHYSICIAN: C. Fraser, MD

\*\*\*\*\*\*\*\*\*\*\*\*\*\*\*\*\*\*\*\*\*\*\*\*\*\*\*\*\*\*\*\*\*\*\*\*\*\*\*\*\*\*\*\*\*\*CHEMISTRY\*\*\*\*\*\*\*\*\*\*\*\*\*\*\*\*\*\*\*\*\*\*\*\*\*\*\*\*\*\*\*\*\*\*\*\*\*\*\*\*\*\*\*\*\*\*

| DAY: | | 1 | 2 | 3 | |
|---|---|---|---|---|---|
| DATE: | | | | | |
| TIME: | | | | | |
| LOCATION: | | | | | |
| | NORMAL | | | | UNITS |
| Albumin | 3.6–5 | 4.9 | 3.8 | | g/dL |
| Total protein | 6–8 | 7.2 | 6.8 | | g/dL |
| Prealbumin | 19–43 | | | | mg/dL |
| Transferrin | 200–400 | | | | mg/dL |
| Sodium | 135–155 | 154 | 148 | 140 | mmol/L |
| Potassium | 3.5–5.5 | 3.2 | 3.7 | 3.7 | mmol/L |
| Chloride | 98–108 | 107 | 104 | 101 | mmol/L |
| $PO_4$ | 2.5–4.5 | 4.0 | 3.5 | 3.6 | mmol/L |
| Magnesium | 1.6–2.6 | 2.2 | 1.9 | 1.8 | mmol/L |
| Osmolality | 275–295 | 319 | 306 | 289 | mmol/kg $H_2O$ |
| Total $CO_2$ | 24–30 | 31 | 28 | 27 | mmol/L |
| Glucose | 70–120 | 71 | 108 | 101 | mg/dL |
| BUN | 8–26 | 22 | 10 | 11 | mg/dL |
| Creatinine | 0.6–1.3 | 1.4 | 0.7 | 0.6 | mg/dL |
| Uric acid | 2.6–6 (women) | 3.6 | | | mg/dL |
| | 3.5–7.2 (men) | | | | |
| Calcium | 8.7–10.2 | 9.1 | | | mg/dL |
| Bilirubin | 0.2–1.3 | 1.1 | | | mg/dL |
| Ammonia ($NH_3$) | 9–33 | | | | $\mu$mol/L |
| SGPT (ALT) | 10–60 | | | | U/L |
| SGOT (AST) | 5–40 | | | | U/L |
| Alk phos | 98–251 | | | | U/L |
| CPK | 26–140 (women) | | | | U/L |
| | 38–174 (men) | | | | |
| LDH | 313–618 | | | | U/L |
| CHOL | 140–199 | | | | mg/dL |
| HDL-C | 40–85 (women) | | | | mg/dL |
| | 37–70 (men) | | | | |
| VLDL | | | | | mg/dL |
| LDL | < 130 | | | | mg/dL |
| LDL/HDL ratio | < 3.22 (women) | | | | |
| | < 3.55 (men) | | | | |
| Apo A | 101–199 (women) | | | | mg/dL |
| | 94–178 (men) | | | | |
| Apo B | 60–126 (women) | | | | mg/dL |
| | 63–133 (men) | | | | |
| TG | 35–160 | | | | mg/dL |
| $T_4$ | 5.4–11.5 | | | | $\mu$g/dL |
| $T_3$ | 80–200 | | | | ng/dL |
| $HbA_{1C}$ | 4.8–7.8 | | | | % |

**U<sub>H</sub>** *UNIVERSITY HOSPITAL*

NAME: Sam Jones                          DOB: 4/13
AGE: 8                                   SEX: M
PHYSICIAN: C. Fraser, MD

\*\*\*\*\*\*\*\*\*\*\*\*\*\*\*\*\*\*\*\*\*\*\*\*\*\*\*\*\*\*\*\*\*\*\*\*\*\*\*\*\*\*\*\*\*\*\*URINALYSIS\*\*\*\*\*\*\*\*\*\*\*\*\*\*\*\*\*\*\*\*\*\*\*\*\*\*\*\*\*\*\*\*\*\*\*\*\*\*\*\*\*\*\*\*\*\*\*

| | NORMAL | DAY: 1 | DAY: 2 | DAY: 3 | UNITS |
|---|---|---|---|---|---|
| DATE: | | | | | |
| TIME: | | | | | |
| LOCATION: | | | | | |
| Coll meth | | First morning | First Morning | First morning | |
| Color | | Amber | Straw | Pale Yellow | |
| Appear | | Cloudy | Slightly Hazy | Clear | |
| Sp grv | 1.003–1.030 | 1.039 | 1.020 | 1.008 | |
| pH | 5–7 | 4.8 | 5.2 | 5.6 | |
| Prot | NEG | NEG | NEG | NEG | mg/dL |
| Glu | NEG | NEG | NEG | NEG | mg/dL |
| Ket | NEG | +1 | NEG | NEG | |
| Occ bld | NEG | NEG | NEG | NEG | |
| Ubil | NEG | NEG | NEG | NEG | |
| Nit | NEG | NEG | NEG | NEG | |
| Urobil | < 1.1 | 0.5 | 0.7 | 0.9 | EU/dL |
| Leu bst | NEG | NEG | NEG | NEG | |
| Prot chk | NEG | NEG | NEG | NEG | |
| WBCs | 0–5 | 2 | 1 | 0 | /HPF |
| RBCs | 0–5 | 1 | 0 | 0 | /HPF |
| EPIs | 0 | 0 | 0 | 0 | /LPF |
| Bact | 0 | 0 | 0 | 0 | |
| Mucus | 0 | 0 | 0 | 0 | |
| Crys | 0 | 0 | 0 | 0 | |
| Casts | 0 | 0 | 0 | 0 | /LPF |
| Yeast | 0 | 0 | 0 | 0 | |

## Case Questions

1.   Define diarrhea. How do osmotic and secretory diarrhea differ? Which does Sam have? What is your rationale?

2.   What are the metabolic consequences of diarrhea?

3.   What signs and symptoms in the physician's interview and physical examination indicate that Sam may be dehydrated?

4.   What are electrolytes?

5.   Evaluate Sam's laboratory values on day 1 of his admission. Use the following table to organize your information. Which values are consistent with diagnosing dehydration? Does Sam have an electrolyte imbalance? Does Sam have an acid–base imbalance? Explain.

| Chemistry | Normal Value | Sam's Value | Function | Significance |
|-----------|-------------|-------------|----------|--------------|
|           |             |             |          |              |
|           |             |             |          |              |
|           |             |             |          |              |
|           |             |             |          |              |
|           |             |             |          |              |
|           |             |             |          |              |
|           |             |             |          |              |

6.  Assess Sam's urinalysis report. Which measures are consistent with his diagnosis?

7.  Assess Sam's height and weight. Which weight would you use— admission weight or usual body weight? What is the most appropriate tool to assess height and weight for an 8-year-old child?

8.  What are Sam's energy and protein needs?

9.  What are Sam's fluid requirements?

10. The physician ordered $D_5W$ ½ NS with 40 mEq KCL @ 100 mL/hr. What is $D_5W$? What is NS?

    a.  How much sodium does this solution provide in 1 liter? In 24 hours?

    b.  How much energy does this solution provide?

    c.  How much potassium does it provide?

11. The physician also ordered Pedialyte 30 cc q hour. What is Pedialyte? Why weren't clear liquids ordered?

12. Are other solutions similar to Pedialyte available?

13. Should Sam have been made NPO? Why or why not?

14. Once Sam's electrolyte imbalances are corrected and diarrhea begins to decrease, what type of diet would you recommend for transition from Pedialyte to solid food?

## Bibliography

American Academy of Pediatrics. Practice parameter: the management of acute gastroenteritis in young children. *Pediatrics.* 1996;97:424–435.

American Dietetic Association. Nutrition management of diarrhea in childhood. In American Dietetic Association. *Manual of Clinical Dietetics,* 6th ed. Chicago: American Dietetic Association, 1996:237–246.

Beyer PL. Medical nutrition therapy for lower gastrointestinal tract disorders. In Mahan LK, Escott-Stump S (eds.), *Krause's Food, Nutrition, and Diet Therapy,* 10th ed. Philadelphia: Saunders, 2000: 671–694.

Burkhart DM. Management of acute gastroenteritis in children. *Am Fam Physician.* 1999;60:2555–2566.

Chan PD, Gennoh JL. Current clinical strategies: pediatrics. Available at www.ccspublishing.com. Accessed September 13, 2000.

Commission on Accreditation for Dietetics Education. Knowledge, skills, and competencies for dietitians. *Accreditation Manual for Dietetics Education Programs,* rev. 4th ed. Chicago: American Dietetic Association, 2000.

Duggan C, Santosham M, Glass RI. The management of acute diarrhea in children: oral rehydration, maintenance, and nutritional therapy. *MMWR.* 1992;41:1–20.

Fayad IM, Hashem M, Duggan C, Refat M, Bakir M, Fontaine O, et al. Comparative efficacy of rice-based and glucose-based oral rehydration salts plus early reintroduction of food. *Lancet.* 1993;342:772–775.

Gavin N, Merrick N, Davidson B. Efficacy of glucose-based oral rehydration therapy. *Pediatrics.* 1996; 98:45–51.

Gould BE. *Pathophysiology for the Health-Related Professions.* Philadephia: Saunders, 1997:70–93.

Hoghton MA, Mittal NK, Sandhu BK, Madhi G. Effects of immediate modified feeding on infantile gastroenteritis. *Br J Gen Pract.* 1996;46:173–175.

Mackenzie A, Barnes G, Shann F. Clinical signs of dehydration in children. *Lancet.* 1989;2(8663):605–607.

Northrup RS, Flanigan TP. Gastroenteritis. *Pediatr Rev.* 1994;15:461–472.

# Celiac Disease

*Introductory Level*

## Objectives

After completing this case, the student will be able to:

1. Apply working knowledge of pathophysiology to interpret diagnostic procedures for celiac disease.
2. Interpret anthropometric and laboratory parameters for nutritional assessment.
3. Determine nutrient requirements for individuals with celiac disease.
4. Demonstrate ability to determine appropriate medical nutrition therapy for individuals with celiac disease.
5. Distinguish classifications of malnutrition.

Mrs. Sandra Guernsey is admitted to University Hospital with severe weight loss, extreme fatigue, and diarrhea. A small bowel biopsy reveals a diagnosis of celiac disease with secondary malabsorption and anemia.

# UNIVERSITY HOSPITAL

## ADMISSION DATABASE

| BED# 1 | DATE: 11/12 | TIME: 0800 | TRIAGE STATUS (ER ONLY): ☐ Red ☐ Yellow ☐ Green ☐ White |
|---|---|---|---|

**Initial Vital Signs**

| TEMP: 98.4 | RESP: 17 | SAO2: |
|---|---|---|

| HT: 5'3" | WT (lb): 92 | B/P: 110/74 | PULSE: 71 |
|---|---|---|---|

| LAST TETANUS 8 years ago | LAST ATE yesterday lunch | LAST DRANK last pm-bedtime |
|---|---|---|

**PRIMARY PERSON TO CONTACT:**
Name: Michael Guernsey
Home #: 512-256-7894
Work #: 512-254-9900

**ORIENTATION TO UNIT:** ☒ Call light ☒ Television/telephone ☒ Bathroom ☒ Visiting ☒ Smoking ☒ Meals ☒ Patient rights/responsibilities

### CHIEF COMPLAINT/HX OF PRESENT ILLNESS

"I have lost a tremendous amount of weight and I have been having terrible diarrhea for awhile now. I don't even have the energy to get off the couch right now."

**PERSONAL ARTICLES:** (Check if retained/describe)
☒ Contacts ☒ R ☒ L     ☐ Dentures ☐ Upper ☐ Lower
☒ Jewelry: wedding band
☐ Other:

### ALLERGIES: Meds, Food, IVP Dye, Seafood: Type of Reaction

NKA

**VALUABLES ENVELOPE:**
☐ Valuables instructions

### PREVIOUS HOSPITALIZATIONS/SURGERIES

2 live births—last child born in September of this year—5 lb 2 oz

full term

**INFORMATION OBTAINED FROM:**
☒ Patient          ☐ Previous record
☒ Family           ☐ Responsible party

Signature   *Sandra Guernsey*

### Home Medications (including OTC)     Codes: A = Sent home     B = Sent to pharmacy     C = Not brought in

| Medication | Dose | Frequency | Time of Last Dose | Code | Patient Understanding of Drug |
|---|---|---|---|---|---|
| prenatal vitamins | 1 | daily | this am | c | yes |
| Kaopectate | 2 T | q 3-4 hours | this pm | c | yes |
|  |  |  |  |  |  |
|  |  |  |  |  |  |
|  |  |  |  |  |  |
|  |  |  |  |  |  |
|  |  |  |  |  |  |
|  |  |  |  |  |  |
|  |  |  |  |  |  |
|  |  |  |  |  |  |

Do you take all medications as prescribed?   ☒ Yes   ☐ No   If no, why?

### PATIENT/FAMILY HISTORY

| | | |
|---|---|---|
| ☐ Cold in past two weeks | ☐ High blood pressure | ☐ Kidney/urinary problems |
| ☐ Hay fever | ☐ Arthritis | ☐ Gastric/abdominal pain/heartburn |
| ☐ Emphysema/lung problems | ☐ Claustrophobia | ☐ Hearing problems |
| ☐ TB disease/positive TB skin test | ☐ Circulation problems | ☐ Glaucoma/eye problems |
| ☐ Cancer | ☐ Easy bleeding/bruising/anemia | ☐ Back pain |
| ☐ Stroke/past paralysis | ☐ Sickle cell disease | ☐ Seizures |
| ☒ Heart attack Father | ☐ Liver disease/jaundice | ☐ Other |
| ☒ Angina/chest pain Father | ☐ Thyroid disease | |
| ☒ Heart problems Father | ☐ Diabetes | |

### RISK SCREENING

Have you had a blood transfusion?   ☐ Yes   ☐ No
Do you smoke?   ☒ Yes   ☐ No
If yes, how many pack(s) 1.5/day for 10 years
Does anyone in your household smoke?   ☒ Yes   ☐ No
Do you drink alcohol?   ☐ Yes   ☒ No
If yes, how often?_____   How much?
When was your last drink?_____/_____/
Do you take any recreational drugs?   ☐ Yes   ☒ No
If yes, type:_____   Route
Frequency:_____   Date last used:_____/_____/

**FOR WOMEN Ages 12–52**
Is there any chance you could be pregnant?   ☐ Yes   ☒ No
If yes, expected date (EDC):
Gravida/Para: 3/2

**ALL WOMEN**
Date of last Pap smear: 10/26/
Do you perform regular breast self-exams?   ☒ Yes   ☐ No

**ALL MEN**
Do you perform regular testicular exams?   ☐ Yes   ☐ No

Additional comments:

✗ *Lynnette Hall, RN, BSN*
Signature/Title

**Client name:** Sandra Guernsey
**DOB:** 3/14
**Age:** 36
**Sex:** Female
**Education:** Bachelor's degree
**Occupation:** Previous secretary for hospital administrator—now at home since recent delivery of son
**Hours of work:** N/A
**Household members:** Husband age 42, son age 4, son age 3 months—all well
**Ethnic background:** Caucasian
**Religious affiliation:** None
**Referring physician:** Roger Smith, MD (gastroenterology)

## Chief complaint:

"I have lost a tremendous amount of weight, and I have been having terrible diarrhea for awhile now. I don't even have the energy to get off the couch right now."

## Patient history:

*Onset of disease:* Patient relates having diarrhea on and off for most of her adult life that she can remember. "It seems that a lot of people in my family have 'funny' stomachs. My mom and grandmother both have problems with diarrhea off and on. Mine got much worse during my most recent pregnancy, and now it is debilitating." She recently delivered a 5 lb 2 oz healthy son @ 39 weeks gestation. She states that she gained 11 lb during this pregnancy but has since lost about 30 lb. She now weighs 92 lb. Her greatest nonpregnant weight was 112 lb just prior to this pregnancy. She describes the diarrhea as foul smelling and doesn't seem affected by what she eats in that she generally has diarrhea no matter what she eats. "I started out breastfeeding my son but stopped about 3 weeks ago because I felt so bad."
*Type of Tx:* None at present
*PMH:* 3 pregnancies—2 live births, 1 miscarriage at 22 weeks. No other significant medical history
*Meds:* Prenatal vitamins, Kaopectate
*Smoker:* Yes
*Family Hx: What?* CAD   *Who?* Father

## Physical exam:

*General appearance:* Thin, pale woman who complains of fatigue, weakness, and diarrhea
*Vitals:* Temp 98.2°F, BP 108/72, HR 78 bpm/normal, RR 17 bpm
*Heart:* Regular rate and rhythm. Heart sounds normal.
*HEENT:*
   *Eyes:* PERRLA sclera pale; fundi benign
   *Throat:* Pharynx clear without postnasal drainage
*Genitalia:* Deferred
*Neurologic:* Intact; alert and oriented
*Extremities:* No edema, strength 5/5
*Skin:* Pale without lesions
*Chest/lungs:* Lungs clear to percussion and auscultation
*Abdomen:* Not distended; diminished bowel sounds

**Nutrition Hx:**

*General:* Patient states that she is hungry all the time. "I do eat but it seems that every time I eat in any large amount that I almost immediately have diarrhea. I do not have nausea or vomiting." Foods that are fried and meat—especially beef—tend to make the diarrhea worse. Relates that she has been relying on chicken noodle soup, crackers, and Sprite for the last several days. Patient states that her greatest nonpregnant weight was prior to her last pregnancy, when she weighed 112 lb. She gained 11 lb with her pregnancy, and her full-term son weighed 5 lb 2 oz.

*Usual dietary intake:*
Likes all foods but has found that she avoids eating because it causes her diarrhea to start.
*24-hr recall* (prior to admission):
AM:      Toast—1 slice, hot tea with sugar.
Lunch:   1 c chicken noodle soup, 2–3 saltine crackers, ½ c applesauce, 12 oz Sprite; throughout
         rest of day, sips of Sprite.

*Food allergies/intolerances/aversions:* NKA
*Previous MNT?* No
*Food purchase/preparation:* Self
*Vit/min intake:* Still taking prenatal vitamins
*Anthropometric measures:* TSF 7.5 mm, MAC 180 mm

**Tx plan:**
24-hour stool collection for direct visual examination; white blood cells; occult blood; Sudan black B fat stain; ova and parasites; electrolytes and osmolality; pH; alkalization; 72-hour fecal fat.
Upper gastrointestinal endoscopy for small bowel biopsy and possible duodenal aspirate.
*Diet:* 100-g fat diet × 3 days

**Hospital course:**
Small bowel biopsy indicates flat mucosa with villus atrophy and hyperplastic crypts—inflammatory infiltrate in lamina propria. Fecal fat indicates steatorrhea and malabsorption. Positive AGA, EMA antibodies

**Dx:**
Celiac disease with secondary malabsorption and anemia

# U<sub>H</sub> *UNIVERSITY HOSPITAL*

NAME: Sandra Guernsey                DOB: 3/14
AGE: 36                              SEX: F
PHYSICIAN: Roger Smith, MD

\*\*\*\*\*\*\*\*\*\*\*\*\*\*\*\*\*\*\*\*\*\*\*\*\*\*\*\*\*\*\*\*\*\*\*\*\*\*\*\*\*\*\*\*\*CHEMISTRY\*\*\*\*\*\*\*\*\*\*\*\*\*\*\*\*\*\*\*\*\*\*\*\*\*\*\*\*\*\*\*\*\*\*\*\*\*\*\*\*\*\*\*

| DAY: | | Admit | 3 | |
|---|---|---|---|---|
| DATE: | | | | |
| TIME: | | | | |
| LOCATION: | | | | |
| | NORMAL | | | UNITS |
| Albumin | 3.6–5 | 2.9 | | g/dL |
| Total protein | 6–8 | 5.5 | | g/dL |
| Prealbumin | 19–43 | 14 | | mg/dL |
| Transferrin | 200–400 | 350 | | mg/dL |
| Sodium | 135–155 | 138 | | mmol/L |
| Potassium | 3.5–5.5 | 3.7 | | mmol/L |
| Chloride | 98–108 | 101 | | mmol/L |
| $PO_4$ | 2.5–4.5 | 2.8 | | mmol/L |
| Magnesium | 1.6–2.6 | 1.6 | | mmol/L |
| Osmolality | 275–295 | 281 | | mmol/kg $H_2O$ |
| Total $CO_2$ | 24–30 | 27 | | mmol/L |
| Glucose | 70–120 | 72 | | mg/dL |
| BUN | 8–26 | 9 | | mg/dL |
| Creatinine | 0.6–1.3 | 0.7 | | mg/dL |
| Uric acid | 2.6–6 (women) | 2.8 | | mg/dL |
| | 3.5–7.2 (men) | | | |
| Calcium | 8.7–10.2 | 8.7 | | mg/dL |
| Bilirubin | 0.2–1.3 | 0.3 | | mg/dL |
| Ammonia ($NH_3$) | 9–33 | 9 | | $\mu$mol/L |
| SGPT (ALT) | 10–60 | 12 | | U/L |
| SGOT (AST) | 5–40 | 8 | | U/L |
| Alk phos | 98–251 | 111 | | U/L |
| CPK | 26–140 (women) | | | U/L |
| | 38–174 (men) | | | |
| LDH | 313–618 | | | U/L |
| CHOL | 140–199 | 128 | | mg/dL |
| HDL-C | 40–85 (women) | | | mg/dL |
| | 37–70 (men) | | | |
| VLDL | | | | mg/dL |
| LDL | < 130 | | | mg/dL |
| LDL/HDL ratio | < 3.22 (women) | | | |
| | < 3.55 (men) | | | |
| Apo A | 101–199 (women) | | | mg/dL |
| | 94–178 (men) | | | |
| Apo B | 60–126 (women) | | | mg/dL |
| | 63–133 (men) | | | |
| TG | 35–160 | | | mg/dL |
| $T_4$ | 5.4–11.5 | | | $\mu$g/dL |
| $T_3$ | 80–200 | + | | ng/dL |
| $HbA_{1C}$ | 4.8–7.8 | + | | % |
| Fecal fat | < 5 | | 11.5 | g |
| AGA antibodies | 0 | + | | — |
| EMA antibodies | 0 | + | | — |

# U**H** *UNIVERSITY HOSPITAL*

NAME: Sandra Guernsey                      DOB: 3/14
AGE: 36                                     SEX: F
PHYSICIAN: Roger Smith, MD

\*\*\*\*\*\*\*\*\*\*\*\*\*\*\*\*\*\*\*\*\*\*\*\*\*\*\*\*\*\*\*\*\*\*\*\*\*\*\*\*\*HEMATOLOGY\*\*\*\*\*\*\*\*\*\*\*\*\*\*\*\*\*\*\*\*\*\*\*\*\*\*\*\*\*\*\*\*\*\*\*\*\*\*\*\*

DAY:                                                        Admit
DATE:
TIME:
LOCATION:

| | NORMAL | | UNITS |
|---|---|---|---|
| WBC | 4.3–10 | 5.2 | $\times 10^3/mm^3$ |
| RBC | 4–5 (women) | 4.9 | $\times 10^6/mm^3$ |
| | 4.5–5.5 (men) | | |
| HGB | 12–16 (women) | 10.5 | g/dL |
| | 13.5–17.5 (men) | | |
| HCT | 37–47 (women) | 35 | % |
| | 40–54 (men) | | |
| MCV | 84–96 | 90 | fL |
| MCH | 27–34 | 27 | pg |
| MCHC | 31.5–36 | 31 | % |
| RDW | 11.6–16.5 | 11.9 | % |
| Plt Ct | 140–440 | 220 | $\times 10^3$ |
| Diff TYPE | | | |
| % GRANS | 34.6–79.2 | 38.6 | % |
| % LYM | 19.6–52.7 | 21.4 | % |
| SEGS | 50–62 | 55 | % |
| BANDS | 3–6 | 4 | % |
| LYMPHS | 25–40 | 28 | % |
| MONOS | 3–7 | 5 | % |
| EOS | 0–3 | 1 | % |
| TIBC | 65–165 (women) | 55 | $\mu$g/dL |
| | 75–175 (men) | | |
| Ferritin | 18–160 (women) | 12 | $\mu$g/dL |
| | 18–270 (men) | | |
| Vitamin $B_{12}$ | 100–700 | 82 | pg/mL |
| Folate | 2–20 | 2 | ng/mL |
| Total T cells | 812–2318 | | $mm^3$ |
| T-helper cells | 589–1505 | | $mm^3$ |
| T-suppressor cells | 325–997 | | $mm^3$ |
| PT | 11–13 | | sec |

## Case Questions

1. What is a 72-hour fecal fat test?

2. Mrs. Guernsey's laboratory report shows her fecal fat was 11.5 gm fat/24 hour. What does this mean?

3. Why was the patient ordered a 100-gm fat diet when her diet history indicates that her symptoms are much worse with fried foods?

4. The small bowel biopsy results state, "flat mucosa with villus atrophy and hyperplastic crypts—inflammatory infiltrate in lamina propria." What do these results tell you about the change in the anatomy of the small intestine?

5. How is celiac disease related to the damage to the small intestine that the endoscopy and biopsy results indicate?

6. What is the etiology of celiac disease? Is anything in Mrs. Guernsey's history typical of patients with celiac disease? Explain.

7. What are AGA and EMA antibodies?

8. Evaluate Mrs. Guernsey's laboratory measures for nutritional significance. Explain the rationale for your evaluation of the laboratory value.

9. Are any symptoms from Mrs. Guernsey's physical examination consistent with her laboratory values? Explain.

10. Calculate the patient's IBW, %IBW, %UBW, and BMI, and explain the nutritional risk associated with each value.

**11.** Evaluate Mrs. Guernsey's other anthropometric measurements. Using the available data, calculate her upper arm muscle area and upper arm fat area. Interpret this information for nutritional signficance.

**12.** Can you diagnose Mrs. Guernsey with malnutrition? If so, what type? What is your rationale?

**13.** Calculate this patient's energy and protein needs.

**14.** Gluten restriction is the major component of the medical nutrition therapy for celiac disease. What is gluten? Where is it found?

**15.** Can patients on a gluten-free diet tolerate oats?

**16.** What sources other than foods might introduce gluten to the patient?

**17.** Are patients with celiac disease also lactose intolerant?

**18.** Are there any other considerations you might need to make when initiating solid food for this patient?

**19.** Mrs. Guernsey's nutritional status is so compromised that she might benefit from high-calorie high-protein supplementation. What would you recommend?

**20.** Would glutamine supplementation help Mrs. Guernsey during the healing process? How could she receive it?

**21.** What result can Mrs. Guernsey expect from restricting all foods with gluten? Will she have to follow this diet for very long?

22.   Evaluate the following excerpt from Mrs. Guernsey's food diary. Identify the foods that might not be tolerated on a gliadin-free diet.

Cornflakes _____

Bologna slices _____

Lean Cuisine—beef tips on rice _____

Skim milk _____

Cheddar cheese spread _____

Green bean casserole (mushroom soup, onions, green beans) _____

Coffee _____

Rice crackers _____

Fruit cocktail _____

Sugar _____

Pudding _____

V8 juice _____

Banana _____

Cola _____

## Bibliography

American Dietetic Association. Celiac disease. In *Manual of Clinical Dietetics,* 6th ed. Chicago: American Dietetic Association, 2000:181–191.

Beyer PL. Medical nutrition therapy for lower gastrointestinal tract disorders. In Mahan LK, Escott-Stump S (eds.), *Krause's Food, Nutrition, and Diet Therapy,* 10th ed. Philadelphia: Saunders, 2000:666–694.

Chartrand LJ, Seidman G. Celiac disease is a lifelong disorder. *Clin Invest Med.* 1996; 19:357.

Commission on Accreditation for Dietetics Education. Knowledge, skills, and competencies for dietitians. *Accreditation Manual for Dietetics Education Programs,* rev. 4th ed. Chicago: American Dietetic Association, 2000:45–49.

Coping with celiac disease. *Tufts University Health & Nutrition Letter;* New York: September 1998.

Feighery C. Celiac disease. *Br Med J.* 1999;319:236–239.

Kemppainen TA, Julkunen RJ, et al. Nutritional status of newly diagnosed celiac disease patients before and after the institution of a celiac disease diet: association with the grade of mucosal villous atrophy. *Am J Clin Nutr.* 1998;67:482.

Russo PA, Chartrand LJ, Seidman E. Comparative analysis of serologic screening tests for initial diagnosis of celiac disease. *Pediatrics.* 1999;104:75–78.

Tso P, Crissinger KD. Overview of digestion and absorption. In Stipanuk MH (ed.), *Biological and Physiological Aspects of Human Nutrition.* Philadelphia: Saunders, 2000:76–90.

Walker JA, Schmitz S, Schmerling J, Visako DH. Revised criteria for diagnosis of coeliac disease. *Arch Dis Child.* 1990;65:900–911.

# Case 15

# Diverticulosis with Incidence of Diverticulitis

*Introductory Level*

## Objectives

After completing this case, the student will be able to:

1. Apply working knowledge of pathophysiology to plan nutrition care.
2. Determine appropriate medical nutrition therapy for diverticulosis.
3. Demonstrate ability to calculate individual nutrient requirements.
4. Interpret the nutrient composition of a dietary history.

5. Modify recipes for individual nutrient needs.

Dr. Greer admitted Mrs. Edna Meyer after she experienced rectal bleeding. An upper GI source of bleeding was ruled out but the colonoscopy noted numerous diverticula. With additional evidence of a lower GI bleed in sigmoid colon, a diagnosis of diverticulosis was determined.

 **UNIVERSITY HOSPITAL**

# ADMISSION DATABASE

Name: Edna Meyer
DOB: 1/17  age 62
Physician: Boyd Greer

| BED#<br>1 | DATE:<br>4/22 | TIME:<br>1400 | TRIAGE STATUS (ER ONLY):<br>☐ Red ☐ Yellow ☐ Green ☐ White |
|---|---|---|---|

**Initial Vital Signs**

| TEMP:<br>98.8 | RESP:<br>15 | | SAO2: |
|---|---|---|---|

| HT:<br>5'1" | WT (lb):<br>155 | B/P:<br>120/82 | PULSE:<br>72 |
|---|---|---|---|

| LAST TETANUS<br>2 years ago | | LAST ATE<br>this am | LAST DRANK<br>this am |
|---|---|---|---|

**CHIEF COMPLAINT/HX OF PRESENT ILLNESS**

"I had a lot of bright red blood in my bowel movement

yesterday morning."

**ALLERGIES: Meds, Food, IVP Dye, Seafood: Type of Reaction**

NKA

**PREVIOUS HOSPITALIZATIONS/SURGERIES**

3 live births

**PRIMARY PERSON TO CONTACT:**
Name: Leonard Meyer
Home #: 613-225-7855
Work #:

**ORIENTATION TO UNIT:** ☒ Call light ☒ Television/telephone
☒ Bathroom ☒ Visiting ☒ Smoking ☒ Meals
☒ Patient rights/responsibilities

**PERSONAL ARTICLES:** (Check if retained/describe)
☐ Contacts ☐ R ☐ L   ☒ Dentures ☒ Upper ☒ Lower
☒ Jewelry: wedding band
☒ Other: glasses

**VALUABLES ENVELOPE:** yes
☒ Valuables instructions

**INFORMATION OBTAINED FROM:**
☐ Patient      ☐ Previous record
☒ Family       ☐ Responsible party

Signature  *Edna Meyer*

| Home Medications (including OTC) | Codes: A=Sent home | | B=Sent to pharmacy | | C=Not brought in |
|---|---|---|---|---|---|
| Medication | Dose | Frequency | Time of Last Dose | Code | Patient Understanding of Drug |
| Prinivil | 5 mg | one time daily | 0800 | c | yes |
|  |  |  |  |  |  |
|  |  |  |  |  |  |
|  |  |  |  |  |  |
|  |  |  |  |  |  |
|  |  |  |  |  |  |
|  |  |  |  |  |  |
|  |  |  |  |  |  |
|  |  |  |  |  |  |
|  |  |  |  |  |  |
|  |  |  |  |  |  |
|  |  |  |  |  |  |

Do you take all medications as prescribed?  ☒ Yes  ☐ No  If no, why?

**PATIENT/FAMILY HISTORY**

| | | |
|---|---|---|
| ☐ Cold in past two weeks | ☒ High blood pressure Patient | ☐ Kidney/urinary problems |
| ☐ Hay fever | ☐ Arthritis | ☐ Gastric/abdominal pain/heartburn |
| ☐ Emphysema/lung problems | ☐ Claustrophobia | ☐ Hearing problems |
| ☐ TB disease/positive TB skin test | ☐ Circulation problems | ☐ Glaucoma/eye problems |
| ☒ Cancer Mother | ☐ Easy bleeding/bruising/anemia | ☐ Back pain |
| ☐ Stroke/past paralysis | ☐ Sickle cell disease | ☐ Seizures |
| ☐ Heart attack | ☐ Liver disease/jaundice | ☐ Other |
| ☐ Angina/chest pain | ☐ Thyroid disease | |
| ☐ Heart problems | ☐ Diabetes | |

**RISK SCREENING**

Have you had a blood transfusion?  ☐ Yes  ☒ No
Do you smoke?  ☐ Yes  ☒ No
If yes, how many pack(s)____ day for____ years
Does anyone in your household smoke?  ☐ Yes  ☒ No
Do you drink alcohol?  ☐ Yes  ☒ No
If yes, how often?_____ How much?____
When was your last drink?_____/_____/_____
Do you take any recreational drugs?  ☐ Yes  ☒ No
If yes, type:_____ Route____
Frequency:_____ Date last used:_____/_____/_____

**FOR WOMEN Ages 12–52**

Is there any chance you could be pregnant?  ☐ Yes  ☐ No
If yes, expected date (EDC):
Gravida/Para:

**ALL WOMEN**

Date of last Pap smear:  /01/this year
Do you perform regular breast self-exams?  ☒ Yes  ☐ No

**ALL MEN**

Do you perform regular testicular exams?  ☐ Yes  ☐ No

Additional comments:

✗ *Betsy Temple, LPN*
Signature/Title

**Client name:** Edna Meyer
**DOB:** 1/17
**Age:** 62
**Sex:** Female
**Education:** High school diploma
**Occupation:** Works at home as a seamstress
**Hours of work:** Varies
**Household members:** Husband age 66, well; granddaughters ages 13, 15
**Ethnic background:** African American
**Religious affiliation:** Baptist
**Referring physician:** Dr. Boyd Greer, family practice

## Chief complaint:

"I had a lot of bright red blood in my bowel movement yesterday morning."

## Patient history:

*Onset of disease:* History of constipation off and on for most of adult life. But recently has also had some episodes of diarrhea with crampy LLQ pain. Presented to MD's office with complaint of blood expelled with bowel movement that morning. Has had 2 other episodes of bleeding in past 24 hours.
*Type of Tx:* None at present
*PMH:* Hypertension Dx 3 years previous
*Meds:* Prinivil (lisinopril) 5 mg daily
*Smoker:* No
*Family Hx: What?* CA    *Who?* Mother died of ovarian cancer; father died of colon cancer

## Physical exam:

*General appearance:* Slightly overweight 62-year-old African American woman in no acute distress; somewhat anxious
*Vitals:* Temp 98.8°F, BP 120/82, HR 72 bpm, RR 15 bpm
*Heart:* S1 and S2 clear; no rub, gallop, or murmur; regular rate
*HEENT:* Unremarkable—normocephalic
  *Eyes:* PERRLA, fundi without lesions
  *Ears:* Clear
  *Nose:* Clear
  *Throat:* Supple, no adenopathy or thyromegaly, no bruits
*Genitalia:* Deferred
*Neurologic:* Alert and oriented ×4; strength 5/5 throughout, DTRs 2+ and symmetrical, sensation intact
*Extremities:* No edema
*Skin:* Warm, dry to touch
*Chest/lungs:* Clear to auscultation and percussion
*Peripheral vascular:* Peripheral pulses palpable
*Abdomen:* Positive bowel sounds throughout, nontender, nondistended

**Nutrition Hx:**

*General:* "My appetite is pretty good—well, probably too good. I like to cook and bake—especially for my granddaughters. I have a problem with being regular, though, and it seems to have gotten worse. I try to drink prune juice or eat prunes, and that helps a little."

*Usual dietary intake:*

| | |
|---|---|
| *Breakfast:* | White toast with butter and jam; fried egg, coffee |
| *Lunch:* | Soup or sandwich—sometimes leftovers from previous day; coffee |
| *Dinner:* | Meat, 1–2 vegetables, rice or potatoes, bread or biscuits, iced tea or coffee |
| *24-hr recall:* | 2 slices white toast with 3 t margarine, 2 t jelly, sliced prunes ½ c, black coffee; |
| | 2 slices white bread with 1 oz ham, 1 T mayonnaise, 2 oz potato chips, coffee; |
| | 2–3 oz pork chop fried, 1 c macaroni and cheese, 1 biscuit, water |
| | 2 slices pound cake with ½ c vanilla ice cream |

*Food allergies/intolerances/aversions:* None
*Previous MNT?* No
*Food purchase/preparation:* Self
*Vit/min intake:* None

**Tx plan:**

NPO
NG to low wall suction
D5NS @ 50 cc/hr
Metronidazole 1 g loading dose, then 500 mg q 6 h; ciprofloxacin 400 mg q 12 h
Strict I/O
Schedule for colonoscopy

**Hospital course:**

NG aspirate heme negative—upper GI source of bleed ruled out. Colonoscopy negative for active bleeding but numerous diverticula noted

**Dx:**

Diverticulosis with evidence of lower GI bleed in sigmoid colon

# U H *UNIVERSITY HOSPITAL*

NAME: Edna Meyer                     DOB: 1/17
AGE: 62                              SEX: F
PHYSICIAN: B. Greer, MD

\*\*\*\*\*\*\*\*\*\*\*\*\*\*\*\*\*\*\*\*\*\*\*\*\*\*\*\*\*\*\*\*\*\*\*\*\*\*\*\*\*\*\*\*\*\*CHEMISTRY\*\*\*\*\*\*\*\*\*\*\*\*\*\*\*\*\*\*\*\*\*\*\*\*\*\*\*\*\*\*\*\*\*\*\*\*\*\*\*\*\*\*\*\*

DAY:                                         1
DATE:
TIME:
LOCATION:

| | NORMAL | | UNITS |
|---|---|---|---|
| Albumin | 3.6–5 | 3.8 | g/dL |
| Total protein | 6–8 | 6.9 | g/dL |
| Prealbumin | 19–43 | | mg/dL |
| Transferrin | 200–400 | | mg/dL |
| Sodium | 135–155 | 138 | mmol/L |
| Potassium | 3.5–5.5 | 4.2 | mmol/L |
| Chloride | 98–108 | 101 | mmol/L |
| $PO_4$ | 2.5–4.5 | 3.2 | mmol/L |
| Magnesium | 1.6–2.6 | 1.9 | mmol/L |
| Osmolality | 275–295 | 285 | mmol/kg $H_2O$ |
| Total $CO_2$ | 24–30 | 25 | mmol/L |
| Glucose | 70–120 | 101 | mg/dL |
| BUN | 8–26 | 12 | mg/dL |
| Creatinine | 0.6–1.3 | 1.1 | mg/dL |
| Uric acid | 2.6–6 (women) | 3.5 | mg/dL |
| | 3.5–7.2 (men) | | |
| Calcium | 8.7–10.2 | 8.9 | mg/dL |
| Bilirubin | 0.2–1.3 | 0.9 | mg/dL |
| Ammonia ($NH_3$) | 9–33 | 9 | $\mu$mol/L |
| SGPT (ALT) | 10–60 | 14 | U/L |
| SGOT (AST) | 5–40 | 8 | U/L |
| Alk phos | 98–251 | 244 | U/L |
| CPK | 26–140 (women) | | U/L |
| | 38–174 (men) | | |
| LDH | 313–618 | | U/L |
| CHOL | 140–199 | 175 | mg/dL |
| HDL-C | 40–85 (women) | 62 | mg/dL |
| | 37–70 (men) | | |
| VLDL | | | mg/dL |
| LDL | < 130 | 111 | mg/dL |
| LDL/HDL ratio | < 3.22 (women) | 1.79 | |
| | < 3.55 (men) | | |
| Apo A | 101–199 (women) | | mg/dL |
| | 94–178 (men) | | |
| Apo B | 60–126 (women) | | mg/dL |
| | 63–133 (men) | | |
| TG | 35–160 | 155 | mg/dL |
| $T_4$ | 5.4–11.5 | | $\mu$g/dL |
| $T_3$ | 80–200 | | ng/dL |
| $HbA_{1C}$ | 4.8–7.8 | | % |

# U<sub>H</sub> *UNIVERSITY HOSPITAL*

NAME: Edna Meyer                          DOB: 1/17
AGE: 62                                   SEX: F
PHYSICIAN: B. Greer, MD

\*\*\*\*\*\*\*\*\*\*\*\*\*\*\*\*\*\*\*\*\*\*\*\*\*\*\*\*\*\*\*\*\*\*\*\*\*\*\*\*\*\*\*\*HEMATOLOGY\*\*\*\*\*\*\*\*\*\*\*\*\*\*\*\*\*\*\*\*\*\*\*\*\*\*\*\*\*\*\*\*\*\*\*\*\*\*\*\*\*\*\*\*

DAY:                                                        1
DATE:
TIME:
LOCATION:

|  | NORMAL |  | UNITS |
|---|---|---|---|
| WBC | 4.3–10 | 8.5 | $\times 10^3/mm^3$ |
| RBC | 4–5 (women) | 4.2 | $\times 10^6/mm^3$ |
|  | 4.5–5.5 (men) |  |  |
| HGB | 12–16 (women) | 12 | g/dL |
|  | 13.5–17.5 (men) |  |  |
| HCT | 37–47 (women) | 37 | % |
|  | 40–54 (men) |  |  |
| MCV | 84–96 | 85 | fL |
| MCH | 27–34 | 28 | pg |
| MCHC | 31.5–36 | 32.5 | % |
| RDW | 11.6–16.5 | 12.2 | % |
| Plt ct | 140–440 |  | $\times 10^3$ |
| Diff TYPE |  |  |  |
| % GRANS | 34.6–79.2 | 55.2 | % |
| % LYM | 19.6–52.7 | 44.6 | % |
| SEGS | 50–62 |  | % |
| BANDS | 3–6 |  | % |
| LYMPHS | 25–40 |  | % |
| MONOS | 3–7 |  | % |
| EOS | 0–3 |  | % |
| TIBC | 65–165 (women) |  | µg/dL |
|  | 75–175 (men) |  |  |
| Ferritin | 18–160 (women) |  | µg/dL |
|  | 18–270 (men) |  |  |
| Vitamin $B_{12}$ | 100–700 |  | pg/mL |
| Folate | 2–20 |  | ng/mL |
| Total T cells | 812–2318 |  | $mm^3$ |
| T-helper cells | 589–1505 |  | $mm^3$ |
| T-suppressor cells | 325–997 |  | $mm^3$ |
| PT | 11–13 |  | sec |

## Case Questions

1.   What are the possible complications of diverticulosis?

2.   What symptoms did Mrs. Meyer indicate in the physician's H&P that are consistent with diverticulosis?

3.   Evaluate Mrs. Meyer's anthropometric data by evaluating UBW, IBW, and BMI. Interpret your calculations.

4.   Estimate Mrs. Meyer's energy and protein requirements. Would you use her IBW or actual body weight in calculating her REE using the Harris-Benedict equation? Explain.

5.   Research indicates that low fiber intake may be related to the risk for the development of diverticulosis. What is the optimal fiber intake for Mrs. Meyer? What guideline would you use to determine this optimal fiber intake?

6.   What are the recommendations for percentage of calories from carbohydrate, protein, and fat for Mrs. Meyer? What guideline did you use and why?

7.   Analyze Mrs. Meyer's 24-hour recall using a computerized dietary analysis program. How does her intake compare to recommendations for kcal, protein, fat, and fiber?

Complete the following table:

| Nutrient | Recommended Amount | Mrs. Meyer's Intake per 24-Hour Recall |
|---|---|---|
| Energy | | |
| Protein | | |
| Carbohydrate | | |
| Fat | | |
| Fiber | | |

8.  Using a computer nutrient analysis, determine whether Mrs. Meyer be concerned about any of her vitamin and mineral intakes.

9.  Mrs. Meyer is currently on clear liquids. What diet would you recommend for her to progress to while she is in the hospital?

10. What should her medical nutrition therapy goal(s) be as the inflammation decreases?

11. Mrs. Meyer tells you she has several recipes for homemade quick breads that she likes to prepare. She wonders if there is a way to increase the fiber intake for the banana bread she makes. Analyze the following recipe for fiber and fat content, and then make recommendations to increase the fiber content in the recipe. Could you make any recommendations to decrease fat content as well?

### Recipe: Edna's Banana Bread

| Ingredient | Fiber Analysis | Fat | Substitution | Fiber Analysis | Fat |
|---|---|---|---|---|---|
| ½ c margarine | | | | | |
| 1 c sugar | | | | | |
| 2 eggs | | | | | |
| 1¾ c all-purpose flour | | | | | |
| 1 t soda | | | | | |
| 1 t baking powder | | | | | |
| ½ t salt | | | | | |
| 3 jars of banana baby food | | | | | |
| 1 t vanilla | | | | | |
| Total before | | | | | |
| Total per serving before (12 slices per loaf) | | | | | |

12. Dr. Greer has suggested that Mrs. Meyer take Fiberall or Metamucil to increase her fiber intake. What are these medications, and how might they help?

## Bibliography

American Dietetic Association. Fiber restricted diet. In *Manual of Clinical Dietetics.* Chicago: American Dietetic Association, 2000:703–707.

American Dietetic Association. High fiber diet. In *Manual of Clinical Dietetics.* Chicago: American Dietetic Association, 2000:709–717.

Beyer PL. Medical nutrition therapy for lower gastrointestinal tract disorders. In Mahan LK, Escott-Stump S (eds.), *Krause's Food, Nutrition, and Diet Therapy,* 10th ed. Philadelphia: Saunders, 2000:666–694.

Commission on Accreditation for Dietetics Education. Knowledge, skills, and competencies for dietitians. *Accreditation Manual for Dietetics Education Programs,* rev. 4th ed. Chicago: American Dietetic Association, 2000.

Engstrom PF, Goosenberg EB. Diverticulosis. In *Diagnosis and Management of Bowel Disease.* PCI; 1999. Available at http://www.medscape.com/PCI/bowel/bowel.ch09/bowel.ch09–01.html. Accessed October 10, 2000.

Escott-Stump S. *Nutrition and Diagnosis Related Care,* 4th ed. Baltimore: Williams & Wilkins, 1998.

Simmang CL, Shires GT. Diverticular disease of the colon. In Feldman M, Sleisenger MH, Scharschmidt BF (eds.), *Gastrointestinal and Liver Disease,* 6th ed. Philadelphia: Saunders, 1998.

# Irritable Bowel Syndrome

*Introductory Level*

## Objectives

After completing this case, the student will be able to:

1. Describe the proposed pathophysiology in irritable bowel syndrome.
2. Identify food and lifestyle patterns that may contribute to symptoms of irritable bowel syndrome.
3. Apply principles of medical nutrition therapy to assist in the medical care of irritable bowel syndrome.
4. Interpret current research to explain the rationale for proposed medical treatment.

5. Determine the nutrient needs of an individual with irritable bowel syndrome.

Mrs. Suzanne Freeman is admitted to University Hospital because of exacerbation of chronic abdominal pain. Extensive evaluation at an outside hospital reveals a normal colonoscopy and endoscopy. She is diagnosed with irritable bowel syndrome (IBS) per established criteria.

 **UNIVERSITY HOSPITAL**

# ADMISSION DATABASE

Name: Suzanne Freeman
DOB: 9/3  age 38
Physician: K. Friedman, MD

| BED# 2 | DATE: 4/3 | TIME: 1200 | TRIAGE STATUS (ER ONLY): ☐ Red ☐ Yellow ☐ Green ☐ White |
|---|---|---|---|

**Initial Vital Signs**

| TEMP: 98 | RESP: 16 | SAO2: | |
|---|---|---|---|

| HT: 5'4" | WT (lb): 109 | B/P: 128/72 | PULSE: 100 |
|---|---|---|---|

| LAST TETANUS ? | LAST ATE 2 hrs ago | LAST DRANK 2 hrs ago |
|---|---|---|

**PRIMARY PERSON TO CONTACT:**
Name: Evan Freeman
Home #: 202-453-7812
Work #: 202-549-7824

**ORIENTATION TO UNIT:** ☒ Call light ☒ Television/telephone ☒ Bathroom ☒ Visiting ☒ Smoking ☒ Meals ☒ Patient rights/responsibilities

### CHIEF COMPLAINT/HX OF PRESENT ILLNESS

"I am here for a workup of my stomach and intestinal problems."

**PERSONAL ARTICLES:** (Check if retained/describe)
☒ Contacts ☒ R ☒ L          ☐ Dentures ☐ Upper ☐ Lower
☒ Jewelry: wedding band
☒ Other: glasses

### ALLERGIES: Meds, Food, IVP Dye, Seafood: Type of Reaction

Milk?

**VALUABLES ENVELOPE:** none
☐ Valuables instructions

### PREVIOUS HOSPITALIZATIONS/SURGERIES

vaginal delivery 8 years previous

**INFORMATION OBTAINED FROM:**
☒ Patient          ☐ Previous record
☒ Family          ☐ Responsible party

Signature  *Suzanne Freeman*

| Home Medications (including OTC) | | Codes: A=Sent home | B=Sent to pharmacy | | C=Not brought in |
|---|---|---|---|---|---|
| Medication | Dose | Frequency | Time of Last Dose | Code | Patient Understanding of Drug |
| Lomotil | 2 T | prn | 3-4 days ago | c | yes |
| | | | | | |
| | | | | | |
| | | | | | |
| | | | | | |
| | | | | | |
| | | | | | |
| | | | | | |
| | | | | | |
| | | | | | |
| | | | | | |
| | | | | | |

Do you take all medications as prescribed?     ☒ Yes     ☐ No     If no, why?

### PATIENT/FAMILY HISTORY

☐ Cold in past two weeks
☐ Hay fever
☐ Emphysema/lung problems
☐ TB disease/positive TB skin test
☐ Cancer
☐ Stroke/past paralysis
☒ Heart attack Paternal grandfather
☒ Angina/chest pain Paternal grandfather
☒ Heart problems Paternal grandfather

☒ High blood pressure Paternal grandfather
☐ Arthritis
☐ Claustrophobia
☐ Circulation problems
☐ Easy bleeding/bruising/anemia
☐ Sickle cell disease
☐ Liver disease/jaundice
☐ Thyroid disease
☐ Diabetes

☐ Kidney/urinary problems
☒ Gastric/abdominal pain/heartburn Patient
☐ Hearing problems
☐ Glaucoma/eye problems
☐ Back pain
☐ Seizures
☐ Other

### RISK SCREENING

Have you had a blood transfusion?     ☐ Yes     ☒ No
Do you smoke?     ☒ Yes     ☐ No
If yes, how many pack(s)  1-2 /day for 20 years
Does anyone in your household smoke?     ☒ Yes     ☐ No
Do you drink alcohol?     ☒ Yes     ☐ No
If yes, how often?  weekly   How much?  1-2 drinks
When was your last drink?  last week /     /
Do you take any recreational drugs?     ☐ Yes     ☒ No
If yes, type:_____     Route
Frequency:_____     Date last used:_____/_____/

**FOR WOMEN Ages 12–52**

Is there any chance you could be pregnant?     ☐ Yes     ☒ No
If yes, expected date (EDC):
Gravida/Para: 1/1

**ALL WOMEN**

Date of last Pap smear:    /   / last year
Do you perform regular breast self-exams?     ☒ Yes     ☐ No

**ALL MEN**

Do you perform regular testicular exams?     ☐ Yes     ☐ No

Additional comments:

**x** *Frank Smithson, RN*
Signature/Title

**Client name:** Suzanne Freeman
**DOB:** 9/3
**Age:** 38
**Sex:** Female
**Education:** College graduate
**Occupation:** Financial analyst
**Hours of work:** M–F 9–6 PM, often takes work home on weekends
**Household members:** Husband age 48, 1 daughter age 8
**Ethnic background:** Caucasian
**Religious affiliation:** Presbyterian
**Referring physician:** Karen Friedman, MD

## Chief complaint:

"I am here for a workup of my stomach and intestinal problems. I have always had a funny stomach, I think. As far back as I can remember I have had times when I had diarrhea and others when I would go for days without going to the bathroom. I have always had to be careful when I eat a large meal or one that is very rich. I almost always have diarrhea then. It is much worse now than it has ever been and is really interfering with my life."

## Patient history:

*Onset of disease:* Patient was admitted because of exacerbation of chronic abdominal pain. The patient describes the pain as lower abdominal cramping associated with several episodes of loose bowel movements a day. Her pain is not affected by defecation. Pain is increased with the frequency of stools that she experiences. These episodes of loose stool are interspersed with normal bowel movements. There is no blood or mucus in stool. No tenesmus or pain on defecation. There is no association between pain and food ingestion. Occasional nausea and nonbloody emesis. No fever or chills. Has experienced some weight loss—she states 10 lbs or so. The patient has been having increasing episodes of abdominal pain and diarrhea since the birth of her child at age 30. She was recently discharged from an outside hospital where she underwent extensive workup because of similar complaints. The evaluation included EGD and colonoscopy, both of which were normal.
*Type of Tx:* None at present
*PMH:* Unremarkable; no previous hospitalizations or significant illness
*Meds:* Over-the-counter antidiarrheal agents
*Smoker:* Yes
*Family Hx: What?* CAD *Who?* Grandfather

## Physical exam:

*General appearance:* Anxious-appearing 38-year-old female
*Vitals:* Temp 98°F, BP 128/72, HR 100 bpm, RR 16 bpm
*Heart:* WNL
*HEENT:* WNL
  *Eyes:* PERRLA
  *Ears:* WNL
  *Nose:* WNL
  *Throat:* WNL

*Genitalia:* Good rectal tone; no masses, tenderness, or blood; brown heme-negative stool
*Neurologic:* Alert and oriented ×3
*Extremities:* WNL
*Skin:* Warm and dry to touch.
*Chest/lungs:* WNL
*Peripheral vascular:* WNL
*Abdomen:* Soft, BS present, no organomegaly or masses. Some lower abdominal tenderness.

**Nutrition Hx:**
*General:* Patient states she rarely cooks because of her schedule and she does not like to cook. Her family eats out 3–4× each week. At home, she uses ready-prepared foods. She never eats breakfast and usually eats fast food for lunch if she eats at all. Lately she has avoided eating in an attempt to control her pain and diarrhea. States there are many foods she does not like. She does not eat fruits or vegetables often. Patient states she occasionally eats cheese but does not drink milk in fear it will cause her to have diarrhea. When questioned more about milk intake, patient states she has never really consumed milk as an adult because of the diarrhea it seems to cause.

*Usual dietary intake:*
*AM:* Coffee
*Lunch:* Burger or sub sandwich, chips or fries, diet soda
*Dinner:* At a variety of restaurants, Italian, Asian, or a buffet/smorgasbord type of restaurant

*Food allergies/intolerances/aversions:* Possibly milk
*Previous MNT?* No
*Anthropometric data:* Ht 5′4″, Wt 109#, UBW 120–125#, TSF 6 mm; MAC 11″
*Food purchase/preparation:* Self and husband
*Vit/min intake:* None

**Tx plan:**
*Admit to GI:* Karen Friedman, MD
*Diagnosis:* R/O irritable bowel syndrome per Rome II criteria
*Condition:* Stable
*Vitals:* Routine
*Diet:* Clear liquids then advance per RD consult
*Activity:* Ad lib
*Lab:* (CBC cd&p, A-8, SMA, CXR, Ca-Mg-PO$_4$, lactose tolerance test)
*Meds:* Begin trial of 5-HT$_3$ antagonist—alosetron; bulking agent—Metamucil, 1 oz in 8 oz of H$_2$O or juice
*PRN Meds:* Tylenol
Nutrition consult

# U<sub>H</sub> _UNIVERSITY HOSPITAL_

NAME: Suzanne Freeman                     DOB: 9/3
AGE: 38                                   SEX: F
PHYSICIAN: K. Friedman, MD

\*\*\*\*\*\*\*\*\*\*\*\*\*\*\*\*\*\*\*\*\*\*\*\*\*\*\*\*\*\*\*\*\*\*\*\*\*\*\*\*\*\*\*\*\*\*\*CHEMISTRY\*\*\*\*\*\*\*\*\*\*\*\*\*\*\*\*\*\*\*\*\*\*\*\*\*\*\*\*\*\*\*\*\*\*\*\*\*\*\*\*\*\*\*\*\*\*\*

DAY:                                                Admit
DATE:
TIME:
LOCATION:

| | NORMAL | | UNITS |
|---|---|---|---|
| Albumin | 3.6–5 | 4.1 | g/dL |
| Total protein | 6–8 | 7.4 | g/dL |
| Prealbumin | 19–43 | | mg/dL |
| Transferrin | 200–400 | | mg/dL |
| Sodium | 135–155 | 138 | mmol/L |
| Potassium | 3.5–5.5 | 3.2 | mmol/L |
| Chloride | 98–108 | 110 | mmol/L |
| $PO_4$ | 2.5–4.5 | 3.7 | mmol/L |
| Magnesium | 1.6–2.6 | 2.1 | mmol/L |
| Osmolality | 275–295 | 285 | mmol/kg $H_2O$ |
| Total $CO_2$ | 24–30 | 27 | mmol/L |
| Glucose | 70–120 | 126 | mg/dL |
| BUN | 8–26 | 7 | mg/dL |
| Creatinine | 0.6–1.3 | 0.5 | mg/dL |
| Uric acid | 2.6–6 (women) | | mg/dL |
| | 3.5–7.2 (men) | | |
| Calcium | 8.7–10.2 | 9.5 | mg/dL |
| Bilirubin | 0.2–1.3 | 0.9 | mg/dL |
| Ammonia ($NH_3$) | 9–33 | | $\mu$mol/L |
| SGPT (ALT) | 10–60 | 27 | U/L |
| SGOT (AST) | 5–40 | 19 | U/L |
| Alk phos | 98–251 | 110 | U/L |
| CPK | 26–140 (women) | | U/L |
| | 38–174 (men) | | |
| LDH | 313–618 | | U/L |
| CHOL | 140–199 | | mg/dL |
| HDL-C | 40–85 (women) | | mg/dL |
| | 37–70 (men) | | |
| VLDL | | | mg/dL |
| LDL | < 130 | | mg/dL |
| LDL/HDL ratio | < 3.22 (women) | | |
| | < 3.55 (men) | | |
| Apo A | 101–199 (women) | | mg/dL |
| | 94–178 (men) | | |
| Apo B | 60–126 (women) | | mg/dL |
| | 63–133 (men) | | |
| TG | 35–160 | | mg/dL |
| $T_4$ | 5.4–11.5 | | $\mu$g/dL |
| $T_3$ | 80–200 | | ng/dL |
| $HbA_{1C}$ | 4.8–7.8 | | % |
| Amylase | 25–125 | 55 | U/L |
| Lipase | 10–140 | 117 | U/L |

**U**H *UNIVERSITY HOSPITAL*

NAME: Suzanne Freeman                    DOB: 9/3
AGE: 38                                   SEX: F
PHYSICIAN: K. Friedman, MD

\*\*\*\*\*\*\*\*\*\*\*\*\*\*\*\*\*\*\*\*\*\*\*\*\*\*\*\*\*\*\*\*\*\*\*\*\*\*\*\*\*\*\*\*HEMATOLOGY\*\*\*\*\*\*\*\*\*\*\*\*\*\*\*\*\*\*\*\*\*\*\*\*\*\*\*\*\*\*\*\*\*\*\*\*\*\*\*\*\*\*\*

DAY:                                              Admit
DATE:
TIME:
LOCATION:

| | NORMAL | | UNITS |
|---|---|---|---|
| WBC | 4.3–10 | 6.91 | $\times\ 10^3/mm^3$ |
| RBC | 4–5 (women) | | $\times\ 10^6/mm^3$ |
| | 4.5–5.5 (men) | | |
| HGB | 12–16 (women) | 12.3 | g/dL |
| | 13.5–17.5 (men) | | |
| HCT | 37–47 (women) | 35.8 | % |
| | 40–54 (men) | | |
| MCV | 84–96 | | fL |
| MCH | 27–34 | | pg |
| MCHC | 31.5–36 | | % |
| RDW | 11.6–16.5 | | % |
| Plt ct | 140–440 | 193 | $\times\ 10^3$ |
| Diff TYPE | | | |
| % GRANS | 34.6–79.2 | | % |
| % LYM | 19.6–52.7 | | % |
| SEGS | 50–62 | 52 | % |
| BANDS | 3–6 | 4 | % |
| LYMPHS | 25–40 | 16 | % |
| MONOS | 3–7 | 5 | % |
| EOS | 0–3 | 1 | % |
| TIBC | 65–165 (women) | | μg/dL |
| | 75–175 (men) | | |
| Ferritin | 18–160 (women) | | μg/dL |
| | 18–270 (men) | | |
| Vitamin $B_{12}$ | 100–700 | | pg/mL |
| Folate | 2–20 | | ng/mL |

# U<sub>H</sub> *UNIVERSITY HOSPITAL*

NAME: Suzanne Freeman                          DOB: 9/3
AGE: 38                                        SEX: F
PHYSICIAN: K. Friedman, MD

```
*****************************************CHEMISTRY*****************************************
```

DAY:                          2
DATE:                         4/4
TIME:
LOCATION:

                                                                              UNITS
```
-------------------------------------------------------------------------------
```
**Lactose Tolerance**                                                         mg/dL
Serum glucose                 70
(1–30 minutes)
Serum glucose                 72
(2–60 minutes)
Serum glucose                 70
(3–120 minutes)

## Case Questions

1. Irritable bowel syndrome (IBS) is a condition that is often undiagnosed and often misunder-stood. What do we currently understand about the etiology of IBS?

2. As Mrs. Freeman's physicians work up a differential diagnosis, what clues point them in the direction of IBS?

3. Her MD notes that diagnosis would be referencing the Rome II criteria. What are these, and how are Mrs. Freeman's symptoms consistent with these criteria?

4. Evaluate the patient's anthropometric data.

   a. Calculate her DBW, percent UBW, and BMI. What can you conclude from this assessment? Explain.

   b. Using the other anthropometric values, MAC and TSF, calculate her upper arm muscle area.

5. Evaluate Mrs. Freeman's laboratory values, and compare her values to the normal values. Are any of nutritional significance? Explain.

6. Mrs. Freeman's lab values indicate normal amylase and lipase. Is this information useful for the physician in making a diagnosis? What do these normal values rule out as a cause of her ab-dominal pain and diarrhea? Explain.

7. How is the lactose tolerance test conducted? What do this patient's results indicate?

8. What foods provide lactose? What should you recommend for treatment of lactase deficiency or lactase nonpersistence? What nutrients do you need to be concerned about for someone who is lactose intolerant?

9.    Evaluate Mrs. Freeman's dietary history, and identify at least four problems that may be aggra-
vating her current condition. Provide a rationale for identifying each.

10.    What are the dietary recommendations for treating irritable bowel syndrome?

11.    The resident working with Dr. Friedman asked the RD to make recommendations for advanc-
ing Mrs. Freeman's diet in the hospital. What diet order would you recommend? Explain.

12.    After you make recommendations to the resident, Mrs. Freeman receives for her first meal:
Cheerios cereal, bran muffin, orange juice, lactose-free milk, and half a banana. She is very
upset and requests to see the RD. She tells you she is very afraid to eat these foods and she
knows she will have diarrhea if she does eat them. What would you tell her?

13.    Using the problems identified in Question 9, outline medical nutrition therapy goals for
Mrs. Freeman. For each goal, identify possible interventions or strategies to meet that goal. You
may use the following table to organize your information.

| Goal | Interventions to Meet Goal |
|---|---|
|  |  |
|  |  |
|  |  |
|  |  |
|  |  |

14.    Mrs. Freeman has lost weight. What would you suggest as her target energy and protein goals?
What standards did you use?

15.    Mrs. Freeman's MD initiated two medications: alosetron and Metamucil. What are these
medications, and what would their role be in her treatment?

# Bibliography

Alderman J. Managing irritable bowel syndrome. *Adv Nurse Prac.* 1999;7(1):40–46.

American Dietetic Association. (2000). Irritable bowel syndrome. In *Manual of Clinical Dietetics,* 6th ed. Chicago: American Dietetic Association, 2000:406–409.

American Gastroenterological Association. Medical position statement: irritable bowel syndrome. *Gastroenterology.* 1997;112:2118–2119.

Beyer PL. Medical nutrition therapy for lower gastrointestinal tract disorders. In Mahan LK, Escott-Stump S (eds.), *Krause's Food, Nutrition, and Diet Therapy,* 10th ed. Philadelphia: Saunders, 2000:666–694.

Browning SM. Constipation, diarrhea, and irritable bowel syndrome. *Primary Care.* 1999;26:113–139.

Camileri M, Northcutt AR, Kong S, Dukes GE, McSorley D, Mangel AW. Efficacy and safety of alosetron in women with irritable bowel syndrome: a randomized, placebo-controlled trial. *Lancet.* 2000;355:1035–1040.

Commission on Accreditation for Dietetics Education. Knowledge, skills, and competencies for dietitians. *Accreditation Manual for Dietetics Education Programs,* rev. 4th ed. Chicago: American Dietetic Association, 2000.

Dixon-Woods M, Critchley S. Medical and lay views of irritable bowel syndrome. *Fam Prac.* 2000;17(2):108–113.

Farthing MJ. Irritable bowel syndrome: new pharmaceutical approaches to treatment. *Baillieres Best Pract Res Clin Gastroenterol.* 1999;13(3):461–471.

Foltz-Gray D. Soothe the savage bowel. *Health.* 2000;14(3):92–98.

Kaplan L. Irritable bowel syndrome: not just a gut feeling. *Nursing Spectrum.* 2000;10:16–18.

Managing irritable bowel syndrome. *Nutr Forum.* 2000;17(3):19–22.

Olden KW. (2000). New insights into irritable bowel syndrome. American College of Gastroenterology 65th Annual Scientific Meeting. Available at http://www.medscape.com/medscape/cno/2000/ACG/Story.cfm?story_id=1713. Accessed October 18, 2000.

Talley NJ. Irritable bowel syndrome: definition, diagnosis, and epidemiology. *Res. Clin. Gastroent.* 1999;13(3):371–384.

Thompson GW, Heaton KW, Smyth GT, Smyth C. Irritable bowel syndrome in general practice: prevalence, characteristics, and referral. *Gut.* 2000;46(1):778–782.

Thompson GW, Irvine JE, Pare P, et al. Comparing Rome I and Rome II criteria for irritable bowel syndrome in a prospective survey of the Canadian population. Program and abstracts of the 65th Annual Scientific Meeting of the American College of Gastroenterology; October 16–18, 2000, New York, NY. Poster 312, p. 461.

# Inflammatory Bowel Disease: Crohn's versus Ulcerative Colitis— Medical and Nutritional Treatment

*Introductory Level*

## Objectives

After completing this case, the student will be able to:

1. Describe the differences between Crohn's disease and ulcerative colitis.
2. Demonstrate ability to interpret the physiological changes resulting from Crohn's disease.
3. Identify the nutritional consequences of the physiological effects of Crohn's disease.
4. Identify potential drug–nutrient interactions.
5. Describe the current medical care for Crohn's disease.
6. Interpret laboratory, dietary, and anthropometric data for assessment of nutritional status.
7. Demonstrate the ability to establish an enteral feeding regimen.

Michael Sims was diagnosed with Crohn's disease 18 months previously. He is now admitted with an acute exacerbation of that disease.

# ADMISSION DATABASE

Name: Michael Sims
DOB: 7/22  age 35
Physician: David Tucker, MD

| BED# 2 | DATE: 9/4 | TIME: 1500 | TRIAGE STATUS (ER ONLY): ☐ Red ☐ Yellow ☐ Green ☐ White |
|---|---|---|---|

**Initial Vital Signs**

| TEMP: 101.5 | RESP: 18 | SAO2: |
|---|---|---|

| HT: 5'9" | WT (lb): 140 | B/P: 125/82 | PULSE: 81 |
|---|---|---|---|

| LAST TETANUS unknown | | LAST ATE this am | LAST DRANK 30 minutes ago |
|---|---|---|---|

**PRIMARY PERSON TO CONTACT:**
Name: Mary Sims
Home #: 614-447-1476
Work #: 614-447-2322

**ORIENTATION TO UNIT:** ☒ Call light  ☒ Television/telephone  ☒ Bathroom  ☒ Visiting  ☒ Smoking  ☒ Meals  ☒ Patient rights/responsibilities

### CHIEF COMPLAINT/HX OF PRESENT ILLNESS

"I was diagnosed with Crohn's disease almost 18 months ago. I have really done OK until recently."

**PERSONAL ARTICLES:** (Check if retained/describe)
☐ Contacts ☐ R ☐ L          ☐ Dentures ☐ Upper ☐ Lower
☐ Jewelry:
☐ Other:

### ALLERGIES: Meds, Food, IVP Dye, Seafood: Type of Reaction

Maybe milk—otherwise NKA.

**VALUABLES ENVELOPE:**
☐ Valuables instructions

### PREVIOUS HOSPITALIZATIONS/SURGERIES

18 months ago—Dx. Crohn's disease

**INFORMATION OBTAINED FROM:**
☒ Patient          ☐ Previous record
☒ Family          ☐ Responsible party

Signature  *Michael Sims*

| Home Medications (including OTC) | Codes: A=Sent home | | B=Sent to pharmacy | | C=Not brought in |
|---|---|---|---|---|---|
| Medication | Dose | Frequency | Time of Last Dose | Code | Patient Understanding of Drug |
| Azulfidine | 500 mg | qid | this am | c | yes |
| multivitamin | 1 | daily | this am | c | yes |
| | | | | | |
| | | | | | |
| | | | | | |
| | | | | | |
| | | | | | |
| | | | | | |
| | | | | | |
| | | | | | |
| | | | | | |

Do you take all medications as prescribed?  ☒ Yes  ☐ No   If no, why?

### PATIENT/FAMILY HISTORY

| | | |
|---|---|---|
| ☐ Cold in past two weeks | ☐ High blood pressure | ☐ Kidney/urinary problems |
| ☐ Hay fever | ☐ Arthritis | ☒ Gastric/abdominal pain/heartburn Patient |
| ☐ Emphysema/lung problems | ☐ Claustrophobia | ☐ Hearing problems |
| ☐ TB disease/positive TB skin test | ☐ Circulation problems | ☐ Glaucoma/eye problems |
| ☐ Cancer | ☐ Easy bleeding/bruising/anemia | ☐ Back pain |
| ☐ Stroke/past paralysis | ☐ Sickle cell disease | ☐ Seizures |
| ☐ Heart attack | ☐ Liver disease/jaundice | ☐ Other |
| ☐ Angina/chest pain | ☐ Thyroid disease | |
| ☐ Heart problems | ☐ Diabetes | |

### RISK SCREENING

Have you had a blood transfusion?  ☐ Yes  ☒ No
Do you smoke?  ☐ Yes  ☒ No
If yes, how many pack(s)      /day for      years
Does anyone in your household smoke?  ☐ Yes  ☒ No
Do you drink alcohol?  ☐ Yes  ☒ No
If yes, how often?      week  How much?
When was your last drink?      /      /
Do you take any recreational drugs?  ☐ Yes  ☒ No
If yes, type:_____      Route
Frequency:_____      Date last used:_____/_____/

**FOR WOMEN Ages 12–52**
Is there any chance you could be pregnant?  ☐ Yes  ☐ No
If yes, expected date (EDC):
Gravida/Para:

**ALL WOMEN**
Date of last Pap smear:
Do you perform regular breast self-exams?  ☐ Yes  ☐ No

**ALL MEN**
Do you perform regular testicular exams?  ☒ Yes  ☐ No

Additional comments:

**✗** *Misty Taylor, LPN*
Signature/Title

**Client name:** Michael Sims
**DOB:** 7/22
**Age:** 35
**Sex:** Male
**Education:** Bachelor's degree
**Occupation:** High school math teacher
**Hours of work:** 8–4:30, some after-school meetings and responsibilities as adviser for school clubs
**Household members:** Wife age 32, well; son age 5, well
**Ethnic background:** Caucasian
**Religious affiliation:** Episcopalian
**Referring physician:** David Tucker, MD (gastroenterology)

## Chief complaint:

"I was diagnosed with inflammatory bowel disease almost 2 years ago. At first they thought I had ulcerative colitis but six months later it was identified as Crohn's disease. I was really sick at that time and was in the hospital for more than two weeks. I have done OK until school started this fall. I've noticed more diarrhea and abdominal pain, but I've tried to keep going since school just started. Now I can't ignore it. There is no way I can work like this. My abdominal pain is unbearable—I seem to have diarrhea constantly, and now I am running a fever."

## Patient history:

*Onset of disease:* Dx. Crohn's disease 18 months previously. Initial diagnostic workup indicated acute disease within last 5–7 cm of jejunum and first 5 cm of ileum—acutely treated with corticosteroids and then maintained on Azulfidine.
*Type of Tx:* Corticosteroids and Azulfidine
*PMH:* Noncontributory
*Meds:* Azulfidine 500 mg qid
*Smoker:* No
*Family Hx:* Noncontributory

## Physical exam:

*General appearance:* Thin, 35-year-old white male in apparent distress.
*Vitals:* Temp 101.5°F, BP 125/82, HR 81 bpm/normal, RR 18 bpm
*Heart:* RRR without murmurs or gallops
*HEENT:*
  *Eyes:* PERRLA, normal fundi
  *Ears:* noncontributory
  *Nose:* noncontributory
  *Throat:* Phayrnx clear
*Rectal:* No evidence of perianal disease
*Neurologic:* Oriented × 4
*Extremities:* No edema, pulses full, no bruits, normal strength, sensation, and DTR
*Skin:* Warm, dry
*Chest/lungs:* Lungs clear to auscultation and percussion
*Abdomen:* Slight distension, extreme tenderness with rebound and guarding; hyperactive bowel sounds

## Nutrition Hx:

*General:* Patient states he has been eating fairly normally for the last year. After first hospitalization, he had lost almost 25 lb, which he regained. He initially ate a low-fiber diet and worked hard to regain the weight he had lost. He drank Boost between meals for several months. His usual weight before his illness was 166–168#. He was at his highest weight (168#) about 6 months ago but now states he has lost most of what he regained.

*Recent dietary intake:*

| | |
|---|---|
| AM: | Cereal, small amount of skim milk, toast or bagel; juice |
| AM snack: | Cola—sometimes crackers or pastry |
| Lunch: | Sandwich (ham or turkey) from home, fruit, chips, cola |
| Dinner: | Meat, pasta or rice, some type of bread. Rarely eats vegetables |
| Bedtime snack: | Cheese and crackers, cookies, cola |
| 24-hr recall: | Has been on clear liquids for past 24 hours since admission |

*Food allergies/intolerances/aversions:* Mr. Sims says he has never liked milk but purposefully avoided it after his diagnosis. He does consume milk products such as cheese, usually without any difficulty.

*Previous MNT?* Yes    *If yes, when:* Last hospitalization

*What?* "The dietitian talked to me about ways to decrease my diarrhea—ways to keep from being dehydrated—and then we worked out a plan to help me regain weight."

*Food purchase/preparation:* Self and wife

*Vit/min intake:* Multivitamin daily

## Tx plan:

R/O acute exacerbation of Crohn's disease vs. infection

CBC/Chem 24

ASCA

CT scan of abdomen and possible esophagogastroduodenoscopy

$D_5W$ NS @ 75 cc/hr

Clear liquids

Nutrition consult

## Hospital course:

CT scan indicated abscess first 2 cm of ileum. Esophagogastroduodenoscopy indicates exacerbation of Crohn's along the first 5–7 cm of ileum. No inflammation of duodenum or jejunum. CDAI score of 210.

*Rx:* $D_5NS$ with 60 mEq KCl @ 100 cc/hr; IV methylprednisolone 60 mg/day; metronidazole 250 mg qid; ciprofloxacin 500 mg bid

NPO

s/p nasoduodenal tube placement

Nutrition consult for enteral nutrition support

U**H** *UNIVERSITY HOSPITAL*

NAME: Michael Sims                         DOB: 7/22
AGE: 35                                    SEX: M
PHYSICIAN: David Tucker, MD

\*\*\*\*\*\*\*\*\*\*\*\*\*\*\*\*\*\*\*\*\*\*\*\*\*\*\*\*\*\*\*\*\*\*\*\*\*\*\*\*\*\*\*\*\*\*CHEMISTRY\*\*\*\*\*\*\*\*\*\*\*\*\*\*\*\*\*\*\*\*\*\*\*\*\*\*\*\*\*\*\*\*\*\*\*\*\*\*\*\*\*\*\*\*

| DAY: | | Admit | 2 | |
|---|---|---|---|---|
| DATE: | | | | |
| TIME: | | | | |
| LOCATION: | | | | |
| | NORMAL | | | UNITS |
| Albumin | 3.6–5 | 2.9 | 2.7 | g/dL |
| Total protein | 6–8 | 5.7 | 5.6 | g/dL |
| Prealbumin | 19–43 | 16 | 16 | mg/dL |
| Transferrin | 200–400 | 220 | 190 | mg/dL |
| Sodium | 135–155 | 138 | 139 | mmol/L |
| Potassium | 3.5–5.5 | 3.5 | 3.6 | mmol/L |
| Chloride | 98–108 | 99 | 101 | mmol/L |
| $PO_4$ | 2.5–4.5 | 3.1 | 3.0 | mmol/L |
| Magnesium | 1.6–2.6 | 1.5 | 1.4 | mmol/L |
| Osmolality | 275–295 | 294 | 285 | mmol/kg $H_2O$ |
| Total $CO_2$ | 24–30 | 27 | 26 | mmol/L |
| Glucose | 70–120 | 79 | 145 | mg/dL |
| BUN | 8–26 | 8 | 9 | mg/dL |
| Creatinine | 0.6–1.3 | 0.6 | 0.7 | mg/dL |
| Uric acid | 2.6–6 (women) | 4.1 | 3.6 | mg/dL |
| | 3.5–7.2 (men) | | | |
| Calcium | 8.7–10.2 | 8.9 | 9.1 | mg/dL |
| Bilirubin | 0.2–1.3 | 0.2 | 0.3 | mg/dL |
| Ammonia ($NH_3$) | 9–33 | 11 | 15 | μmol/L |
| SGPT (ALT) | 10–60 | 12 | 15 | U/L |
| SGOT (AST) | 5–40 | 15 | 22 | U/L |
| Alk phos | 98–251 | 101 | 99 | U/L |
| CPK | 26–140 (women) | | | U/L |
| | 38–174 (men) | | | |
| LDH | 313–618 | | | U/L |
| CHOL | 140–199 | 121 | 130 | mg/dL |
| HDL-C | 40–85 (women) | | | mg/dL |
| | 37–70 (men) | | | |
| VLDL | | | | mg/dL |
| LDL | < 130 | | | mg/dL |
| LDL/HDL ratio | < 3.22 (women) | | | |
| | < 3.55 (men) | | | |
| Apo A | 101–199 (women) | | | mg/dL |
| | 94–178 (men) | | | |
| Apo B | 60–126 (women) | | | mg/dL |
| | 63–133 (men) | | | |
| TG | 35–160 | 72 | | mg/dL |
| $T_4$ | 5.4–11.5 | | | μg/dL |
| $T_3$ | 80–200 | | | ng/dL |
| $HbA_{1C}$ | 4.8–7.8 | | | % |
| ANCA | negative | – | | U/L |
| ASCA | negative | + | | U/L |

# $U_H$ _UNIVERSITY HOSPITAL_

NAME: Michael Sims                     DOB: 7/22
AGE: 35                                SEX: M
PHYSICIAN: David Tucker, MD

\*\*\*\*\*\*\*\*\*\*\*\*\*\*\*\*\*\*\*\*\*\*\*\*\*\*\*\*\*\*\*\*\*\*\*\*\*\*\*\*\*\*HEMATOLOGY\*\*\*\*\*\*\*\*\*\*\*\*\*\*\*\*\*\*\*\*\*\*\*\*\*\*\*\*\*\*\*\*\*\*\*\*\*\*\*\*\*\*

| DAY: | | Admit | 2 | |
|------|--------|-------|---|-------|
| DATE: | | | | |
| TIME: | | | | |
| LOCATION: | | | | |
| | NORMAL | | | UNITS |
| WBC | 4.3–10 | 15.5 | | $\times\ 10^3/mm^3$ |
| RBC | 4–5 (women) | 4.9 | | $\times\ 10^6/mm^3$ |
| | 4.5–5.5 (men) | | | |
| HGB | 12–16 (women) | 12.9 | | g/dL |
| | 13.5–17.5 (men) | | | |
| HCT | 37–47 (women) | 38 | | % |
| | 40–54 (men) | | | |
| MCV | 84–96 | 87 | | fL |
| MCH | 27–34 | 30 | | pg |
| MCHC | 31.5–36 | 33 | | % |
| RDW | 11.6–16.5 | 13.2 | | % |
| Plt ct | 140–440 | 452 | | $\times\ 10^3$ |
| Diff TYPE | | | | |
| % GRANS | 34.6–79.2 | | | % |
| % LYM | 19.6–52.7 | | | % |
| SEGS | 50–62 | | | % |
| BANDS | 3–6 | | | % |
| LYMPHS | 25–40 | | | % |
| MONOS | 3–7 | | | % |
| EOS | 0–3 | | | % |
| TIBC | 65–165 (women) | 219 | | μg/dL |
| | 75–175 (men) | | | |
| Ferritin | 18–160 (women) | 16 | | μg/dL |
| | 18–270 (men) | | | |
| Vitamin $B_{12}$ | 100–700 | 300 | | pg/mL |
| Folate | 2–20 | 3 | | ng/mL |
| Total T cells | 812–2318 | | | $mm^3$ |
| T-helper cells | 589–1505 | | | $mm^3$ |
| T-suppressor cells | 325–997 | | | $mm^3$ |
| PT | 11–13 | | | sec |

## Case Questions

1.  What is inflammatory bowel disease? What does current medical literature indicate regarding its etiology?

2.  What does a CDAI score of 210 indicate?

3.  Mr. Sims was initially diagnosed with ulcerative colitis and then diagnosed with Crohn's. How could this happen? What are the similarities and differences between Crohn's disease and ulcerative colitis?

4.  What did you find in Mr. Sims's history and physical that is consistent with his diagnosis of Crohn's? Explain.

5.  What are the potential nutritional consequences of Crohn's disease? Are there any specific concerns when there is disease in the jejunum and ileum?

6.  Mr. Sims had an esophagogastroduodenoscopy as part of his diagnostic workup. What is this procedure?

7.  Evaluate Mr. Sims's anthropometric data by evaluating UBW, IBW, and BMI. Interpret your calculations.

8.  Identify any significant laboratory measurements from both his admission hematology and his chemistry labs. How might they be related to his admitting diagnosis?

| Lab | Patient's Value | Normal Value | Relationship to Diagnosis |
|-----|-----------------|--------------|---------------------------|
|     |                 |              |                           |
|     |                 |              |                           |
|     |                 |              |                           |
|     |                 |              |                           |
|     |                 |              |                           |
|     |                 |              |                           |
|     |                 |              |                           |
|     |                 |              |                           |
|     |                 |              |                           |
|     |                 |              |                           |
|     |                 |              |                           |

9.  Dr. Tucker started Mr. Sims on methylprednisolone 60 mg/day; metronidazole 250 mg qid; ciprofloxacin 500 mg bid. What are these medications? What are the potential drug–nutrient interactions?

10. Determine Mr. Sims's energy and protein requirements.

11. Dr. Tucker asked for a nutrition consult for enteral feeding. Do you agree that Mr. Sims should begin on enteral feeding? Why or why not?

12. Plan an enteral feeding regimen for Mr. Sims via a nasogastric feeding using a continuous drip administration. Determine formula choice, initiation rate, and goal rate. Give your rationale for your formula choice.

13. Look at Mr. Sims's labs on day 2 of his admission.

   a. Why do you think his albumin has decreased to 2.7 mg/dL? What are the limitations of evaluating albumin as a measure of short-term nutritional status?


   b. Mr. Sims's glucose increased to 145 mg/dL. Why do you think this level is now abnormal? What should be done about it?


14. As Mr. Sims's RD, what would you monitor to assure his tolerance to the enteral feeding?


15. Mr. Sims's symptoms now are much improved. He is now being weaned from the IV methyl-prednisolone and will be started on a regimen of 6-mercaptopurine and oral corticosteroids preparing for discharge. He will also start on an oral diet. What nutritional recommendations can you make for this transition to an oral diet? When would you recommend his enteral feeding be discontinued?

# Bibliography

American Society for Parenteral and Enteral Nutrition. First treatment approved for Crohn's disease. *FDA Consumer.* 1998;32(6):3.

Belluzzi A, Brignola C, Campieri M, et al. Effect of an enteric-coated fish-oil preparation on relapses in Crohn's disease. *N Engl J Med.* 1996;334:1557–1560.

Best WR, Becktel JM, Singleton JW. Rederived values of the eight coefficients of the Crohn's disease activity index (CDAI). *Gastroenterology.* 1979;77:843–846.

Beyer PL. (2000). Medical nutrition therapy for lower gastrointestinal tract disorders. In Mahan LK, Escott-Stump S (eds.), *Krause's Food, Nutrition, and Diet Therapy,* 10th ed. Philadelphia: Saunders, 2000:666–694.

Brandt LJ, Bernstein LH, Boley SJ, et al. Metronidazole therapy for perineal Crohn's disease: a follow-up study. *Gastroenterology.* 1982;83:383–387.

Buzby GP, Williford WO, Peterson OL, et al. A randomized clinical trial of total parenteral nutrition in malnourished surgical patients: the rationale and impact of previous clinical trials and pilot study on protocol design. *Am J Clin Nutr.* 1988;47:357–365.

Chan AT, Fleming R, O'Fallon WM, et al. Estimated versus measured basal energy requirements in patients with Crohn's disease. *Gastroenterology.* 1986;91:75–78.

Christie PM, Hill GL. Effect of intravenous nutrition on nutrition and function in acute attacks of inflammatory bowel disease. *Gastroenterology.* 1990;99:730–736.

Commission on Accreditation for Dietetics Education. Knowledge, skills, and competencies for dietitians. *Accreditation Manual for Dietetics Education Programs,* rev. 4th ed. Chicago: American Dietetic Association, 2000:29–45.

Den HE, Hiele M, Peeters M, et al. Effect of long-term oral glutamine supplements on small intestinal permeability in patients with Crohn's disease. *J Parenter Enteral Nutr.* 1999;23:7–11.

Fernandez-Banares F, Cabre E, Gonzalez-Huix F, Gassull MA. Enteral nutrition as primary therapy in Crohn's disease. *Gut.* 1994;35(1)Suppl:S55–59.

Gonzalez-Huix F, de Leon R, Fernandez-Banares F, et al. Polymeric enteral diets as primary treatment of active Crohn's disease: a prospective steroid controlled trial. *Gut.* 1993;34:778–782.

Greenberg GR, Fleming CR, Jeejeebhoy KN, et al. Controlled trial of bowel rest and nutritional support in the management of Crohn's disease. *Gut.* 1988;29:1309–1315.

Griffiths AM. Enteral nutrition: the neglected primary therapy of active Crohn's disease. *J Pediatr Gastroenterol Nutr.* 2000;31:3–5.

Griffiths AM, Ohlsson A, Sherman PM, et al. Meta-analysis of enteral nutrition as a primary treatment of active Crohn's disease. *Gastroenterology.* 1995;108:1056–1067.

Guidelines for the use of parenteral and enteral nutrition in adult and pediatric patients. *J Parenter Enteral Nutr.* 1993;17:7SA–11SA.

Heuschkel RB. Enteral nutrition in children with Crohn's disease. *J Pediatr Gastroenterol Nutr.* 2000;31:575.

Kelly DG, Fleming CR. Nutritional considerations in inflammatory bowel diseases. *Gastroenterol Clin North Am.* 1995;24:597–611.

Korelitz BI, Present DH. Favorable effects of 6-mercaptopurine in fistulae of Crohn's disease. *Dig Dis Sci.* 1985;30:58–64.

Kornbluth A, Salomon P, Sachar DB. Crohn's disease. In Sleisenger MH, Fordtran JS (eds.), *Gastrointestinal Disease.* Philadelphia: Saunders, 1994:1270–1304.

Lewis JD, Fisher RL. Nutrition support in inflammatory bowel disease. *Med Clin North Am.* 1994;78:1443–1456.

Lochs H, Steinhardt HJ, Klaus-Wentz B, et al. Comparison of enteral nutrition and drug treatment in active Crohn's disease. *Gastroenterology.* 1991;101:881–888.

Mansfield JC, Giaffer MH, Holdsworth CD. Controlled trial of oligopeptide versus amino acid diet in treatment of active Crohn's disease. *Gut.* 1995;36:60–66.

O'Brien CJ, Giaffer MH, Cann PA, et al. Elemental diet in steroid-dependent and steroid-refractory Crohn's disease. *Am J Gastroenterol.* 1991;86:1614–1618.

O'Morain C, Segal AW, Levi AJ. Elemental diet as primary treatment of acute Crohn's disease: a controlled trial. *BMJ.* 1984;288:1859–1862.

Rayhorn N. Understanding inflammatory bowel disease. *Nursing.* 1999;29:1257–1261.

Rigaud D, Angel LA, Cerf M, et al. Mechanisms of decreased food intake during weight loss in adult Crohn's disease patients without obvious malabsorption. *Am J Clin Nutr.* 1994;60:775–781.

Rigaud D, Cosnes J, Quintrec YL, et al. Controlled trial comparing two types of enteral nutrition in treatment of active Crohn's disease: elemental versus polymeric diet. *Gut.* 1991;32:1492–1497.

Riordan AM, Hunter JO, Cowan RE, et al. Treatment of active Crohn's disease by exclusion diet: East Anglian Multicentre controlled trial. *Lancet.* 1993;342:1131–1134.

Royall D, Greenberg GR, Allard JP, et al. Total enteral nutrition support improves body composition of patients with active Crohn's disease. *J Parenter Enteral Nutr.* 1995;19:95–99.

Royall D, Jeejeebhoy KN, Baker JP, et al. Comparison of amino acid versus peptide based enteral diets in active Crohn's disease: clinical and nutritional outcome. *Gut.* 1994;35:783–787.

Shepherd H, Barr G, Jewell D. Use of an intravenous steroid regimen in the treatment of acute Crohn's disease. *J Clin Gastroenterol.* 1986;8:154–159.

Soderholm JD, Olaison G, Lindberg E, et al. Different intestinal permeability patterns in relatives and spouses of patients with Crohn's disease: an inherited defect in mucosal defense? *Gut.* 1999;44:96–100.

Souba WW, Wilmore DW. Diet and nutrition in the care of the patient with surgery, trauma, and sepsis. In Shils ME, Olson JA, Shike M (eds.), *Modern Nutrition in Health and Disease.* Philadelphia: Lea & Febiger, 1994:1207–1240.

Stokes MA, Hill GL. Total energy expenditure in patients with Crohn's disease: measurement by the combined body scan technique. *J Parenter Enteral Nutr.* 1993;17:3–7.

# Inflammatory Bowel Disease: Crohn's versus Ulcerative Colitis—Surgical Treatment Resulting in Short Bowel Syndrome

*Introductory Level*

## Objectives

After completing this case, the student will be able to:

1. Identify the nutritional consequences of surgical resection of the small intestine.
2. Calculate parenteral nutrition formulation.
3. Determine the appropriate methods to monitor nutritional support.
4. Interpret laboratory, nutritional, and anthropometric information for nutritional assessment.
5. Develop a transitional feeding plan.

Michael Sims is readmitted 2½ years after his initial diagnosis of Crohn's disease. He undergoes a resection of 200 cm of jejunum and proximal ileum with placement of jejunostomy because of a small bowel obstruction.

**UNIVERSITY HOSPITAL**

# ADMISSION DATABASE

Name: Michael Sims
DOB: 7/22 age 36
Physician: David Tucker, MD

| BED#<br>2 | DATE:<br>12/15 | TIME:<br>1000 | TRIAGE STATUS (ER ONLY):<br>☐ Red  ☐ Yellow  ☐ Green  ☐ White |
|---|---|---|---|

**Initial Vital Signs**

| TEMP:<br>102 | RESP:<br>17 | SAO2: | |
|---|---|---|---|

| HT:<br>5'9" | WT (lb):<br>137 | B/P:<br>110/78 | PULSE:<br>82 |
|---|---|---|---|

| LAST TETANUS<br>unknown | | LAST ATE<br>this am | LAST DRANK<br>30 minutes ago |
|---|---|---|---|

**PRIMARY PERSON TO CONTACT:**
Name: Mary Sims
Home #: 614-447-1476
Work #: 614-447-2322

**ORIENTATION TO UNIT:** ☒ Call light ☒ Television/telephone
☒ Bathroom ☒ Visiting ☒ Smoking ☒ Meals
☒ Patient rights/responsibilities

### CHIEF COMPLAINT/HX OF PRESENT ILLNESS

"I am here to see if I need surgery for my Crohn's."

**PERSONAL ARTICLES:** (Check if retained/describe)
☐ Contacts ☐ R ☐ L  ☐ Dentures ☐ Upper ☐ Lower
☐ Jewelry:
☐ Other:

### ALLERGIES: Meds, Food, IVP Dye, Seafood: Type of Reaction

maybe milk—otherwise NKA

**VALUABLES ENVELOPE:**
☐ Valuables instructions

### PREVIOUS HOSPITALIZATIONS/SURGERIES

2½ years ago–Dx. Crohn's disease

September–Abscess and acute exacerbation of Crohn's

**INFORMATION OBTAINED FROM:**
☒ Patient   ☐ Previous record
☒ Family   ☐ Responsible party

Signature *Michael Sims*

### Home Medications (including OTC)   Codes: A=Sent home   B=Sent to pharmacy   C=Not brought in

| Medication | Dose | Frequency | Time of Last Dose | Code | Patient Understanding of Drug |
|---|---|---|---|---|---|
| 6-mercaptopurine | | daily | | c | yes |
| multivitamin | 1 | daily | | c | yes |
| | | | | | |
| | | | | | |
| | | | | | |
| | | | | | |
| | | | | | |
| | | | | | |
| | | | | | |
| | | | | | |

Do you take all medications as prescribed?  ☒ Yes  ☐ No   If no, why?

### PATIENT/FAMILY HISTORY

| | | |
|---|---|---|
| ☐ Cold in past two weeks | ☐ High blood pressure | ☐ Kidney/urinary problems |
| ☐ Hay fever | ☐ Arthritis | ☒ Gastric/abdominal pain/heartburn Patient |
| ☐ Emphysema/lung problems | ☐ Claustrophobia | ☐ Hearing problems |
| ☐ TB disease/positive TB skin test | ☐ Circulation problems | ☐ Glaucoma/eye problems |
| ☐ Cancer | ☐ Easy bleeding/bruising/anemia | ☐ Back pain |
| ☐ Stroke/past paralysis | ☐ Sickle cell disease | ☐ Seizures |
| ☐ Heart attack | ☐ Liver disease/jaundice | ☐ Other |
| ☐ Angina/chest pain | ☐ Thyroid disease | |
| ☐ Heart problems | ☐ Diabetes | |

### RISK SCREENING

Have you had a blood transfusion?  ☐ Yes  ☒ No
Do you smoke?  ☐ Yes  ☒ No
If yes, how many pack(s) ___ /day for ___ years
Does anyone in your household smoke?  ☐ Yes  ☒ No
Do you drink alcohol?  ☐ Yes  ☒ No
If yes, how often? ___ How much? ___
When was your last drink?  ___/___/___
Do you take any recreational drugs?  ☐ Yes  ☒ No
If yes, type:_____ Route ___
Frequency:_____ Date last used:_____/_____/_____

**FOR WOMEN Ages 12–52**

Is there any chance you could be pregnant?  ☐ Yes  ☐ No
If yes, expected date (EDC):
Gravida/Para:

**ALL WOMEN**

Date of last Pap smear:
Do you perform regular breast self-exams?  ☐ Yes  ☐ No

**ALL MEN**

Do you perform regular testicular exams?  ☒ Yes  ☐ No

Additional comments:

✗ *Rosie Martin, RN*
Signature/Title

**Client name:** Michael Sims
**DOB:** 7/22
**Age:** 36
**Sex:** Male
**Education:** Bachelor's degree
**Occupation:** High school math teacher
**Hours of work:** 8–4:30, some after-school meetings and responsibilities as adviser for school clubs
**Household members:** Wife age 33, well; son age 6, well
**Ethnic background:** Caucasian
**Religious affiliation:** Episcopalian
**Referring physician:** David Tucker, MD (gastroenterologist)

## Chief complaint:

"I was diagnosed with inflammatory bowel disease more than 3 years ago and with Crohn's six months later. I have had several episodes of the Crohn's over the last 2½ years. I was hospitalized and started on new meds, which seemed to be working. Now, obviously, it seems I am really sick again. My doctor says l may have to have surgery."

## Patient history:

*Onset of disease:* Dx Crohn's disease 2½ years previous. Regimens have included Azulfidine, cortico-steroids, antibiotics, and most recently 6-mercaptopurine.
*Type of Tx:* 6-MP
*PMH:* Noncontributory—no previous surgeries
*Meds:* 6-mercaptopurine
*Smoker:* No
*Family Hx:* Noncontributory

## Physical exam:

*General appearance:* Thin, 36-year-old white male in apparent distress
*Vitals:* Temp 102°F, BP 110/78, HR 82 bpm/normal, RR 17 bpm
*Heart:* RRR without murmurs or gallops
*HEENT:*
    *Eyes:* PERRLA, normal fundi
    *Ears:* Noncontributory
    *Nose:* Noncontributory
    *Throat:* Phayrnx clear
*Rectal:* No evidence of perianal disease
*Neurologic:* Oriented × 4
*Extremities:* No edema, pulses full, no bruits; normal strength, sensation, and DTR
*Skin:* Warm, dry
*Chest/lungs:* Lungs clear to auscultation and percussion
*Abdomen:* Distention, extreme tenderness with rebound and guarding; minimal bowel sounds
*Anthropometrics:* Ht 5'9", Adm Wt 137#, UBW 166–168#, Wt 15 months ago: 140#

**Nutrition Hx:**

*General:* Patient states that he has been eating OK but not really back to normal since his last hospitalization. "I tried to drink Boost again between meals." Patient weighed 140# at last hospitalization and gained only about 6 lb, which he later lost. His usual weight before his illness was 166–168#. He was at his highest weight (168#) about 21 months ago.

*Recent dietary intake:*

| | |
|---|---|
| *Breakfast:* | Cereal, small amount of skim milk, toast or bagel, juice |
| *AM snack:* | Cola—sometimes crackers or pastry |
| *Lunch:* | Sandwich (ham or turkey) from home, fruit, chips, cola |
| *Dinner:* | Meat, pasta, or rice, some type of bread. Rarely eats vegetables |
| *Bedtime snack:* | Cheese and crackers, cookies, cola |
| *24-hr recall:* | Has been on clear liquids for past 24 hours since admission |

*Food allergies/intolerances/aversions:* Mr. Sims says he has never liked milk but purposefully avoided it after his diagnosis. He does consume milk products usually without any difficulty.

*Previous MNT?* Yes   *If yes, when:* Last hospitalization.

*What?* "The dietitian worked with me to move from eating a low-fiber diet to one that is much higher in fiber. We went back over the ways to help me regain weight. I know what to do—it is just that the pain and diarrhea make my appetite so bad. It is really hard for me to eat."

*Food purchase/preparation:* Self and wife

*Vit/min intake:* Multivitamin daily

**Tx plan:**

R/O small bowel obstruction versus acute exacerbation of Crohn's disease
CBC/Chemistries
ASCA
CT scan of abdomen
NPO
$D_5W$ with 60 mEq KCl @ 125 cc/hr
Surgical consult
Nutrition support consult

**Hospital course:**

CT scan and esophagogastroduodenoscopy indicated bowel obstruction and continued exacerbation of Crohn's disease. Mr. Sims underwent resection of 200 cm of jejunum and proximal ileum with placement of jejunostomy. The ileocecal valve was preserved. Mr. Sims did not have an ileostomy, and his entire colon remains intact. Mr. Sims was placed on parenteral nutrition support immediately postoperatively, and a nutrition support consult was ordered.

# U H  *UNIVERSITY HOSPITAL*

NAME: Michael Sims                    DOB: 7/22
AGE: 36                               SEX: M
PHYSICIAN: D. Tucker, MD

\*\*\*\*\*\*\*\*\*\*\*\*\*\*\*\*\*\*\*\*\*\*\*\*\*\*\*\*\*\*\*\*\*\*\*\*\*\*\*\*\*\*\*\*\*\*\*CHEMISTRY\*\*\*\*\*\*\*\*\*\*\*\*\*\*\*\*\*\*\*\*\*\*\*\*\*\*\*\*\*\*\*\*\*\*\*\*\*\*\*\*\*\*

DAY:                                              1
DATE:
TIME:
LOCATION:

| | NORMAL | | UNITS |
|---|---|---|---|
| Albumin | 3.6–5 | 3.2 | g/dL |
| Total protein | 6–8 | 5.5 | g/dL |
| Prealbumin | 19–43 | 11 | mg/dL |
| Transferrin | 200–400 | 180 | mg/dL |
| Sodium | 135–155 | 136 | mmol/L |
| Potassium | 3.5–5.5 | 3.7 | mmol/L |
| Chloride | 98–108 | 101 | mmol/L |
| $PO_4$ | 2.5–4.5 | 2.9 | mmol/L |
| Magnesium | 1.6–2.6 | 1.8 | mmol/L |
| Osmolality | 275–295 | 280 | mmol/kg $H_2O$ |
| Total $CO_2$ | 24–30 | 26 | mmol/L |
| Glucose | 70–120 | 82 | mg/dL |
| BUN | 8–26 | 11 | mg/dL |
| Creatinine | 0.6–1.3 | 0.8 | mg/dL |
| Uric acid | 2.6–6 (women) | 5.1 | mg/dL |
| | 3.5–7.2 (men) | | |
| Calcium | 8.7–10.2 | 9.1 | mg/dL |
| Bilirubin | 0.2–1.3 | 0.8 | mg/dL |
| Ammonia ($NH_3$) | 9–33 | 11 | $\mu$mol/L |
| SGPT (ALT) | 10–60 | 35 | U/L |
| SGOT (AST) | 5–40 | 22 | U/L |
| Alk phos | 98–251 | 123 | U/L |
| CPK | 26–140 (women) | | U/L |
| | 38–174 (men) | | |
| LDH | 313–618 | | U/L |
| CHOL | 140–199 | 149 | mg/dL |
| HDL-C | 40–85 (women) | 38 | mg/dL |
| | 37–70 (men) | | |
| VLDL | | | mg/dL |
| LDL | < 130 | 111 | mg/dL |
| LDL/HDL ratio | < 3.22 (women) | 2.92 | |
| | < 3.55 (men) | | |
| Apo A | 101–199 (women) | | mg/dL |
| | 94–178 (men) | | |
| Apo B | 60–126 (women) | | mg/dL |
| | 63–133 (men) | | |
| TG | 35–160 | 85 | mg/dL |
| $T_4$ | 5.4–11.5 | | $\mu$g/dL |
| $T_3$ | 80–200 | | ng/dL |
| $HbA_{1C}$ | 4.8–7.8 | | % |

## UNIVERSITY HOSPITAL

Name: Michael Sims
Physician: D. Tucker, MD

# PATIENT CARE SUMMARY SHEET

Date: 12/20 | Room: 315 | Wt. Yesterday: 138 lb | Today: 138.25 lb | Postdialysis     lb

| Temp °F | NIGHTS | | | | | | | | DAYS | | | | | | | | EVENINGS | | | | | | | |
|---|---|---|---|---|---|---|---|---|---|---|---|---|---|---|---|---|---|---|---|---|---|---|---|---|
| | 00 | 01 | 02 | 03 | 04 | 05 | 06 | 07 | 08 | 09 | 10 | 11 | 12 | 13 | 14 | 15 | 16 | 17 | 18 | 19 | 20 | 21 | 22 | 23 |
| 105 | | | | | | | | | | | | | | | | | | | | | | | | |
| 104 | | | | | | | | | | | | | | | | | | | | | | | | |
| 103 | | | | | | | | | | | | | | | | | | | | | | | | |
| 102 | | | | | | | | | | | | | | | | | | | | | | | | |
| 101 | | | | | | | | | | | | | | | | | | | | | | | | |
| 100 | | × | | | | | | | | | | | | | | | | | | | | | | |
| 99 | | | | | | | | | | | | × | | | | | | | | | × | | | |
| 98 | | | | | | | | | | | | | | | | | | | | | | | | |
| 97 | | | | | | | | | | | | | | | | | | | | | | | | |
| 96 | | | | | | | | | | | | | | | | | | | | | | | | |
| Pulse | | 77 | | | | | | | | | 81 | | | | | | | | 79 | | | | | |
| Respiration | | 18 | | | | | | | | | 17 | | | | | | | | 17 | | | | | |
| BP | | 101/73 | | | | | | | | | 92/70 | | | | | | | | 110/75 | | | | | |
| Blood Glucose | | 142 | | | | | | | | | 121 | | | | | | | | 132 | | | | | |
| Appetite/Assist | | | | | | | | | | | | | | | | | | | | | | | | |
| INTAKE | | | | | | | | | | | | | | | | | ` | | | | | | | |
| Oral | | | | | | | | | | | | | | | | | | | | | | | | |
| IV | 85 | 85 | 85→ | | | | | | → | | | | | | | | → | | | | | | | |
| TF Formula/Flush | | | | | | | | | | | | | | | | | | | | | | | | |
| Shift Total | 680 | | | | | | | | 680 | | | | | | | | 680 | | | | | | | |
| OUTPUT | | | | | | | | | | | | | | | | | | | | | | | | |
| Void | | | | | | | | | | | | | | | | | | | | | | | | |
| Cath. | | | 520 | | | | | | | 250 | | | 250 | | | | 350 | | | 220 | | | | |
| Emesis | | | | | | | | | | | | | | | | | | | | | | | | |
| BM | | | | | | | | | | | | | | | | | | | | | | | | |
| Drains | | | | | | | | | | | | | | | | | | | | | | | | |
| Shift Total | 520 | | | | | | | | 500 | | | | | | | | 570 | | | | | | | |
| Gain | +160 | | | | | | | | +180 | | | | | | | | +110 | | | | | | | |
| Loss | | | | | | | | | | | | | | | | | | | | | | | | |
| Signatures | Angela Phelps, RN | | | | | | | | Leslie Snyder, RN | | | | | | | | D. Magee, RN | | | | | | | |
| | | | | | | | | | | | | | | | | | | | | | | | | |
| | | | | | | | | | | | | | | | | | | | | | | | | |

## Case Questions

1.  Mr. Sims has had a 200-cm resection of his jejunum and proximal ileum. How long is the small intestine, and how significant is this resection?

2.  What nutrients are normally digested and absorbed in the portion of the small intestine that has been resected?

3.  What would be the optimal method of determining Mr. Sims's energy and protein needs?

4.  Evaluate the anthropometric data available for Mr. Sims.

5.  Evaluate Mr. Sims's admission labs for nutritional significance.

6.  The members of the nutrition support team note that his serum phosphorus and serum magnesium are at the low end of the normal range. Why might that be of concern?

7.  What is refeeding syndrome? Is Mr. Sims at risk for this syndrome? How can it be prevented?

8.  Calculate Mr. Sims's energy requirements. Compare the Harris-Benedict equation with appropriate activity and stress factors to calculations with the Ireton-Jones equation.

9.  Which numbers would you use to target Mr. Sims's nutrition support? Explain.

10.  What would you estimate Mr. Sims's protein requirements to be?

11.  The surgeon notes that Mr. Sims probably will not resume eating by mouth for at least 7–10 days. What information would the nutrition support team evaluate in deciding the route for nutrition support?

12. Mr. Sims was started on parenteral nutrition postoperatively. Initially, he was prescribed to receive 200 g dextrose/L, 42.5 g amino acids/L, 30 g lipid/L. His parenteral nutrition was initiated at 50 cc/hr with a goal rate of 85 cc/hr. Do you agree with the team's decision to initiate parenteral nutrition? Will this meet his estimated nutritional needs? Explain.

13. Indirect calorimetry revealed the following information:

| Data | Mr. Sims's Data |
| --- | --- |
| Oxygen consumption (mL/min) | 295 |
| CO$_2$ production (mL/min) | 261 |
| RQ | 0.88 |
| RMR | 2022 |

What does this information tell you about Mr. Sims?

14. Would you make any changes in his prescribed nutrition support? What should be monitored to assure adequacy of his nutrition support? Explain.

15. What should the nutrition support team monitor daily? What should be monitored weekly? Explain your answers.

16. Evaluate the following 24-hour urine data: 24-hour urinary nitrogen for 12/20: 18.4 grams. By using the daily nursing record that records the amount of PN received, calculate Mr. Sims's nitrogen balance on postoperative day 4. How would you interpret this information? Should you be concerned? Are there problems with the accuracy of nitrogen balance studies? Explain.

17. Is Mr. Sims a likely candidate for short bowel syndrome? Define *short bowel syndrome*, and provide a rationale for your answer.

18. What type of adaptation can the small intestine make after resection?

**19.** What classic symptoms of short bowel syndrome should Mr. Sims's health care team monitor for?

**20.** Mr. Sims is being evaluated for participation in a clinical trial with the new drug ALX-0600. What is this drug, and how might it help Mr. Sims?

**21.** On postop day 10, Mr. Sims's team notes that he has had bowel sounds for the previous 48 hours and had his first bowel movement. The nutrition support team recommends consideration of an oral diet. What should Mr. Sims be allowed to try first? What would you monitor for tolerance? If successful, when can the parenteral nutrition be weaned?

## Bibliography

American Society for Parenteral and Enteral Nutrition. Gottschlich M., ed., *The Science and Practice of Nutrition Support.* Dubuque, IA: Kendall/Hunt, 2001.

Beyer PL. Medical nutrition therapy for lower gastrointestinal tract disorders. In Mahan LK, Escott-Stump S (eds.), *Krause's Food, Nutrition, and Diet Therapy,* 10th ed. Philadelphia: Saunders, 2000:666–694.

Commission on Accreditation for Dietetics Education. Knowledge, skills, and competencies for dietitians. *Accreditation Manual for Dietetics Education Programs,* rev. 4th ed. Chicago: American Dietetic Association, 2000:45–49.

NPS pharmaceuticals receives orphan drug status for ALX-0600 as treatment for short bowel syndrome. *NPS Pharmaceuticals.* Available at http://www.npsp.com. Accessed March 1, 2001.

Solomon SM, Kirby DF. The refeeding syndrome: a review. *J Parenter Enteral Nutr.* 1990;14:90–97.

Vanderhoof JA, Langnas AN. Short bowel syndrome in children and adults. *Gastroenterology.* 1997;113:1767.

Zeman F. The intestinal tract and accessory organs. In *Clinical Nutrition and Dietetics.* New York: Macmillan, 1990:218–279.

# MEDICAL NUTRITION THERAPY FOR PANCREATIC AND LIVER DISORDERS

## Introduction

The liver and pancreas are often called "ancillary" organs of digestion; however, the term *ancillary* does little to describe their importance in digestion, absorption, and metabolism of carbohydrate, protein, and lipid. The cases in this section portray common conditions affecting these organs and dramatically outline their effects on nutritional status.

Pancreatic and liver disease are commonly caused by toxicity from excessive alcohol ingestion and infections such as those caused by hepatitis viruses. The cases in this section focus on those etiologies. The incidence of malnutrition is very high in these disease states. It is often difficult for the practitioner to simultaneously treat the disease with appropriate medical nutrition therapy and prevent malnutrition.

Generalized symptoms of these diseases center around interruption of normal metabolism in these organs. Jaundice, anorexia, fatigue, abdominal pain, steatorrhea, and malabsorption are signs or symptoms of hepatobiliary disease. These symptoms respond to medical nutrition therapy but also interfere with maintenance of an adequate nutritional status.

The incidence of hepatobiliary disease has significantly increased over the last several decades. Cirrhosis of the liver is the fifth leading cause of death in the United States. The most common cause of cirrhosis is chronic alcohol ingestion; the second most common cause is viral hepatitis. This case focuses on hepatitis C viral (HCV), which is the most common hepatic viral infection in the United States. It is diagnosed in approximately 300,000 Americans each year. Almost 4 million Americans have the disease today; more than 90% of them have chronically infected livers, and of those, over 1 million will progress to cirrhosis or hepatic carcinoma. Researchers estimate that between the years 2010 and 2019, health care costs for HCV will approach $81.8 billion. The medical profession hopes that new treatments using interferon and ribavirin will prevent the natural progression of this disease.

Treatment of cirrhosis is primarily supportive. The only cure is a liver transplant. Therefore medical nutrition therapy is crucial for preventing protein-calorie malnutrition, minimizing the symptoms of the disease, and maintaining quality of life.

# Chronic Pancreatitis Secondary to Chronic Alcoholism

*Introductory Level*

## Objectives

After completing this case, the student will be able to:

1. Describe the anatomic features and physiologic function of the pancreas.
2. Explain etiology and risk factors for development of chronic pancreatitis.
3. Apply working knowledge of the pathophysiology of chronic pancreatitis.
4. Collect pertinent information, and use nutrition assessment techniques to determine baseline nutritional status.
5. Calculate parenteral nutrition formulations (develop an appropriate parenteral nutrition regimen).
6. Evaluate standard parenteral nutrition formulations.
7. Develop behavioral outcomes for the patient.
8. Identify appropriate MNT goals.
9. Design nutrition education for the patient with alcoholism and chronic pancreatitis.
10. Demonstrate ability to communicate in the medical record.

Ms. Elena Jordan is admitted for evaluation of her recurring epigastric pain accompanied by nausea, vomiting, diarrhea, and weight loss. Dr. Paula Bennett diagnoses Ms. Jordan with pancreatitis, probably secondary to chronic alcohol ingestion.

 **UNIVERSITY HOSPITAL**

## ADMISSION DATABASE

Name: E. Jordan
DOB: 10/7  age 30
Physician: P. Bennett

| BED#<br>1 | DATE:<br>4/22 | TIME:<br>1532 | TRIAGE STATUS (ER ONLY):<br>☐ Red  ☐ Yellow  ☐ Green  ☐ White |
|---|---|---|---|

**Initial Vital Signs**

| TEMP:<br>100.8 | RESP:<br>80 | | SAO2: |
|---|---|---|---|

| HT:<br>5'8" | WT (lb):<br>112 | B/P:<br>125/76 | PULSE:<br>114 |
|---|---|---|---|

| LAST TETANUS<br>10+ years | LAST ATE<br>over 24 hr ago | LAST DRANK<br>5 hr ago |
|---|---|---|

**PRIMARY PERSON TO CONTACT:**
Name: Michele Jordan-mother
Home #: 555-3847
Work #: same

**CHIEF COMPLAINT/HX OF PRESENT ILLNESS**

severe abdominal pain

**ORIENTATION TO UNIT:** ☒ Call light  ☒ Television/telephone
☒ Bathroom  ☒ Visiting  ☒ Smoking  ☒ Meals
☒ Patient rights/responsibilities

**PERSONAL ARTICLES:** (Check if retained/describe)
☒ Contacts  ☒ R  ☒ L          ☐ Dentures  ☐ Upper  ☐ Lower
☒ Jewelry: watch
☐ Other:

**ALLERGIES: Meds, Food, IVP Dye, Seafood: Type of Reaction**

none known

**VALUABLES ENVELOPE:**
☐ Valuables instructions

**PREVIOUS HOSPITALIZATIONS/SURGERIES**

**INFORMATION OBTAINED FROM:**
☒ Patient          ☐ Previous record
☐ Family          ☐ Responsible party

Signature  *Elena Jordan*

| Home Medications (including OTC) | | Codes: A=Sent home | | B=Sent to pharmacy | | C=Not brought in |
|---|---|---|---|---|---|---|
| Medication | Dose | Frequency | Time of Last Dose | Code | Patient Understanding of Drug |
| | | | | | |
| | | | | | |
| | | | | | |
| | | | | | |
| | | | | | |
| | | | | | |
| | | | | | |
| | | | | | |
| | | | | | |
| | | | | | |

Do you take all medications as prescribed?  ☐ Yes  ☐ No  If no, why?

**PATIENT/FAMILY HISTORY**

| | | |
|---|---|---|
| ☐ Cold in past two weeks | ☐ High blood pressure | ☐ Kidney/urinary problems |
| ☒ Hay fever Patient | ☐ Arthritis | ☐ Gastric/abdominal pain/heartburn |
| ☐ Emphysema/lung problems | ☒ Claustrophobia Mother | ☐ Hearing problems |
| ☐ TB disease/positive TB skin test | ☐ Circulation problems | ☐ Glaucoma/eye problems |
| ☐ Cancer | ☐ Easy bleeding/bruising/anemia | ☐ Back pain |
| ☐ Stroke/past paralysis | ☒ Sickle cell disease Maternal grandmother | ☐ Seizures |
| ☐ Heart attack | ☐ Liver disease/jaundice | ☐ Other |
| ☐ Angina/chest pain | ☐ Thyroid disease | |
| ☐ Heart problems | ☐ Diabetes | |

**RISK SCREENING**

Have you had a blood transfusion?  ☐ Yes  ☒ No
Do you smoke?  ☐ Yes  ☒ No
If yes, how many pack(s) _____ /day for _____ years
Does anyone in your household smoke?  ☐ Yes  ☐ No
Do you drink alcohol?  ☒ Yes  ☐ No
If yes, how often?  daily  How much?  2-3 drinks
When was your last drink?  __4/20__/____
Do you take any recreational drugs?  ☒ Yes  ☐ No
If yes, type: marijuana  Route inhale
Frequency: <1/month  Date last used: don't remember

**FOR WOMEN Ages 12–52**

Is there any chance you could be pregnant?  ☐ Yes  ☒ No
If yes, expected date (EDC):
Gravida/Para: 0/0

**ALL WOMEN**

Date of last Pap smear: 2/15/
Do you perform regular breast self-exams?  ☒ Yes  ☐ No

**ALL MEN**

Do you perform regular testicular exams?  ☐ Yes  ☐ No

Additional comments:

✗ *Miriam Link, RN*
Signature/Title

**Client name:** Elena Jordan
**DOB:** 10/7
**Age:** 30
**Sex:** Female
**Education:** Bachelor's degree
**Occupation:** Pharmaceutical sales rep
**Hours of work:** Varies—usually 50+ hours/week
**Household members:** Lives alone
**Ethnic background:** Biracial
**Religious affiliation:** Agnostic
**Referring physician:** Paula Bennett, MD (gastroenterology)

## Chief complaint:

"I'm tired of hurting so much. I've had this terrible pain in my stomach for the past 2 days. I took a client out to dinner the other night, but I couldn't eat much. This has been happening off and on for the past 9 months, but the pain has never gone around to my back before."

## Patient history:

Ms. Jordan is a 30-yo woman who has been well until 12 months ago when she began to experience bouts of epigastric pain. Most recently the pain has started to radiate to her back and lasts from 4 hours to 3 days. She c/o poor appetite and a recent, unintentional weight loss of 10 pounds. She reports two loose stools per day for the past 4 months. She says they are foul smelling. As of late, she c/o anorexia and nausea.
*Onset of disease:* 12 months ago
*Type of Tx:* Antacids
*PMH:* Currently weighs 112 pounds; weighed 140 a year ago
*Meds:* Ortho-Tri-Cyclen, 28-day cycle
*Smoker:* No
*Family Hx:* No family history of GI disease

## Physical exam:

*General appearance:* Thin, 30-yo female with temporal muscle wasting who appears to be in a moderate amount of discomfort
*Vitals:* Temp 100.8°F, BP 125/76 mm Hg, HR 114 bpm, RR 80 bpm
*Heart:* Regular rate and rhythm, heart sounds normal
*HEENT:*
   *Eyes:* PERRLA
   *Ears:* Noncontributory
   *Nose:* Noncontributory
   *Throat:* Noncontributory
*Genitalia:* Normal female
*Neurologic:* Alert, oriented times three
*Extremities:* Noncontributory
*Skin:* Smooth, warm, and dry, slightly tented, no edema
*Chest/lungs:* Lungs are clear
*Peripheral vascular:* Pulse 4+ bilaterally, warm, no edema
*Abdomen:* Flat, bowel sounds normal, tenderness in epigastric region, liver and spleen not enlarged

**Nutrition Hx:**

*General:* Pt reports that her appetite has usually been good, but for the last 6–9 months, she's had difficulty eating due to nausea. Ms. Jordan is a pharmaceutical rep who travels outside the area two weeks every month. Because a large part of her job entails entertaining clients, she eats many of her meals in restaurants and consumes 2–3 alcoholic beverages per night. When questioned about her history of alcohol intake, Ms. Jordan stated that she started drinking in high school on the weekends when her friends had parties. She drank only beer and had only 1–2 cans per night (total of 2–4 beers per weekend). When she entered college, Ms. Jordan continued to drink on the weekends with her friends. She would drink beer in the college bars and at house parties—often consuming 5 or more drinks in one evening. After graduation from college, she was hired as a pharmaceutical representative, which entailed eating most of her meals (while on the road) in restaurants. When she entertains clients—usually men—she feels she has to match them drink for drink to help land business. Ms. Jordan usually drinks wine or beer when out with clients because she believes only people with alcohol problems drink "hard liquor."

*Usual dietary intake:*
*At home:*

| | |
|---|---|
| *Breakfast:* | Dry bagel, 1 c black coffee |
| *Lunch:* | Diet Coke, Lean Cuisine—usually Swedish meatballs (with noodles) |
| *Dinner:* | 5 oz white wine while preparing dinner |
| | Grilled salmon—usually 2–3 ounces, seasoned with salt and pepper |
| | Baked potato—medium sized, with butter, sour cream, and chives |
| | 2 stalks steamed broccoli with cheese sauce (made from Cheez Whiz) |
| | 2 glasses (5 oz) white wine with dinner |
| *HS Snack:* | Bowl (∼ 1½ c) Häagen-Dazs rum raisin ice cream |

*On the road:*

| | |
|---|---|
| *Breakfast:* | ¾ c dry cereal (varies) with 1½ c 2% milk, 1 c orange juice, 1 c black coffee |
| *Lunch:* | (Often doesn't eat lunch, but when she does) McDonald's fruit and yogurt parfait, medium Diet Coke |
| *Dinner:* | Usually some type of appetizer—most likely fried mushrooms |
| | Spinach salad with hot bacon dressing |
| | Fettuccine Alfredo or small (6 oz) filet mignon with garlic mashed potatoes |
| | 2–3 glasses of wine (6-oz glasses) |
| *After-dinner drink:* | Usually sherry |

*Food allergies/intolerances/aversions:* None
*Previous MNT?* No
*Food purchase/preparation:* Self, eats in restaurants often (2 weeks of each month)
*Vit/min intake:* None
*Current diet order:* NPO

**Tx plan:**
Pregnancy test
CBC
Chemistry with liver and pancreatic enzymes

Urinalysis
Upper GI w/ small bowel series
CT scan of abdomen and pelvis
1 liter NS bolus, then $D_5NS$ @ 150 cc/h
Demerol 25 mg IM q 4–6 h
72-hr stool collection for fecal fat
NPO
Chlordiazepoxide 25 mg IV q 6h × 3d
Thiamin 100 mg IV q d × 3d
Folic acid 1 mg IV q d × 3d
Multivitamins 1 amp in first liter of IV fluids

# U<sub>H</sub> UNIVERSITY HOSPITAL

NAME: E. Jordan                          DOB: 10/7
AGE: 30                                  SEX: Female
PHYSICIAN: P. Bennett, MD

\*\*\*\*\*\*\*\*\*\*\*\*\*\*\*\*\*\*\*\*\*\*\*\*\*\*\*\*\*\*\*\*\*\*\*\*\*\*\*\*\*\*\*\*\*\*CHEMISTRY\*\*\*\*\*\*\*\*\*\*\*\*\*\*\*\*\*\*\*\*\*\*\*\*\*\*\*\*\*\*\*\*\*\*\*\*\*\*\*\*\*\*\*\*\*

DAY:                                     Admit
DATE:
TIME:
LOCATION:

| | NORMAL | | UNITS |
|---|---|---|---|
| Albumin | 3.6–5 | 3.6 | g/dL |
| Total protein | 6–8 | 6 | g/dL |
| Prealbumin | 19–43 | 20.5 | mg/dL |
| Transferrin | 200–400 | 155 | mg/dL |
| Sodium | 135–155 | 145 | mEq/L |
| Potassium | 3.5–5.5 | 4.6 | mEq/L |
| Chloride | 98–108 | 105 | mEq/L |
| $PO_4$ | 2.5–4.5 | 3.3 | mEq/L |
| Magnesium | 1.6–2.6 | 2.1 | mEq/L |
| Osmolality | 275–295 | 296 | mmol/kg $H_2O$ |
| Total $CO_2$ | 24–30 | 27 | mmol/L |
| Glucose | 70–120 | 130 | mg/dL |
| BUN | 8–26 | 18 | mg/dL |
| Creatinine | 0.6–1.3 | 0.75 | mg/dL |
| Uric acid | 2.6–6 (women) | 4.7 | mg/dL |
| | 3.5–7.2 (men) | | |
| Calcium | 8.7–10.2 | 9.3 | mg/dL |
| Bilirubin | 0.2–1.3 | 1.5 | mg/dL |
| Ammonia ($NH_3$) | 9–33 | 27 | $\mu$mol/L |
| SGPT (ALT) | 10–60 | 70 | U/L |
| SGOT (AST) | 5–40 | 50 | U/L |
| Alk phos | 98–251 | 178 | U/L |
| CPK | 26–140 (women) | 140 | U/L |
| | 38–174 (men) | | |
| LDH | 313–618 | 323 | U/L |
| Lipase | 0–417 | 521 | U/L |
| Amylase | 25–125 | 750 | U/L |
| CHOL | 140–199 | 225 | mg/dL |
| HDL-C | 40–85 (women) | 40 | mg/dL |
| | 37–70 (men) | | |
| VLDL | | | mg/dL |
| LDL | <130 | 129 | mg/dL |
| LDL/HDL ratio | <3.22 (women) | | |
| | <3.55 (men) | | |
| Apo A | 101–199 (women) | | mg/dL |
| | 94–178 (men) | | |
| Apo B | 60–126 (women) | | mg/dL |
| | 63–133 (men) | | |
| TG | 35–160 | 250 | mg/dL |
| $T_4$ | 5.4–11.5 | | $\mu$g/dL |
| $T_3$ | 80–200 | | ng/dL |
| HbA$_{1C}$ | 4.8–7.8 | 7.5 | % |

# U H *UNIVERSITY HOSPITAL*

NAME: E. Jordan                          DOB: 10/7
AGE: 30                                  SEX: Female
PHYSICIAN: P. Bennett, MD

************************************************HEMATOLOGY*********************************************

DAY:                                              Admit
DATE:
TIME:
LOCATION:

| | NORMAL | | UNITS |
|---|---|---|---|
| WBC | 4.3–10 | 14.5 | $\times\ 10^3/mm^3$ |
| RBC | 4–5 (women) | 4.7 | $\times\ 10^6/mm^3$ |
| | 4.5–5.5 (men) | | |
| HGB | 12–16 (women) | 11.6 | g/dL |
| | 13.5–17.5 (men) | | |
| HCT | 37–47 (women) | 35.7 | % |
| | 40–54 (men) | | |
| MCV | 84–96 | 101.5 | fL |
| MCH | 27–34 | 29 | pg |
| MCHC | 31.5–36 | 33.4 | % |
| RDW | 11.6–16.5 | 13.6 | % |
| Plt ct | 140–440 | 359 | $\times\ 10^3$ |
| Diff TYPE | | | |
| % GRANS | 34.6–79.2 | 84.2 | % |
| % LYM | 19.6–52.7 | 51 | % |
| SEGS | 50–62 | 59 | % |
| BANDS | 3–6 | 4 | % |
| LYMPHS | 25–40 | 34 | % |
| MONOS | 3–7 | 5 | % |
| EOS | 0–3 | 2 | % |
| TIBC | 65–165 (women) | 148 | $\mu$g/dL |
| | 75–175 (men) | | |
| Ferritin | 18–160 (women) | 17.9 | $\mu$g/dL |
| | 18–270 (men) | | |
| Vitamin B$_{12}$ | 100–700 | | pg/mL |
| Folate | 2–20 | | ng/mL |
| Total T cells | 812–2318 | | mm$^3$ |
| T-helper cells | 589–1505 | | mm$^3$ |
| T-suppressor cells | 325–997 | | mm$^3$ |
| PT | 11–13 | | sec |

 UNIVERSITY HOSPITAL

Name: E. Jordan

Physician: P. Bennett

# PATIENT CARE SUMMARY SHEET

Date: 4/22  Room: 1  Wt Yesterday:  Today: 112 lb  Postdialysis  lb

| Temp °F | NIGHTS | | | | | | | | DAYS | | | | | | | | EVENINGS | | | | | | | |
|---|---|---|---|---|---|---|---|---|---|---|---|---|---|---|---|---|---|---|---|---|---|---|---|
| | 00 | 01 | 02 | 03 | 04 | 05 | 06 | 07 | 08 | 09 | 10 | 11 | 12 | 13 | 14 | 15 | 16 | 17 | 18 | 19 | 20 | 21 | 22 | 23 |
| 105 | | | | | | | | | | | | | | | | | | | | | | | | |
| 104 | | | | | | | | | | | | | | | | | | | | | | | | |
| 103 | | | | | | | | | | | | | | | | | | | | | | | | |
| 102 | | | | | | | | | | | | | | | | | | | | | | | | |
| 101 | | | | | | | | | | ✕ | | | | | | | | | | | | | | |
| 100 | | | | | | | | | | | | | | | | | | | | | | | | |
| 99 | | | | | | | | | | | | | | | | | | | | | | | | |
| 98 | | | | | | | | | | | | | | | | | | | | | | | | |
| 97 | | | | | | | | | | | | | | | | | | | | | | | | |
| 96 | | | | | | | | | | | | | | | | | | | | | | | | |
| Pulse | | | | | | | | | | 114 | | | | | | | | | | | | | | |
| Respiration | | | | | | | | | | 80 | | | | | | | | | | | | | | |
| BP | | | | | | | | | | 125/76 | | | | | | | | | | | | | | |
| Blood Glucose | | | | | | | | | | | | | | | | | | | | | | | | |
| Appetite/Assist | | | | | | | | | | NPO | | | | | | | | | | | | | | |
| INTAKE | | | | | | | | | | | | | | | | | | | | | | | | |
| Oral | | | | | | | | | | | | | | | | | | | | | | | | |
| IV | | | | | | | | | | 125→ | | | | | | | | | | | | | | |
| TF Formula/Flush | | | | | | | | | | | | | | | | | | | | | | | | |
| Shift Total | | | | | | | | | | | | | | | | | | | | | | | | |
| OUTPUT · | | | | | | | | | | | | | | | | | | | | | | | | |
| Void | | | | | | | | | | | | | | | | | | | | | | | | |
| Cath. | | | | | | | | | | | | | | | | | | | | | | | | |
| Emesis | | | | | | | | | | | | | | | | | | | | | | | | |
| BM | | | | | | | | | | | | | | | | | | | | | | | | |
| Drains | | | | | | | | | | | | | | | | | | | | | | | | |
| Shift Total | | | 875 | | | | | | | | | | | | | | | | | | | | | |
| Gain | | | | | | | | | | | | | | | | | | | | | | | | |
| Loss | | | | | | | | | | | | | | | | | | | | | | | | |
| Signatures | | | *Miriam Link, RN* | | | | | | | | | | | | | | | | | | | | | |

## Case Questions

1. The pancreas is an exocrine and endocrine gland. Describe exocrine and endocrine functions of the pancreas.

2. Factors that influence pancreatic secretion during a meal can be subdivided into three phases (cephalic, gastric, and intestinal). Describe the action of the pancreas within each phase.

3. Dr. Bennett makes a diagnosis of chronic pancreatitis. Define *chronic pancreatitis*.

4. Go to Ms. Jordan's lab reports. Which are important in diagnosing pancreatitis? What are her values? What other labs are consistent with her diagnosis?

5. What physical symptoms in the physician's history and physical are consistent with Ms. Jordan's diagnosis?

6. What is the most common etiology for pancreatitis? Explain the rationale for the development of pancreatitis.

7. Explain how alcohol is metabolized.

8. Women absorb and metabolize alcohol differently from men, making them more vulnerable than men to alcohol-related organ damage. Describe how alcohol is metabolized in women.

9. Overall, women are more vulnerable to the effects of alcohol. List several health problems associated with alcohol consumption.

10. One year ago, Ms. Jordan weighed 140 lb. On admission, she weighed 112 lb. Calculate her percent weight loss.

**11.** Discuss how you would interpret Ms. Jordan's weight loss.

**12.** Dr. Bennett specifically wanted to see Ms. Jordan's blood glucose level. Why?

**13.** When Dr. Bennett admitted Ms. Jordan, she evaluated the severity of her pancreatitis using Ranson's criteria. What are Ranson's criteria, and how is this test scored?

**14.** Why were thiamin, folic acid, and a multivitamin supplement ordered on admission?

**15.** Ms. Jordan's mean corpuscular volume (MCV) was elevated on admission. What might cause this?

**16.** What do the U.S. Dietary Guidelines indicate regarding alcohol intake?

**17.** Define "drink" (beer, wine, liquor) as indicated in the U.S. Dietary Guidelines.

**18.** How would you respond to Ms. Jordan's comment that she mostly drinks wine and beer because only people who drink "hard liquor" develop alcohol problems?

**19.** Estimate Ms. Jordan's usual dietary intake for the following (show your calculations):

|  | At Home | On the Road |
|---|---|---|
| Alcohol | g | g |
| Alcohol kcal | kcal | kcal |
| Total energy | kcal | kcal |
| % energy as alcohol | % | % |
| Protein | g | g |
| **At home (calculations):** | | |
| **On the road (calculations):** | | |

**20.** What assessment would you make in each of the following areas?

| Parameter | Assessment |
|---|---|
| Anthropometric data | |
| Biochemical data | |
| Clinical data | |
| Dietary data | |

**21.** Using Harris-Benedict equation (HBE), estimate Ms. Jordan's energy needs with her current weight and with her usual body weight.

**22.** Calculate Ms. Jordan's protein needs.

**23.** Hospital day 2: Pt remains stable on IV fluid. Her pain has been somewhat controlled with parenteral analgesics, but she is still unable to eat. Dr. Bennett consults you to evaluate the parenteral nutrition she has suggested: a dextrose-based parenteral solution with 4.25% amino acids, 25% dextrose, along with electrolytes, vitamins, and trace elements at a rate of 85 cc/hr in addition to 500 cc/day of 10% lipids. Will this meet the patient's energy and protein needs?

**24.** When developing parenteral regimens during pancreatitis, you may find that patients have difficulty with high-dextrose solutions as well as lipid emulsions. What guidelines exist for these issues? What labs would you monitor? Would you recommend any changes in the prescribed parenteral regimen? Explain.

**25.** Why do you think Dr. Bennett ordered parenteral nutrition rather than enteral nutrition support? What is the current standard of practice?

**26.** Outline nutritional management for an underweight patient with pancreatitis when the patient is able to eat again.

**27.** Write an initial nutrition-screening progress note:

S:

O:

A:

P:

**28.** What will be the key factors in preventing exacerbations of Ms. Jordan's pancreatitis in the future?

# Bibliography

American Dietetic Association, *Manual of Clinical Dietetics,* 5th ed. Chicago: American Dietetic Association, 1996.

American Dietetic Association and Morrison Health Care. *Medical Nutrition Therapy Across the Continuum of Care.* Chicago: American Dietetic Association, 1998.

Center for Science in the Public Interest. *Fact sheet: women and alcohol.* Available at http://www.cspinet.org/booze/women.htm. Accessed November 21, 2000.

Chronic pancreatitis. *The Merck Manual.* Available at http://www.merck.com. Accessed May 21, 2001.

Commission on Accreditation for Dietetics Education. Knowledge, skills, and competencies for dietitians. *Accreditation Manual for Dietetics Education Programs,* rev. 4th ed. Chicago: American Dietetic Association, 2000.

Davis MA. Acute pancreatitis. Library of the National Medical Society. Available at http://www.medical-library.org. Accessed September 22, 2000.

Escott-Stump S. *Nutrition and Diagnosis-Related Care,* 4th ed. Baltimore: Williams & Wilkins, 1998.

Fischbach F. *Manual of Laboratory and Diagnostic Tests,* 6th ed. Philadelphia: Lippincott, 2000.

Hebuterne X, Hastier P, Peroux JL, et al. Resting energy expenditure in patients with alcoholic chronic pancreatitis. *Dig Dis Sci.* 1996; 41:533–539.

Mann K, Batra A, Gunther A, Schroth G. Do women develop alcoholic brain damage more readily than men? *Alcohol Clin Exp Res.* 1992;16(6):1052–1056.

MedlinePlus Health information. Available at http://www.nlm.nih.gov/medlineplus/druginformation.html. Accessed September 8, 2000.

Mezey E, Kolman CJ, Diehl AM, Mitchell MC, Herlong HF. Alcohol and dietary intake in the development of chronic pancreatitis and liver disease in alcoholism. *Am J Clin Nutr.* 1988;48(1):148–151.

National Institute on Alcohol Abuse and Alcoholism. Are women more vulnerable to alcohol effects? Alcohol Alert No. 46 (1999). Rockville, MD: U.S. Department of Health and Human Services. Available at http://www.nih.gov. Accessed April 21, 2001.

National Institute on Alcohol Abuse and Alcoholism. Alcohol and the liver. Alcohol Alert No. 42 (1998). Rockville, MD: U.S. Department of Health and Human Services. Available at http://www.nih.gov. Accessed July 24, 2000.

Seidner DL, Fuhrman MP. Nutrition support in pancreatitis. In American Society for Parenteral, Enteral Nutrition (eds.), *The Science and Practice of Nutrition Support: A Case Based Core Curriculum.* Dubuque, IA: Kendall/Hunt, 2000.

Smith-Warner SA, Spiegelman D, Yaun S, van den Brandt PA, Folsom AR, Goldbohn RA, Graham S, Holmbert L, Howe GR, Marshall JR, Miller AB, Potter JD, Speizer FE, Willett WC, Wolk A, Hunber DJ. Alcohol and breast cancer in women: a pooled analysis of cohort studies. *JAMA.* 1998;279(7):535–540.

Stern R. Spotlight on alcohol. In Insel P, Turner RE, Ross D. (eds.), *Nutrition.* Boston: Jones and Bartlett, 2001.

Swearingen PL, Ross DG. *Manual of Medical-Surgical Nursing Care,* 4th ed. St. Louis: Mosby, 1999.

Urbano-Marquez A, Estruch R, Fernandez-Sola J, Nicolas JM, Pare JC, Rubin E. The greater risk of alcoholic cardiomyopathy and myopathy in women compared with men. *JAMA.* 1995;274(2):149–154.

# Case 20

# Acute Hepatitis

*Introductory Level*

## Objectives

After completing this case, the student will be able to:

1. Demonstrate working knowledge of the pathophysiology of hepatitis.
2. Collect pertinent information and use nutrition assessment techniques to determine baseline nutritional status.
3. Develop behavioral outcomes for the patient.
4. Identify appropriate MNT goals.

5. Identify key components of nutrition education for the patient with hepatitis.
6. Communicate in the medical record.

Dr. Teresa Wilcox is a 48-year-old woman who is admitted with complaint of fatigue, vague upper quadrant pain, nausea, and anorexia. Elevated liver enzymes lead to further evaluation of possible infection with hepatitis C.

# ADMISSION DATABASE

| BED#<br>1 | DATE:<br>5/17 | TIME:<br>1400 | TRIAGE STATUS (ER ONLY):<br>☐ Red  ☐ Yellow  ☐ Green  ☐ White |
|---|---|---|---|

**PRIMARY PERSON TO CONTACT:**
Name: Dennis Gustat
Home #: 555-3947
Work #: same

### Initial Vital Signs

| TEMP:<br>99.6 | RESP:<br>20 | SAO2: |
|---|---|---|

| HT:<br>5'6" | WT (lb):<br>145 | B/P:<br>100/60 | PULSE:<br>75 |
|---|---|---|---|

**ORIENTATION TO UNIT:** ☒ Call light  ☒ Television/telephone
☒ Bathroom  ☒ Visiting  ☒ Smoking  ☒ Meals
☒ Patient rights/responsibilities

| LAST TETANUS<br>5 yrs ago | LAST ATE<br>yesterday | LAST DRANK<br>this morning |
|---|---|---|

### CHIEF COMPLAINT/HX OF PRESENT ILLNESS

Fatigue, malaise

Anorexia, N/V

**PERSONAL ARTICLES:** (Check if retained/describe)
☒ Contacts ☒ R ☒ L          ☐ Dentures ☐ Upper ☐ Lower
☒ Jewelry: ring
☒ Other: watch

### ALLERGIES: Meds, Food, IVP Dye, Seafood: Type of Reaction

Penicillin

**VALUABLES ENVELOPE:**
☐ Valuables instructions

### PREVIOUS HOSPITALIZATIONS/SURGERIES

None

**INFORMATION OBTAINED FROM:**
☒ Patient          ☐ Previous record
☐ Family          ☐ Responsible party

Signature  *Terri Wilcox*

### Home Medications (including OTC)      Codes: A=Sent home          B=Sent to pharmacy          C=Not brought in

| Medication | Dose | Frequency | Time of Last Dose | Code | Patient Understanding of Drug |
|---|---|---|---|---|---|
| Prempro | 0.65/5 | qd | this morning | c | yes |
| Allegra | 60 mg | qd | this morning | c | yes |
| Wellbutrin | 75 mg | bid | this morning | c | yes |
| | | | | | |
| | | | | | |
| | | | | | |
| | | | | | |
| | | | | | |
| | | | | | |
| | | | | | |

Do you take all medications as prescribed?  ☒ Yes  ☐ No  If no, why?

### PATIENT/FAMILY HISTORY

| | | |
|---|---|---|
| ☐ Cold in past two weeks | ☒ High blood pressure Mother | ☐ Kidney/urinary problems |
| ☐ Hay fever | ☒ Arthritis Mother | ☒ Gastric/abdominal pain/heartburn Father |
| ☐ Emphysema/lung problems | ☐ Claustrophobia | ☐ Hearing problems |
| ☐ TB disease/positive TB skin test | ☐ Circulation problems | ☒ Glaucoma/eye problems Mother |
| ☒ Cancer Maternal grandmother | ☐ Easy bleeding/bruising/anemia | ☐ Back pain |
| ☐ Stroke/past paralysis | ☐ Sickle cell disease | ☐ Seizures |
| ☐ Heart attack | ☒ Liver disease/jaundice Paternal grandfather | ☒ Other ALS, Paternal grandmother |
| ☐ Angina/chest pain | ☐ Thyroid disease | |
| ☐ Heart problems | ☒ Diabetes Father | |

### RISK SCREENING

Have you had a blood transfusion?  ☐ Yes  ☒ No
Do you smoke?  ☐ Yes  ☒ No
If yes, how many pack(s) ____/day for ____ years
Does anyone in your household smoke?  ☐ Yes  ☒ No
Do you drink alcohol?  ☒ Yes  ☐ No
If yes, how often? socially   How much? 1-2 glasses of wine/week
When was your last drink? __5/15/__
Do you take any recreational drugs?  ☐ Yes  ☒ No
If yes, type:_____ Route
Frequency:_____ Date last used:_____/_____/

**FOR WOMEN Ages 12–52**

Is there any chance you could be pregnant?  ☐ Yes  ☒ No
If yes, expected date (EDC):
Gravida/Para: 0/0

**ALL WOMEN**

Date of last Pap smear: 4/01/
Do you perform regular breast self-exams?  ☒ Yes  ☐ No

**ALL MEN**

Do you perform regular testicular exams?  ☐ Yes  ☐ No

Additional comments:

✗ *Val Koetting, RN*
Signature/Title

**Client name:** Teresa (Terri) Wilcox
**DOB:** 3/5
**Age:** 48
**Sex:** Female
**Education:** Doctoral degree
**Occupation:** Professor at the university
**Hours of work:** Varies—afternoon office hours, some afternoon classes, and some evening classes during the semester; often works on weekends
**Household members:** Significant other (59 yo), hyperlipidemia
**Ethnic background:** European American
**Religious affiliation:** Buddhist
**Referring physician:** Phillip Horowitz, MD (gastroenterology)

**Chief complaint:**
"I just feel so tired. I can hardly move, my joints ache so much, and my muscles feel sore."

**Patient history:**
*Onset of disease:* Dr. Wilcox is a 48-yo postmenopausal female who has been living with a significant other for 20 years. In addition to c/o fatigue, aches, and pains, she complains of vague right upper quadrant pain, nausea, and anorexia. She has been in relatively good health for the past 30 years. Gravida 0 / para 0. On admission to the hospital, all laboratory tests proved negative except for liver enzymes.
*Type of Tx:* Rule out hepatitis C; test for anti-HCV and HCV RNA, nutrition consult to determine appropriate medical nutrition therapy; abstain from alcohol
*PMH:* Seasonal allergies treated with antihistamines.
*Meds:* Prempro, 0.625/5 1 tab po daily; Allegra, 60 mg po qd; Wellbutrin, 75 mg po qd
*Smoker:* No
*Family Hx: What?* Mother (living)—HTN, diverticulitis, cholecystitis, carpal tunnel syndrome; father (deceased)—diabetes mellitus, peptic ulcer disease; maternal grandmother—cholecystitis, bilateral breast cancer; maternal grandfather—leukemia; paternal grandfather—cirrhosis; paternal grandmother—amyotrophic lateral sclerosis

**Physical exam:**
*General appearance:* Tired-looking middle-aged female
*Vitals:* Temp 99.6°F, BP 100/60 mm Hg, HR 75 bpm, RR 20 bpm
*Heart:* Regular rate and rhythm, no gallops or rubs, point of maximal impulse at the fifth intercostal space in the midclavicular line
*HEENT:*
   *Head:* normocephalic
   *Eyes:* Wears contact lenses to correct myopia, PERRLA
   *Ears:* Tympanic membranes w/out lesions
   *Nose:* Dry mucous membranes w/out lesions
   *Throat:* Normal mucosa w/out exudates or lesions
*Genitalia:* Normal female
*Neurologic:* Alert and oriented × 3
*Extremities:* Normal muscular tone, normal ROM
*Skin:* Warm and dry

*Chest/lungs:* Respirations normal, no crackles, rhonchi, wheezes, or rubs noted
*Peripheral vascular:* Pulse 3+ bilaterally
*Abdomen:* Tattoo on lower abdomen close to pubis; upper right abdomen guarding

## Nutrition Hx:

*General:* Appetite is usually good, but has not had an appetite for the past few weeks. She eats cereal and orange juice for breakfast most mornings (orange juice every AM). Lunch is usually in the office (carry-out) or at a restaurant. Dinner at home, but may be carry-out. If carry-out, it's usually Chinese food.

*Usual dietary intake:*

| | |
|---|---|
| Breakfast: | Sugar Frosted Flakes or Frosted Mini-Wheats—1½ c, 2% milk—about ½ c |
| | Occasionally a banana sliced on top of cereal; strawberries or raspberries in season |
| | Calcium-fortified orange juice—8 oz |
| AM: | Unsweetened, flavored hot or iced tea during morning in office |
| Lunch: | Cheeseburger—1 oz American cheese, 3 oz beef patty, bun, lettuce, tomato slice, dill pickle spear |
| | 12 oz Diet Coke |
| | Half-order waffle-cut french fries with ketchup |
| Dinner: | Cashew shrimp: 3 oz shrimp, 1½ c vegetables (baby corn, water chestnuts, sliced carrots, pea pods), ¼ c cashews, 1 c steamed rice |
| | 5 oz wine, usually white |
| HS snack: | 3–4 small Famous Amos chocolate chip with pecans cookies |
| | Ice water |
| 24-hr recall: | Sips of orange juice, hot tea; 4 saltine crackers; 5 c Jello; 12 oz Sprite; .75 c cream of chicken soup; hot tea |

*Food allergies/intolerances/aversions:* Does not like liver or lima beans; NKA
*Previous MNT?* No
*Food purchase/preparation:* Self and/or significant other
*Vit/min intake:* 400 mg vitamin E, 500 mg calcium multivitamin/mineral q d
*Current diet order:* high kcal, high protein

## Tx plan:

Anti-HCV and HCV RNA tests
Chemistry, CBC I
Vitamin B complex supplement
High-kcal, high-protein diet (per dietitian)
Bed rest
Prempro, 0.625/5 1 tab po daily
Allegra, 60 mg po qd
Wellbutrin, 75 mg po qd

# U<sub>H</sub> *UNIVERSITY HOSPITAL*

NAME: T. Wilcox                          DOB: 3/5
AGE: 48                                  SEX: Female
PHYSICIAN: P. Horowitz, MD

\*\*\*\*\*\*\*\*\*\*\*\*\*\*\*\*\*\*\*\*\*\*\*\*\*\*\*\*\*\*\*\*\*\*\*\*\*\*\*\*\*\*CHEMISTRY\*\*\*\*\*\*\*\*\*\*\*\*\*\*\*\*\*\*\*\*\*\*\*\*\*\*\*\*\*\*\*\*\*\*\*\*\*\*\*\*\*

DAY:                                          Admit
DATE:
TIME:
LOCATION:

| | NORMAL | | UNITS |
|---|---|---|---|
| Albumin | 3.6–5 | 4.2 | g/dL |
| Total protein | 6–8 | 7.1 | g/dL |
| Prealbumin | 19–43 | 36 | mg/dL |
| Transferrin | 200–400 | 325 | mg/dL |
| Sodium | 135–155 | 143 | mEq/L |
| Potassium | 3.5–5.5 | 4.8 | mEq/L |
| Chloride | 98–108 | 102 | mEq/L |
| PO$_4$ | 2.5–4.5 | 3.6 | mEq/L |
| Magnesium | 1.6–2.6 | 2.1 | mEq/L |
| Osmolality | 275–295 | 280 | mmol/kg H$_2$O |
| Total CO$_2$ | 24–30 | 28 | mmol/L |
| Glucose | 70–120 | 110 | mg/dL |
| BUN | 8–26 | 20 | mg/dL |
| Creatinine | 0.6–1.3 | 0.9 | mg/dL |
| Uric acid | 2.6–6 (women) | 5.9 | mg/dL |
| | 3.5–7.2 (men) | | |
| Calcium | 8.7–10.2 | 9.3 | mg/dL |
| Bilirubin | 0.2–1.3 | 1.5 | mg/dL |
| Ammonia (NH$_3$) | 9–33 | 28 | $\mu$mol/L |
| SGPT (ALT) | 10–60 | 340 | U/L |
| SGOT (AST) | 5–40 | 500 | U/L |
| Alk phos | 98–251 | 302 | U/L |
| CPK | 26–140 (women) | 138 | U/L |
| | 38–174 (men) | | |
| LDH | 313–618 | 695 | U/L |
| CHOL | 140–199 | 199 | mg/dL |
| HDL–C | 40–85 (women) | 50 | mg/dL |
| | 37–70 (men) | | |
| VLDL | | | mg/dL |
| LDL | <130 | 125 | mg/dL |
| LDL/HDL ratio | <3.22 (women) | | |
| | <3.55 (men) | | |
| Apo A | 101–199 (women) | | mg/dL |
| | 94–178 (men) | | |
| Apo B | 60–126 (women) | | mg/dL |
| | 63–133 (men) | | |
| TG | 35–160 | 152 | mg/dL |
| T$_4$ | 5.4–11.5 | | $\mu$g/dL |
| T$_3$ | 80–200 | | ng/dL |
| HbA$_{1C}$ | 4.8–7.8 | 4.9 | % |

# U<sub>H</sub> *UNIVERSITY HOSPITAL*

```
NAME: T. Wilcox                          DOB: 3/5
AGE: 48                                  SEX: Female
PHYSICIAN: P. Horowitz, MD
```

******************************************URINALYSIS******************************************

| DAY: | | Admit | 2 | d/c | |
|------|------|-------|---|-----|------|
| DATE: | | | | | |
| TIME: | | | | | |
| LOCATION: | | | | | |
| | NORMAL | | | | UNITS |
| Coll meth | | First morning | First morning | First morning | |
| Color | | Pale yellow | Pale yellow | Pale yellow | |
| Appear | | Clear | Clear | Clear | |
| Sp grv | 1.003–1.030 | | | | |
| pH | 5–7 | | | | |
| Prot | NEG | 1+ | | | mg/dL |
| Glu | NEG | | | | mg/dL |
| Ket | NEG | | | | |
| Occ bld | NEG | | | | |
| Ubil | NEG | | | | |
| Nit | NEG | | | | |
| Urobil | < 1.1 | 1.8 | | | EU/dL |
| Leu bst | NEG | | | | |
| Prot chk | NEG | | | | |
| WBCs | 0–5 | | | | /HPF |
| RBCs | 0–5 | | | | /HPF |
| EPIs | 0 | | | | /LPF |
| Bact | 0 | | | | |
| Mucus | 0 | | | | |
| Crys | 0 | | | | |
| Casts | 0 | | | | /LPF |
| Yeast | 0 | | | | |

## Case Questions

1. Several specific viruses are responsible for hepatitis symptoms. Describe the following characteristics of each virus.

|  | Hepatitis A HAV | Hepatitis B HBV | Hepatitis C HCV | Hepatitis D HDV | Hepatitis E HEV |
|---|---|---|---|---|---|
| Likely mode of transmission |  |  |  |  |  |
| Symptoms |  |  |  |  |  |
| Population most often affected |  |  |  |  |  |
| Means of reducing exposure |  |  |  |  |  |
| Treatment |  |  |  |  |  |

2. Describe hepatitis C to Dr. Wilcox.

3. What signs and symptoms does Dr. Wilcox have?

**4.**    Are there any other signs or symptoms of hepatitis that Dr. Wilcox does not have?

**5.**    Examine the patient's chemistry report. What values would steer Dr. Horowitz to the patient's diagnosis? What do these values measure, and what is their relationship to liver disease?

**6.**    The results of the anti-HCV and HCV RNA tests that Dr. Horowitz ordered were positive. What does this mean?

**7.**    Dr. Wilcox is devastated by the diagnosis. She tells Dr. Horowitz that she's never had a blood transfusion, never been exposed to blood products or experienced a needle stick. She has never used IV drugs, inhaled cocaine, or been in a relationship with anyone who has. She has had only one sexual partner in her lifetime. The only time she has come into contact with any kind of needles was during her undergraduate college years when she was tattooed. What is the likelihood that Dr. Wilcox contracted hepatitis C from her tattoo?

**8.**    Describe how the symptoms of hepatitis are related to the pathophysiology of this disease. Include at least five symptoms in your discussion.

**9.**    The course of infection usually follows four phases. What are these? Which one is Dr. Wilcox in? What is your rationale?

**10.**    Once the diagnosis of hepatitis C is made, the physician orders 3-MU interferon alfa-2b sq q d and Rebetol 200 mg po bid. What are these medications, and what do they do?

**11.**    What are the nutritional side effects of interferon and ribavirin?

**12.**    Given these side effects, what can the dietetic professional do to help the patient maintain positive nutritional status?

**13.**    As you assess Dr. Wilcox's nutritional status, what are your concerns?

14. Because resting energy expenditure varies in liver disease, indirect calorimetry is recommended. You do not have access to this means of measurement. How would you estimate Dr. Wilcox's energy and protein requirements? What would that estimate be?

15. Dr. Horowitz requested your consultation to order the patient's diet. What do you recommend?

16. How will you be able to determine if this diet prescription is appropriate?

17. Dr. Wilcox tells you that a friend suggested she use milk thistle to help fight the hepatitis virus. What would you tell her?

18. Using the patient's usual dietary intake, help her plan a menu.

| Usual Diet | Suggested Substitutions | Kcal | Pro (g) |
|---|---|---|---|
| 1½ c Sugar Frosted Flakes | | | |
| ½ c 2% milk | | | |
| 1 banana | | | |
| 1 c calcium-fortified orange juice | | | |
| Iced tea | | | |
| Cheeseburger | | | |
| 12 oz Diet Coke | | | |
| French fries | | | |
| 3 oz shrimp | | | |
| 1½ c vegetables | | | |
| 1 c steamed rice | | | |
| Water | | | |
| 4 cookies | | | |

19. What behavioral goals would you plan with the patient?

# Bibliography

Agency for Healthcare Research and Quality. Milk thistle: effects on liver disease and cirrhosis and clinical adverse effects. Evidence Report/Technology Assessment: Number 21. Available at http://www.ahrq.gov/clinic/milktsum.htm. Accessed May 16, 2001.

American Dietetic Association, *Manual of Clinical Dietetics*, 6th ed. Chicago: American Dietetic Association, 2000.

American Dietetic Association and Morrison Health Care. *Medical Nutrition Therapy Across the Continuum of Care*. Chicago: American Dietetic Association, 1998.

Centers for Disease Control, National Center for Infectious Diseases. Viral hepatitis C fact sheet. Available at http://www.cdc.gov/ncidod/diseases/hepatitis/c/fact.htm. Accessed May 16, 2001.

Commission on Accreditation for Dietetics Education. Knowledge, skills, and competencies for dietitians. *Accreditation Manual for Dietetics Education Programs*, rev. 4th ed. Chicago: American Dietetic Association, 2000.

Hepatitis. *The Merck Manual*. Available at http://www.merck.com. Accessed May 15, 2001.

Hepatitis Foundation International. Hepatitis C (revised 11/00). Available at http://www.hepfi.org. Accessed May 15, 2001.

Fischbach F. *Manual of Laboratory and Diagnostic Tests*, 6th ed. Philadelphia: Lippincott, 2000.

MedlinePlus Health information. Available at http://www.nlm.nih.gov/medlineplus/druginformation.html. Accessed September 8, 2000.

National Institute of Diabetes, Digestive and Kidney Diseases. Chronic hepatitis C: current disease management. Available at http://intelihealth.com. Accessed May 15, 2001.

Swearingen PL, Ross DG. Endocrine disorders. *Manual of Medical-Surgical Nursing Care*, 4th ed. St. Louis: Mosby, 1999.

Teran JC, McCullough AJ. Nutrition in liver disease. In American Society for Parenteral, Enteral Nutrition (ed.), *The Science and Practice of Nutrition Support: A Case Based Core Curriculum*. Dubuque, IA: Kendall/Hunt, 2000.

World Health Organization. Hepatitis C, fact sheet no. 164, revised October 2000. Available at http://www.who.int. Accessed May 16, 2001.

# Cirrhosis of the Liver with Resulting Hepatic Encephalopathy

*Advanced Practice*

## Objectives

After completing this case, the student will be able to:

1. Integrate knowledge of the pathophysiology of cirrhosis with development of medical nutrition therapy.
2. Collect pertinent information and use nutrition assessment techniques to determine baseline nutritional status.
3. Develop behavioral outcomes for the patient.
4. Identify appropriate MNT goals.
5. Determine key components of nutrition education for the patient with cirrhosis.

Dr. Teresa Wilcox, introduced in Case 20, is admitted to University Hospital with increasing symptoms of liver disease 3½ years after being diagnosed with acute hepatitis. Liver biopsy and CT scan confirm her diagnosis of cirrhosis of the liver secondary to chronic hepatitis C infection.

**UNIVERSITY HOSPITAL**

# ADMISSION DATABASE

| BED# 1 | DATE: 12/19 | TIME: 1400 | TRIAGE STATUS (ER ONLY): ☐ Red ☐ Yellow ☐ Green ☐ White |
|---|---|---|---|

**PRIMARY PERSON TO CONTACT:**
Name: Dennis Gustat
Home #: 555-3947
Work #: same

### Initial Vital Signs

| TEMP: 96.9 | RESP: 19 | SAO2: |
|---|---|---|

| HT: 5'6" | WT (lb): 125 | B/P: 102/65 | PULSE: 72 |
|---|---|---|---|

**ORIENTATION TO UNIT:** ☒ Call light ☒ Television/telephone ☒ Bathroom ☒ Visiting ☒ Smoking ☒ Meals ☒ Patient rights/responsibilities

| LAST TETANUS 8 yrs ago | LAST ATE am | LAST DRANK am |
|---|---|---|

### CHIEF COMPLAINT/HX OF PRESENT ILLNESS

N/V, anorexia

Fatigue, weakness

**PERSONAL ARTICLES:** (Check if retained/describe)
☒ Contacts ☒ R ☒ L  ☐ Dentures ☐ Upper ☐ Lower
☒ Jewelry:
☐ Other:

### ALLERGIES: Meds, Food, IVP Dye, Seafood: Type of Reaction

Penicillin

**VALUABLES ENVELOPE:**
☒ Valuables instructions

### PREVIOUS HOSPITALIZATIONS/SURGERIES

3 years ago for hepatitis C

**INFORMATION OBTAINED FROM:**
☒ Patient  ☐ Previous record
☐ Family  ☐ Responsible party

Signature  *Terri Wilcox*

### Home Medications (including OTC)    Codes: A=Sent home    B=Sent to pharmacy    C=Not brought in

| Medication | Dose | Frequency | Time of Last Dose | Code | Patient Understanding of Drug |
|---|---|---|---|---|---|
| Prempro | 0.65/5 | qd | this am | c | yes |
| Allegra | 60 | bid | this am | c | yes |
| Wellbutrin | 75 mg | bid | this am | c | yes |
| | | | | | |
| | | | | | |
| | | | | | |
| | | | | | |
| | | | | | |
| | | | | | |
| | | | | | |

Do you take all medications as prescribed?  ☒ Yes  ☐ No  If no, why?

### PATIENT/FAMILY HISTORY

| | | |
|---|---|---|
| ☐ Cold in past two weeks | ☒ High blood pressure Mother | ☐ Kidney/urinary problems |
| ☐ Hay fever | ☒ Arthritis Mother | ☒ Gastric/abdominal pain/heartburn Father |
| ☐ Emphysema/lung problems | ☐ Claustrophobia | ☐ Hearing problems |
| ☐ TB disease/positive TB skin test | ☐ Circulation problems | ☒ Glaucoma/eye problems Mother |
| ☒ Cancer Maternal grandmother | ☐ Easy bleeding/bruising/anemia | ☐ Back pain |
| ☐ Stroke/past paralysis | ☐ Sickle cell disease | ☐ Seizures |
| ☐ Heart attack | ☒ Liver disease/jaundice Patient | ☐ Other ALS, Paternal grandmother |
| ☐ Angina/chest pain | ☐ Thyroid disease | |
| ☐ Heart problems | ☒ Diabetes Father | |

### RISK SCREENING

Have you had a blood transfusion?  ☐ Yes  ☒ No
Do you smoke?  ☐ Yes  ☒ No
If yes, how many pack(s) ___/day for ___ years
Does anyone in your household smoke?  ☐ Yes  ☒ No
Do you drink alcohol?  ☐ Yes  ☒ No
If yes, how often? ___ How much? ___
When was your last drink?  ___/___/___
Do you take any recreational drugs?  ☐ Yes  ☒ No
If yes, type:_____  Route _____
Frequency:_____  Date last used:___/___/___

**FOR WOMEN Ages 12–52**

Is there any chance you could be pregnant?  ☐ Yes  ☒ No
If yes, expected date (EDC):
Gravida/Para: 0/0

**ALL WOMEN**

Date of last Pap smear: 5/12/
Do you perform regular breast self-exams?  ☒ Yes  ☐ No

**ALL MEN**

Do you perform regular testicular exams?  ☐ Yes  ☐ No

Additional comments:

x *Connie Bussard, RN*
Signature/Title

**Client name:** Teresa (Terri) Wilcox
**DOB:** 3/5
**Age:** 51
**Sex:** Female
**Education:** Doctoral degree
**Occupation:** Professor at the university
**Hours of work:** Varies—afternoon office hours, some afternoon classes, and some evening classes during the semester
**Household members:** Significant other (age 62), hyperlipidemia
**Ethnic background:** European American
**Religious affiliation:** Buddhist
**Referring physician:** Phillip Horowitz, MD (gastroenterology)

## Chief complaint:

"It just seems as if I can't get enough rest. I feel so weak. Sometimes I'm so tired, I can't go to campus to teach my classes. Does my skin look yellow to you?"

## Patient history:

Pt is a 51-yo postmenopausal wf who was in relatively good health until 3 years ago when she was Dx with hepatitis C. Origin of the hepatitis is thought to be a tattoo pt obtained as a college student over 30 years ago. Currently, she c/o fatigue, anorexia, N/V, and weakness. She has lost 10# since her last office visit, which was 6 mos ago. She also reports that she has noticed more bruising than usual lately.
*Type of Tx:* Rule out cirrhosis
*PMH:* Hepatitis C Dx 3 years ago; seasonal allergies treated with antihistamines
*Meds:* Prempro, 0.625/5 1 tab po daily; Allegra, 60 mg po qd; Wellbutrin, 75 mg po qd
*Smoker:* No
*Family Hx:* What? Mother (living)—HTN, diverticulitis, cholecystitis, carpal tunnel syndrome; father (deceased)—diabetes mellitus, peptic ulcer disease; maternal grandmother—cholecystitis, bilateral breast cancer; maternal grandfather—leukemia; paternal grandfather—cirrhosis; paternal grandmother—amyotrophic lateral sclerosis

## Physical exam:

*General appearance:* Tired-looking middle-aged female
*Vitals:* Temp 96.9°F, BP 102/65 mm Hg, HR 72 bpm, RR 19 bpm
*Heart:* Regular rate and rhythm, no gallops or rubs, point of maximal impulse at the fifth intercostal space in the midclavicular line
*HEENT:*
  *Head:* Normocephalic
  *Eyes:* Wears contact lenses to correct myopia, PERRLA
  *Ears:* Tympanic membranes w/out lesions
  *Nose:* Dry mucous membranes w/out lesions
  *Throat:* Enlarged esophageal veins
*Genitalia:* Normal female
*Neurologic:* Alert and oriented × 3
*Extremities:* Normal muscular tone, normal ROM; no edema; no asterixis noted
*Skin:* Warm and dry; bruising noted on lower arms and legs; telangiectasias noted on chest

*Chest/lungs:* Respirations normal, no crackles, rhonchi, wheezes, or rubs noted
*Peripheral vascular:* Pulse 3+ bilaterally
*Abdomen:* Tattoo on lower abdomen close to pubis; mild distension, splenomegaly w/out hepatomegaly

**Nutrition Hx:**
*General:* Has not had an appetite for the past few weeks. She drinks calcium-fortified orange juice for breakfast most mornings. Lunch is usually soup and crackers with a Diet Coke. Dinner at home, but may be carry-out. If carry-out, it's usually Chinese or Italian food.
*Usual dietary intake:* Sips of water, juice, and Diet Coke only. Has not eaten for the past 2 days.

*Food allergies/intolerances/aversions:* Does not like liver or lima beans
*Previous MNT?* Small, frequent meals, plenty of liquids. 3 yrs ago
*Food purchase/preparation:* Self and/or significant other
*Vit/min intake:* 400 mg vitamin E, 500 mg calcium multivitamin/mineral qd
*Current diet order:* soft, high kcal, high protein

**Tx plan:**
Prempro, 0.625/5 1 tab po daily
Allegra, 60 mg po qd
Wellbutrin, 75 mg po qd
CT scan of liver and biopsy
Endoscopy
Test stool for occult blood
Daily I/O
Spironolactone 25 mg qid
Propranolol 40 mg bid
Soft, high-kcal, high-protein diet—small, frequent meals
Multivitamin/mineral supplement
Bed rest

# U̲H̲ *UNIVERSITY HOSPITAL*

NAME: T. Wilcox                        DOB: 3/5
AGE: 51                                SEX: Female
PHYSICIAN: P. Horowitz, MD


\*\*\*\*\*\*\*\*\*\*\*\*\*\*\*\*\*\*\*\*\*\*\*\*\*\*\*\*\*\*\*\*\*\*\*\*\*\*\*\*\*\*CHEMISTRY\*\*\*\*\*\*\*\*\*\*\*\*\*\*\*\*\*\*\*\*\*\*\*\*\*\*\*\*\*\*\*\*\*\*\*\*\*\*\*\*\*

| DAY: | | Admit | d/c | |
| --- | --- | --- | --- | --- |
| DATE: | | | | |
| TIME: | | | | |
| LOCATION: | | | | |
| | NORMAL | | | UNITS |
| Albumin | 3.6–5 | 2.1 | | g/dL |
| Total protein | 6–8 | 5.4 | | g/dL |
| Prealbumin | 19–43 | 17 | | mg/dL |
| Transferrin | 200–400 | 187 | | mg/dL |
| Sodium | 135–155 | 135 | | mEq/L |
| Potassium | 3.5–5.5 | 5.0 | | mEq/L |
| Chloride | 98–108 | 102 | | mEq/L |
| $PO_4$ | 2.5–4.5 | 3.6 | | mEq/L |
| Magnesium | 1.6–2.6 | 2.1 | | mEq/L |
| Osmolality | 275–295 | 284 | | mmol/kg $H_2O$ |
| Total $CO_2$ | 24–30 | 28 | | mmol/L |
| Glucose | 70–120 | 115 | | mg/dL |
| BUN | 8–26 | 24 | | mg/dL |
| Creatinine | 0.6–1.3 | 1.2 | | mg/dL |
| Uric acid | 2.6–6 (women) | 5.9 | | mg/dL |
| | 3.5–7.2 (men) | | | |
| Calcium | 8.7–10.2 | 9.3 | | mg/dL |
| Bilirubin | 0.2–1.3 | 3.7 | | mg/dL |
| Ammonia ($NH_3$) | 9–33 | 33 | | $\mu$mol/L |
| SGPT (ALT) | 10–60 | 92 | | U/L |
| SGOT (AST) | 5–40 | 230 | | U/L |
| Alk phos | 98–251 | 275 | | U/L |
| CPK | 26–140 (women) | 138 | | U/L |
| | 38–174 (men) | | | |
| LDH | 313–618 | 658 | | U/L |
| CHOL | 140–199 | 199 | | mg/dL |
| HDL–C | 40–85 (women) | 50 | | mg/dL |
| | 37–70 (men) | | | |
| VLDL | | | | mg/dL |
| LDL | <130 | 125 | | mg/dL |
| LDL/HDL ratio | <3.22 (women) | | | |
| | <3.55 (men) | | | |
| Apo A | 101–199 (women) | | | mg/dL |
| | 94–178 (men) | | | |
| Apo B | 60–126 (women) | | | mg/dL |
| | 63–133 (men) | | | |
| TG | 35–160 | 256 | | mg/dL |
| $T_4$ | 5.4–11.5 | | | $\mu$g/dL |
| $T_3$ | 80–200 | | | ng/dL |
| $HbA_{1C}$ | 4.8–7.8 | 4.9 | | % |

U_H *UNIVERSITY HOSPITAL*

NAME: T. Wilcox                          DOB: 3/5
AGE: 51                                  SEX: Female
PHYSICIAN: P. Horowitz, MD

*************************************HEMATOLOGY*********************************************

DAY:                                              Admit
DATE:
TIME:
LOCATION:

| | NORMAL | | UNITS |
|---|---|---|---|
| WBC | 4.3–10 | 4.0 | $\times\ 10^3/mm^3$ |
| RBC | 4–5 (women) | 3.9 | $\times\ 10^6/mm^3$ |
| | 4.5–5.5 (men) | | |
| HGB | 12–16 (women) | 10.9 | g/dL |
| | 13.5–17.5 (men) | | |
| HCT | 37–47 (women) | 35.9 | % |
| | 40–54 (men) | | |
| MCV | 84–96 | 102 | fL |
| MCH | 27–34 | 29 | pg |
| MCHC | 31.5–36 | 35.4 | % |
| RDW | 11.6–16.5 | 12.4 | % |
| Plt ct | 140–440 | 342 | $\times\ 10^3$ |
| Diff TYPE | | | |
| % GRANS | 34.6–79.2 | 54.2 | % |
| % LYM | 19.6–52.7 | 20.6 | % |
| SEGS | 50–62 | 51 | % |
| BANDS | 3–6 | 4.2 | % |
| LYMPHS | 25–40 | 30 | % |
| MONOS | 3–7 | 4.2 | % |
| EOS | 0–3 | 2.8 | % |
| TIBC | 65–165 (women) | 160 | μg/dL |
| | 75–175 (men) | | |
| Ferritin | 18–160 (women) | 160 | μg/dL |
| | 18–270 (men) | | |
| Vitamin $B_{12}$ | 100–700 | 700 | pg/mL |
| Folate | 2–20 | 20 | ng/mL |
| Total T cells | 812–2318 | | $mm^3$ |
| T-helper cells | 589–1505 | | $mm^3$ |
| T-suppressor cells | 325–997 | | $mm^3$ |
| PT | 11–13 | 14.5 | sec |

# U<sub>H</sub> _UNIVERSITY HOSPITAL_

```
NAME: T. Wilcox                        DOB: 3/5
AGE: 51                                SEX: Female
PHYSICIAN: P. Horowitz, MD
```

```
     *****************************************URINALYSIS*****************************************
```

| DAY: | | Admit | 2 | d/c | |
|---|---|---|---|---|---|
| DATE: | | | | | |
| TIME: | | | | | |
| LOCATION: | | | | | |
| | NORMAL | | | | UNITS |
| Coll meth | | Random Specimen | | | |
| Color | | Dark | | | |
| Appear | | Slightly hazy | | | |
| Sp grv | 1.003-1.030 | 1.025 | | | |
| pH | 5-7 | 5.9 | | | |
| Prot | NEG | 1+ | | | mg/dL |
| Glu | NEG | NEG | | | mg/dL |
| Ket | NEG | NEG | | | |
| Occ bld | NEG | NEG | | | |
| Ubil | NEG | 1+ | | | |
| Nit | NEG | NEG | | | |
| Urobil | < 1.1 | 1.8 | | | EU/dL |
| Leu bst | NEG | NEG | | | |
| Prot chk | NEG | NEG | | | |
| WBCs | 0-5 | 3.8 | | | /HPF |
| RBCs | 0-5 | 2.7 | | | /HPF |
| EPIs | 0 | 0 | | | /LPF |
| Bact | 0 | 0 | | | |
| Mucus | 0 | 0 | | | |
| Crys | 0 | 0 | | | |
| Casts | 0 | 0 | | | /LPF |
| Yeast | 0 | 0 | | | |

## Case Questions

1. Nearly all blood that leaves the stomach and intestines must pass through the liver. In addition to acting as the body's "largest chemical factory," the liver has thousands of functions, many related to nutrition. List at least six functions related to nutrition.

2. The CT scan and liver biopsy confirm the diagnosis of cirrhosis. What is cirrhosis?

3. What is the etiology of cirrhosis?

4. What is the cause of this patient's cirrhosis?

5. List the signs and symptoms of cirrhosis, and describe the cause of each.

6. What signs and symptoms did this patient have?

7. Hypoglycemia is an additional symptom that cirrhotic patients may experience. What is the physiological basis for this? Is this a potential problem? Explain.

8. What are the treatments for cirrhosis?

9. Esophageal varices are visualized on the endoscopy. What are esophageal varices, and what causes them?

10. Examine the patient's chemistry values. Which labs support the diagnosis of cirrhosis? Explain their connection to the diagnosis.

11. Examine the patient's hematology values. Which are abnormal, and why? Does she have any physical symptoms consistent with your findings?

12. If the patient's cirrhosis is not treated at this time, how would you expect the disease to progress?

13. Protein-calorie malnutrition is commonly associated with cirrhosis. What are potential causes of malnutrition in cirrhosis? Explain each cause.

14. How does malnutrition affect the prognosis of cirrhosis?

15. Some nutrition assessment indices are not applicable to patients with cirrhosis. List those indices used in nutrition assessment to determine whether the measurement is appropriate or not. Briefly give your rationale.

16. Dr. Horowitz notes that the patient has lost 10 lb since her last exam. How would you interpret this weight loss?

17. What signs and/or symptoms would you monitor to determine further liver decompensation?

18. Dr. Horowitz prescribes two medications to assist with the patient's symptoms. What is the rationale for these medications, and what are the pertinent nutritional implications of each?

| Rationale for Rx | Nutritional Implications |
|---|---|
| Spironolactone | |
| Propranolol | |

19. If the patient's condition worsens (for example, acute varices bleeding, progresses to hepatic encephalopathy), the following medications could be used. Describe each drug classification and its mechanism.

| Drug | Classification | Mechanism |
|---|---|---|
| Vasopressin | | |
| Lactulose | | |
| Neomycin | | |
| Ferrous sulfate | | |
| Bisacodyl | | |
| Docusate | | |
| Diphenhydramine | | |

20. Why was a soft, 4-gm Na diet ordered?

21. Calculate the patient's energy and protein needs. What guidelines did you use and why?

**22.**   Outline medical nutrition therapy for the following stages of cirrhosis.

| | |
|---|---|
| Cirrhosis w/out encephalopathy | |
| Cirrhosis w/ acute encephalopathy | |
| Cirrhosis w/ chronic encephalopathy | |

**23.**   What is the recommendation regarding alcohol intake even though cirrhosis is caused by hepatitis C virus?

**24.**   Dr. Wilcox is discharged on a soft, 4-g Na diet with a 2-L fluid restriction. Do you agree with this decision?

**25.**   This patient asks if she can use a salt substitute at home. What would you tell her?

**26.**   What suggestions might you make to assist with compliance for the fluid restriction?

**27.**   Over the next 6 months, the patient's cirrhosis worsens. She is evaluated and found to be a good candidate for a liver transplant. She is placed on a transplant list and 20 weeks later receives a transplant. After the liver transplant, what diet and nutritional recommendations will the patient need before discharge? For long-term?

|  | Immediate posttransplant (first 2 mos) | Long-term posttransplant |
|---|---|---|
| Kcal |  |  |
| Protein |  |  |
| Fat |  |  |
| CHO |  |  |
| Sodium |  |  |
| Fluid |  |  |
| Calcium |  |  |
| Vitamins |  |  |

# Bibliography

American Dietetic Association, *Manual of Clinical Dietetics,* 6th ed. Chicago: American Dietetic Association, 2000.

American Dietetic Association and Morrison Health Care. *Medical Nutrition Therapy Across the Continuum of Care.* Chicago: American Dietetic Association, 1998.

Centers for Disease Control, National Center for Infectious Diseases. Viral hepatitis C fact sheet. Available at http://www.cdc.gov/ncidod/diseases/hepatitis/c/fact.htm. Accessed May 16, 2001.

Commission on Accreditation for Dietetics Education. Knowledge, skills, and competencies for dietitians. *Accreditation Manual for Dietetics Education Programs,* rev. 4th ed. Chicago: American Dietetic Association, 2000.

Escott-Stump S. *Nutrition and Diagnosis-Related Care,* 4th ed. Baltimore: Williams & Wilkins, 1998.

Family Doctor. Portal hypertension. Available at http://familydoctor.org/handouts/188.html. Accessed July 24, 2000.

Fischbach F. *Manual of Laboratory and Diagnostic Tests,* 6th ed. Philadelphia: Lippincott, 2000.

Ghalib R. Hepatic diseases. In Carey CF, Lee HH, Woeltje KF (eds.), *The Washington Manual of Medical Therapeutics,* 29th ed. Philadelphia: Lippincott, Williams & Wilkins, 1998.

Harvard Medical School. Hepatitis C: cirrhosis. Available at http://intelihealth.com. Accessed May 15, 2001.

Hasse JM, Matarese LE. Medical nutrition therapy for liver, biliary system, and exocrine pancreas disorders. In Mahan LK, Escott-Stump S (eds.), *Krause's Food, Nutrition, and Diet Therapy,* 10th ed. Philadelphia: Saunders, 2000.

Hepatic & biliary disorders. *The Merck Manual.* Available at http://www.merck.com. Accessed May 22, 2001.

Hepatitis Central. Cirrhosis. Available at http://hepatitis-central.com/hcv/liver/causes.html. Accessed May 15, 2001.

Hepatitis Central. Cirrhosis, many causes. Available at http://hepatitis-central.com/hcv/liver/causes.html. Accessed May 15, 2001.

Hepatitis Central. Cirrhosis symptoms and signs. Available at http://hepatitis-central.com/hcv/liver/causes.html. Accessed May 15, 2001.

Hepatitis Central. Complications of cirrhosis. Available at http://hepatitis-central.com/hcv/liver/causes.html. Accessed May 15, 2001.

Hepatitis Foundation International. Hepatitis C (revised November 2000). Available at http://www.hepfi.org. Accessed May 15, 2001.

MedlinePlus Health information. Available at http://www.nlm.nih.gov/medlineplus/druginformation.html. Accessed September 8, 2000.

Miller MJ, Bottcher J, Selberg O, Weselmann S, Boker KHW, Schwarze M, von zur Mihlen A, Manns MP. Hypermetabolism in clinically stable patients with liver cirrhosis. *Am J Clin Nutr.* 1999;69:1194–1201.

National Institute of Diabetes, Digestive and Kidney Diseases. Cirrhosis of the liver. Available at http://www.niddk.nih.gov/health/digest/pubs/cirrhosis/cirrhosis.htm. Accessed May 22, 2001.

Richardson RA, Davidson HI, Hinds A, Cowan S, Rae P, Garden OJ. Influence of the metabolic sequelae of liver cirrhosis on nutritional intake. *Am J Clin Nutr.* 1999;69:331–337.

Swearingen PL, Ross DG. Endocrine disorders. *Manual of Medical-Surgical Nursing Care,* 4th ed. St. Louis: Mosby, 1999.

Teran JC, McCullough AJ. Nutrition in liver disease. In American Society for Parenteral and Enteral Nutrition (ed.), *The Science and Practice of Nutrition Support: A Case Based Core Curriculum.* Dubuque, IA: Kendall/Hunt, 2000.

Worman HJ. Cirrhosis. Available at http://cpmcnet.columbia.edu/dept/gi/cirrhosis.html. Accessed May 22, 2001.

# MEDICAL NUTRITION THERAPY FOR NEUROLOGICAL AND PSYCHIATRIC DISORDERS

## Introduction

Pharmacological treatment of both neurological and psychiatric conditions is common. Significant drug–nutrient interactions have been identified with several classes of these medications. Monoamine oxidase inhibitors such as Nardil or Parnate have long been recognized as a source of potential drug–nutrient interaction and require restriction of foods high in tyramine. Another example of drug–nutrient interaction in psychiatric conditions is the effects of antidepressants on appetite. Many, such as Elavil, increase appetite and result in weight gain. Others, such as Prozac, may actually suppress appetite.

The first case in this unit examines not only drug–nutrient interaction but also the use of alternative therapy. The interest in complementary and alternative medicine has significantly increased in the United States over the previous two decades, during which more than 40% of Americans have used some form of alternative therapy. Of particular interest to nutrition professionals is the use of herbal therapies. Educa-

tion and research on their use is important in the nutrition profession. This case combines both situations: herbal therapy and drug–nutrient interaction.

The second case in this unit approaches neurological conditions through one of the most common diagnoses, Parkinson's disease. Nutrition issues in neurological conditions may begin at impairment in ability to obtain food. Symptoms such as impaired vision or ambulation may result in the inability to shop or prepare adequate meals. The next stage of impairment may include physical symptoms interfering with the ability to eat. Most neurological conditions can potentially interfere with chewing, swallowing, or feeding oneself. These problems are often not easily identified or easily solved. Each situation is highly individualized and requires a comprehensive nutrition assessment. The role of medical nutrition therapy is crucial throughout the course of the disease to maintain nutritional status and quality of life.

# Depression: Drug–Nutrient Interaction

*Introductory Level*

## Objectives

After completing this case, the student will be able to:

1. Describe alternative nutrition and herbal therapies.
2. Demonstrate working knowledge of pharmacology: nutrient–nutrient and drug–nutrient interactions.
3. Analyze nutrition assessment data to determine baseline nutritional status.
4. Identify potential drug–nutrient interactions and appropriate nutrition interventions for the prevention or treatment of drug–nutrient interactions.
5. Assess dietary intake for nutritional adequacy.
6. Identify health behaviors and educational needs.
7. Analyze sociocultural and ethnic food consumption issues and trends.
8. Interpret current research.

Ivanna Geitl, a 20-year-old college student, is admitted from University Health Services for evaluation of acute depression.

 **UNIVERSITY HOSPITAL**

## ADMISSION DATABASE

Name: Ivanna Geitl
DOB: 9/17  age 20
Physician: J. L. Byrd, MD

| BED# | DATE: | TIME: | TRIAGE STATUS (ER ONLY): |
|---|---|---|---|
| 2 | 11/1 | 1700 | ☐ Red ☐ Yellow ☐ Green ☐ White |

**Initial Vital Signs**

| TEMP: | RESP: | | SAO2: |
|---|---|---|---|
| 97.4 | 16 | | |

| HT: | WT (lb): | B/P: | PULSE: |
|---|---|---|---|
| 5'11" | 160 | 75/92 | 82 |

| LAST TETANUS | LAST ATE | LAST DRANK |
|---|---|---|
| 2 years ago | last pm | this am |

**PRIMARY PERSON TO CONTACT:**
Name: Mary Brown
Home #: 554-1242
Work #:

**ORIENTATION TO UNIT:** ☒ Call light ☒ Television/telephone ☒ Bathroom ☒ Visiting ☒ Smoking ☒ Meals ☒ Patient rights/responsibilities

### CHIEF COMPLAINT/HX OF PRESENT ILLNESS

"All I want to do is sleep. I haven't been able to eat--I don't feel hungry. I can hardly make myself go to classes. And when I go, I can't concentrate on what the professors are saying."

### ALLERGIES: Meds, Food, IVP Dye, Seafood: Type of Reaction

NKA

### PREVIOUS HOSPITALIZATIONS/SURGERIES

**PERSONAL ARTICLES:** (Check if retained/describe)
☐ Contacts ☐ R ☐ L       ☐ Dentures ☐ Upper ☐ Lower
☒ Jewelry: ring
☐ Other:

**VALUABLES ENVELOPE:**
☐ Valuables instructions

**INFORMATION OBTAINED FROM:**
☒ Patient       ☐ Previous record
☐ Family        ☐ Responsible party

Signature  *Ivanna Geitl*

**Home Medications (including OTC)**  Codes: A = Sent home   B = Sent to pharmacy   C = Not brought in

| Medication | Dose | Frequency | Time of Last Dose | Code | Patient Understanding of Drug |
|---|---|---|---|---|---|
| oral contraceptives | | | | | |
| | | | | | |
| | | | | | |
| | | | | | |
| | | | | | |
| | | | | | |
| | | | | | |
| | | | | | |
| | | | | | |
| | | | | | |

Do you take all medications as prescribed? ☒ Yes ☐ No  If no, why?

### PATIENT/FAMILY HISTORY

| | | |
|---|---|---|
| ☐ Cold in past two weeks | ☐ High blood pressure | ☐ Kidney/urinary problems |
| ☐ Hay fever | ☐ Arthritis | ☐ Gastric/abdominal pain/heartburn |
| ☐ Emphysema/lung problems | ☐ Claustrophobia | ☐ Hearing problems |
| ☐ TB disease/positive TB skin test | ☐ Circulation problems | ☐ Glaucoma/eye problems |
| ☐ Cancer | ☐ Easy bleeding/bruising/anemia | ☐ Back pain |
| ☐ Stroke/past paralysis | ☐ Sickle cell disease | ☐ Seizures |
| ☐ Heart attack | ☐ Liver disease/jaundice | ☐ Other |
| ☐ Angina/chest pain | ☐ Thyroid disease | |
| ☐ Heart problems | ☐ Diabetes | |

### RISK SCREENING

Have you had a blood transfusion? ☐ Yes ☐ No
Do you smoke? ☒ Yes ☐ No
If yes, how many pack(s) ½ /day for 2 years
Does anyone in your household smoke? ☒ Yes ☐ No
Do you drink alcohol? ☒ Yes ☐ No
If yes, how often? weekends   How much? 4–5 drinks
When was your last drink? 10/31/
Do you take any recreational drugs? ☐ Yes ☒ No
If yes, type:   Route
Frequency:_____   Date last used:

**FOR WOMEN Ages 12–52**
Is there any chance you could be pregnant? ☐ Yes ☒ No
If yes, expected date (EDC):
Gravida/Para:

**ALL WOMEN**
Date of last Pap smear: 8/12 / this year
Do you perform regular breast self-exams? ☒ Yes ☐ No

**ALL MEN**
Do you perform regular testicular exams? ☐ Yes ☐ No

Additional comments:

✗ *Mary Elizabeth Jenkins, RN*
Signature/Title

**Client name:** Ivanna Geitl
**DOB:** 9/17
**Age:** 20
**Sex:** Female
**Education:** International college student—junior year
**Occupation:** Student
**Hours of work:** Varies
**Household members:** Lives with roommates—Erika, 20; Veronica, 21; and Carol, 21—all are in excellent health. At home in Germany, she lives with her mother (age 45), father (age 50), and grandmother (age 70).
**Ethnic background:** German
**Religious affiliation:** Lutheran
**Referring physician:** J. L. Byrd, MD (family medicine, University Health Service)

## Chief complaint:

"All I want to do is sleep. I haven't been able to eat—I don't feel hungry. I can hardly make myself go to classes. And when I go, I can't concentrate on what the professors are saying. My roommates tell me that I've been really moody. I'm so lonely being here in the U.S. without my family. But I wouldn't want them to see how poorly I'm doing in school. It's just hopeless—I'll never meet their expectations. Maybe it would be better for everyone—me, my roommates, and my family—if I just wasn't around."

## Patient history:

Ms. Geitl is a 20-yo female university student from Germany. She seeks medical attention at the urging of her roommates, who report that her mood has become increasingly depressed over the past two semesters. She has become withdrawn and moody—a significant change from her affect since first coming to the U.S. to attend college. She is an otherwise healthy young woman. Ms. Geitl reports a 5-pound weight loss in the past 3 months. She takes birth control pills for contraception and regulation of menses. Her mother has been treated for depression with St. John's wort by the family physician for the past 10 years.
*Onset of disease:* 9 months ago
*Type of Tx:* Antidepressant medication and counseling therapy
*PMH:* Noncontributory
*Meds:* Oral contraceptives
*Smoker:* Yes. ½ pack per day
*Family Hx: What?* Mother regularly takes *Hypericum perforatum* for mood swings.

## Physical exam:

*General appearance:* Thin and pale
*Vitals:* Temp 97.4°F, BP 75/92 mm Hg, HR 80 bpm, RR 20 bpm
*Heart:* Regular rate and rhythm, no gallops or rubs, point of maximal impulse at fifth intercostal space in midclavicular line
*HEENT:*
  *Head:* Normocephalic, equal carotid pulses
  *Eyes:* PERRLA
  *Ears:* Membranes dry

*Nose:* Moist, pink mucous membranes

*Throat:* Tissue pink and smooth w/out exudates or lesions

*Genitalia:* Normal female

*Neurologic:* Lethargic and oriented to person, place, and time

*Extremities:* Noncontributory

*Skin:* Soft, smooth, dry, and flexible

*Chest/lungs:* Respirations normal; no crackles, rhonchi, wheezes, or rubs noted

*Peripheral vascular:* Normal pulse (3+) bilaterally, warm, no edema

*Abdomen:* Nontender, no guarding, normal bowel sounds

**Nutrition Hx:**

*General:* Appetite has been poor.

*Usual dietary intake:*

*Breakfast:*   Black coffee—2 cups

*Lunch:*        No time for lunch—maybe a diet cola

*PM snack:*    Low-fat frozen yogurt—approx. ½ cup of chocolate or strawberry

*Dinner:*       (usually with roommates—they take turns cooking)

              3″ square Stouffer's lasagna—ate half

              1 cup steamed broccoli (plain, only with salt and pepper)

              1 breadstick

              Diet cola soft drink

*HS snack:*    Air-popped popcorn sprayed with butter-flavored Pam and salted

              Diet cola soft drink

*24-hr recall:*

*Breakfast:*   1 cup black coffee

              1 slice dry whole wheat toast

*Lunch:*        2 cups chicken and noodle soup

              2 saltine crackers

              ½ cup strawberry gelatin

              1 12-oz can diet cola

*Dinner:*       2 peach halves

              1 cup cottage cheese (low fat)

              1 cup black coffee

*Food allergies/intolerances/aversions (specify):* None

*Previous MNT?* No

*Food purchase/preparation:* Self and roommates

*Vit/min intake:* None

*Current diet order:* Regular, select

**Tx plan:**

Zoloft, 50 mg qd

Referral to house psychologist for counseling

Nutrition consult re: Poor eating habits

## Case Questions

1. What is depression?

2. Dr. Byrd has decided to treat Ms. Geitl with Zoloft, a selective serotonin reuptake inhibitor (SSRI). Are there any pertinent nutritional considerations when using this medication?

3. How do serotonin reuptake inhibitors (SSRIs) work?

4. During the diet history, you ask Ms. Geitl if she uses any OTC vitamins, minerals, or herbal supplements. She tells you her mother suggested she try *Hypericum perforatum* (St. John's wort) because in Germany it is prescribed to treat depression. Ms. Geitl did as her mother suggested, as it is available without prescription in the United States. What is St. John's wort?

5. How is St. John's wort used in the United States?

6. How does St. John's wort work as an antidepressant?

7. Does St. John's wort have any side effects?

8. How is St. John's wort regulated in the United States?

9. How is St. John's wort used in Europe?

10. Why do you think people are interested in alternative medicine and herbal treatments?

11. Because Ms. Geitl is ambulatory, you are able to measure her height and weight. She is 5'11" tall and weighs 160 pounds. You also determine that she is of medium frame. Because Ms. Geitl is from Germany, she is used to reporting her weight in kilograms and her height in centimeters. Convert her height and weight to metric numbers.

**12.** Is Ms. Geitl's recent weight loss anything to be worried about?

**13.** Because Ms. Geitl is alert and cooperative, you ask her to complete a Patient-Generated Subjective Global Assessment (PG-SGA) of Nutritional Status. How would you score her? (See Appendix I.)

| Sections | Score |
|---|---|
| Box 1 | |
| Box 2 | |
| Box 3 | |
| Box 4 | |
| Weight loss section (Table 1) | |
| Disease section (Table 2) | |
| Metabolic section (Table 3) | |
| Physical section (Table 4) | |
| **Total** | |

**SGA Rating**

Select one:

| ☐ A = well nourished | ☐ B = moderately (or suspected of being) malnourished | ☐ C = severely malnourished |
|---|---|---|

**14.** How would you triage nutritional intervention?

**15.** What methods are available to estimate Ms. Geitl's energy needs?

**16.** Calculate Ms. Geitl's basal energy needs using one of the methods you listed in Question 15.

**17.** What is Ms. Geitl's estimated energy expenditure?

**18.** Evaluate her diet history and her 24-hour recall. Is she meeting her energy needs?

**19.** What would you advise?

**20.** List each factor from your nutritional assessment and then determine an expected outcome from each.

| Assessment Factor | Expected Outcome |
|---|---|
|  |  |
|  |  |
|  |  |
|  |  |

**21.** What is your immediate concern regarding this patient's use of St. John's wort?

**22.** Review the initial nutrition note written for this patient. Is this progress note appropriate? Is it complete? Any errors? Any omissions?

11/1/00    Nutrition Note
1028    S    Pt reports decreased appetite for duration of 6 mo, & wt loss of 5#. Diet history indicates a usual intake of ~600 kcal/day & 31 g protein. 24-hr recall indicates intake of ~625 kcal and 46 g protein. Pt takes St. John's wort & oral contraceptives. No vitamins or minerals reported. Smokes ½ pack cigarettes/day.
        O    20 yo wf college student, 5'11", 160#, UBW 165#
            Estimated nutrient needs ~ 2600 kcal and 58 g protein (0.8 g/d)
            Meds: Zoloft.
            PG-SGA score—B (suspected of becoming malnourished)
        A    Potential drug–herb interaction between Zoloft & oral contraceptives with St. John's wort. Pt may be at risk of malnutrition based on PG-SGA score, and she is not meeting her energy or protein needs with her oral intake.
        P    Contact MD verbally regarding potential drug–herb interaction between Zoloft & oral contraceptives with St. John's wort. Recommend d/c St. John's wort. Pt's likes and dislikes noted. Adjust meals accordingly. Work with Pt to assist with increasing kcal and protein intake on d/c.

            Darren Arela, RD

# Bibliography

American Dietetic Association. *Manual of Clinical Dietetics,* 6th ed. Chicago: American Dietetic Association, 2000.

American Dietetic Association and Morrison Health Care. *Medical Nutrition Therapy Across the Continuum of Care.* Chicago: American Dietetic Association, 1998.

American Herbal Pharmacopoeia and Therapeutic Compendium. St. John's wort (*Hypericum perforatum*) monograph. *Herbalgram. J Am Bot Council and Herb Res Found.* 1997;40:1–16.

Bennett DA. Neuropharmacology of St. John's wort (*Hypericum*). *Ann Pharmacotherapy.* 1998;32:1201–1208.

Blumenthal M, Busse WR, Goldbert A, Gruenwald J, Hall T, Riggins CW, Rister RS (eds.). *The Complete German Commission E Monographs. Therapeutic Guide to Herbal Medicines.* Austin, TX: American Botanical Council. Published in cooperation with Integrative Medicine Communications (Boston, MA), 1998.

Commission on Accreditation for Dietetics Education. Knowledge, skills, and competencies for dietitians. *Accreditation Manual for Dietetics Education Programs,* rev. 4th ed. Chicago: American Dietetic Association, 2000.

Cupp MJ. Herbal remedies: adverse effects and drug interactions. *American Family Physician.* 1999;March 1; 59(5):1239–1245.

Depression (unipolar disorder). *The Merck Manual.* Available at http://www.merck.com. Accessed November 15, 2000.

Escott-Stump S. *Nutrition and Diagnosis-Related Care,* 4th ed. Baltimore: Williams & Wilkins, 1998.

Food and Agriculture Organization (FAO), World Health Organization (WHO). *Energy and Protein Requirements.* Technical Report Series 724. Geneva: WHO, 1985.

Fischbach F. *Manual of Laboratory and Diagnostic Tests,* 6th ed. Philadelphia: Lippincott, 2000.

Fugh-Berman A. Herb–drug interactions. *Lancet.* 2000; 355(9198, January 8):134–138.

Henney JE. Risk of drug interactions with St. John's wort. *JAMA.* April 5, 2000. Available at http://jama.ama-assn.org. Accessed November 14, 2000.

Lee RD, Nieman DC. *Nutritional Assessment,* 2nd ed. St. Louis: Mosby, 1996.

Library of the National Medical Society. Physicians' Drug Resource. Available at http://www.medical-library.org/index.htm. Accessed November 16, 2000.

Linde K, Ramirez G, Murlow CD, Weidenhammer W, Melchart D. St. John's wort for depression: an overview and metaanalysis of randomized clinical trials. *Br Med J.* 1996;313(7052):253–258.

*Mayo Clinic Health Oasis.* Available at http://www.mayohealth.org. Accessed November 17, 2000.

MedlinePlus Health information. Available at http://www.nlm.nih.gov/medlineplus/druginformation.html. Accessed September 8, 2000.

Moore LB, Goodwin B, Jones SA, Wisely GB, Serabjit-Singh CJ, Wilson TM, Collins JL, Kliewer SA. St. John's wort induces hepatic drug metabolism through activation of the pregnane X receptor. *Proc Natl Acad Sci USA.* 2000;97(13), June 20:7500–7502.

Murray M. Common questions about St. John's wort extracts. *Am J Natural Med.* 1997;4(7):14–19.

National Center for Complementary and Alternative Medicine, National Institutes of Health. *St. John's wort.* Available at http://nccam.nih.gov/nccam. Accessed November 14, 2000.

National Institute of Mental Health. *Depression.* Available at http://www.nimh.nih.gov/publicat/depression.cfm. Accessed November 14, 2000.

National Institute of Mental Health. *Medications.* Available at http://www.nimh.nih.gov/publicat/medicate.cfm. Accessed November 14, 2000.

National Institute of Mental Health. *Questions and Answers About St. John's Wort.* Available at http://www.nimh.nih.gov/publicat/stjohnqa.cfm. Accessed November 14, 2000.

National Mental Health Association. *Clinical Depression: Learning to Recognize Clinical Depression.* Available at http://www.nmha.org. Accessed November 14, 2000.

*Physicians' Desk Reference for Herbal Medicines.* Montvale, NJ: Medical Economics Company, 1998.

Pressman A, Shelley D (eds.). *Integrative Medicine: The Patient's Essential Guide to Conventional and Complementary Treatments for More Than 300 Common Disorders.* New York: St. Martin's Press, 2000.

Pronsky ZM. *Powers and Moore's Food–Medication Interactions,* 11th ed. Birchrunville, PA: Food–Medication Interactions, 2000.

# Parkinson's Disease with Dysphagia

*Introductory Level*

## Objectives

After completing this case, the student will be able to:

1. Demonstrate working knowledge of the pathophysiology of Parkinson's disease.
2. Analyze anthropometric, biochemical, clinical, and dietary data to complete comprehensive nutrition assessment.
3. Assess and identify nutritional health risks in dysphagia.
4. Determine appropriate nutrient requirements.
5. Use strategies to assess need for adaptive feeding techniques.
6. Interpret laboratory parameters relating to nutrition.
7. Design nutrition recommendations that will promote pleasurable eating.

Mrs. Ruth Leaming, a 77-year-old woman in an advanced stage of Parkinson's disease is admitted to University Hospital. She is experiencing coughing, choking, and difficulty eating.

 **UNIVERSITY HOSPITAL**

## ADMISSION DATABASE

Name: R. Leaming
DOB: 4/5  age 77
Physician: S. Goldman

| BED#<br>1 | DATE:<br>11/10 | TIME:<br>0930 | TRIAGE STATUS (ER ONLY):<br>☐ Red  ☐ Yellow  ☐ Green  ☐ White |
|---|---|---|---|

**Initial Vital Signs**

| TEMP:<br>98.6 | RESP:<br>18 | SAO2: |
|---|---|---|

| HT :<br>5'0" | WT (lb):<br>90 | B/P:<br>135/85 | PULSE:<br>80 |
|---|---|---|---|

| LAST TETANUS<br>20 years ago | LAST ATE<br>0730 | LAST DRANK<br>0730 |
|---|---|---|

PRIMARY PERSON TO CONTACT:
Name: Phil Leaming
Home #: 555-1234
Work #: 555-4321

ORIENTATION TO UNIT: ☒ Call light  ☒ Television/telephone
☒ Bathroom  ☒ Visiting  ☒ Smoking  ☒ Meals
☒ Patient rights/responsibilities

### CHIEF COMPLAINT/HX OF PRESENT ILLNESS

Trouble swallowing; chokes on food

PERSONAL ARTICLES: (Check if retained/describe)
☐ Contacts ☐ R ☐ L      ☐ Dentures ☐ Upper ☐ Lower
☐ Jewelry
☒ Other: glasses

### ALLERGIES: Meds, Food, IVP Dye, Seafood: Type of Reaction

VALUABLES ENVELOPE:
☐ Valuables instructions

### PREVIOUS HOSPITALIZATIONS/SURGERIES

INFORMATION OBTAINED FROM:
☒ Patient          ☐ Previous record
☒ Family           ☐ Responsible party

Signature  *Phil Leaming*

| Home Medications (including OTC) | | Codes: A=Sent home | B=Sent to pharmacy | | C=Not brought in |
|---|---|---|---|---|---|
| Medication | Dose | Frequency | Time of Last Dose | Code | Patient Understanding of Drug |
| Sinemet | 50/200 mg | bid | 0700 | c | yes |
| | | | | | |
| | | | | | |
| | | | | | |
| | | | | | |
| | | | | | |
| | | | | | |
| | | | | | |
| | | | | | |
| | | | | | |
| | | | | | |

Do you take all medications as prescribed?    ☒ Yes    ☐ No    If no, why?

### PATIENT/FAMILY HISTORY

| | | |
|---|---|---|
| ☐ Cold in past two weeks | ☐ High blood pressure | ☐ Kidney/urinary problems |
| ☐ Hay fever | ☐ Arthritis | ☐ Gastric/abdominal pain/heartburn |
| ☐ Emphysema/lung problems | ☐ Claustrophobia | ☐ Hearing problems |
| ☐ TB disease/positive TB skin test | ☐ Circulation problems | ☐ Glaucoma/eye problems |
| ☐ Cancer | ☐ Easy bleeding/bruising/anemia | ☐ Back pain |
| ☐ Stroke/past paralysis | ☐ Sickle cell disease | ☐ Seizures |
| ☐ Heart attack | ☐ Liver disease/jaundice | ☒ Other Parkinson's disease |
| ☐ Angina/chest pain | ☐ Thyroid disease | |
| ☐ Heart problems | ☐ Diabetes | |

### RISK SCREENING

Have you had a blood transfusion?    ☐ Yes  ☒ No
Do you smoke?    ☐ Yes  ☒ No
If yes, how many pack(s)    /day for    years
Does anyone in your household smoke?    ☐ Yes  ☒ No
Do you drink alcohol?    ☐ Yes  ☒ No
If yes, how often?        How much?
When was your last drink?    /   /
Do you take any recreational drugs?    ☐ Yes  ☒ No
If yes, type:    Route
Frequency:    Date last used:

**FOR WOMEN Ages 12–52**

Is there any chance you could be pregnant?    ☐ Yes    ☐ No
If yes, expected date (EDC):
Gravida/Para:

**ALL WOMEN**

Date of last Pap smear:   /   /
Do you perform regular breast self-exams?    ☐ Yes    ☒ No

**ALL MEN**

Do you perform regular testicular exams?    ☐ Yes    ☐ No

Additional comments:

✗ *Carol Boushey, RN, BSN*
Signature/Title

**Client name:** Ruth Leaming
**DOB:** 4/5
**Age:** 77
**Sex:** Female
**Education:** High school diploma
**Occupation:** Retired hairdresser
**Hours of work:** N/A
**Household members:** Widowed, lives with son, age 55, and his family
**Ethnic background:** European American
**Religious affiliation:** Protestant
**Referring physician:** Sol Goldman, MD

## Chief complaint:

"Every time I eat, something gets stuck in my throat. I cough and feel like I'm choking every time I eat. I'm afraid to eat any more."

## Patient history:

Mrs. Leaming is a 77-yo white female who is a retired hairdresser. She lives with her son (55 yo) and his wife. She is in the advanced stages of Parkinson's disease and is being treated with Sinemet. Mrs. Leaming has lost 10 pounds in the past 6 months. Her son and daughter-in-law report that she has been eating less and less at meals, mainly because of coughing and choking episodes.
*Onset of disease:* Parkinson's—10 years ago
*Type of Tx:* R/o dysphagia
*PMH:* Parkinson's disease, 10 years duration
*Meds:* Sinemet: carbidopa/levodopa, 50/200 mg controlled-release tablet bid
*Smoker:* No
*Family Hx:* Noncontributory

## Physical exam:

*General appearance:* Slight, elderly female in no apparent distress
*Vitals:* Temp 98.6°F, BP 135/85 mm Hg, HR 80 bpm, RR 18 bpm
*Heart:* Regular rate and rhythm, no gallops or rubs, point of maximal impulse at the fifth intercostal space in the midclavicular line
*HEENT:*
   *Head:* Normocephalic; dry, dull hair; sunken cheeks
   *Eyes:* Wears glasses for myopia, bilateral redness, fissured eyelid corners
   *Ears:* Tympanic membranes normal
   *Nose:* Dry mucous membranes w/out lesions
   *Throat:* Slightly dry mucous membranes w/out exudates or lesions
*Genitalia:* Normal w/out lesions
*Neurologic:* Alert and oriented × 3; decreased blink reflex; positive palmomental; diminished postural relexes
*Extremities:* Reduced strength, koilonychias, slight tremor in hands
*Skin:* Warm and dry, poor turgor, angular stomatitis and cheilosis noted on lips
*Chest/lungs:* Respirations normal, no crackles, rhonchi, wheezes, or rubs noted
*Peripheral vascular:* Bilateral, 3+ pedal pulses
*Abdomen:* Hypoactive bowel sounds, distended, w/out masses or organomegaly

**Nutrition Hx:**
*General:* Mrs. Leaming states that she has been having trouble swallowing solid foods. She complains of something "stuck in her throat."

*Usual dietary intake:*
Mostly liquids, because solids are difficult to swallow. Before developing difficulty in swallowing, Mrs. Leaming usually ate the following:

| | |
|---|---|
| *Breakfast:* | Orange juice |
| | Cream of wheat or raisin bran with 2% milk |
| | 1 banana |
| | Coffee with 2% milk and sweetener |
| *Lunch:* | Chicken tortellini soup (cheese tortellini cooked in chicken broth) |
| | Saltine crackers—about 8 |
| | Canned pears—2 halves |
| | Iced tea with sweetener |
| *Dinner:* | Baked chicken—usually ½ breast |
| | Baked potato—1 medium |
| | Steamed broccoli—approx. 1 cup with 1 tsp margarine |
| | Canned peaches—6–8 slices |
| | Iced tea with sweetener |
| *HS snack:* | Popcorn—4 cups popped with melted butter w/ Coca Cola or |
| | 1 cup Häagen Dazs ice cream |
| *24-hr recall:* | N/A |

*Food allergies/intolerances/aversions:* Food, especially dry or sticky food, gets stuck in pt's throat.
*Previous MNT?* N/A
*Food purchase/preparation:* Son and daughter-in-law
*Vit/min intake:* Multivitamin/mineral supplement daily, 500 mg calcium 3 ×/day
*Current diet order:* NPO

**Tx plan:**
Bedside swallowing assessment
Fiberoptic endoscopy
Modified barium swallow
Evaluation by speech-language pathologist and dietitian
Staged dysphagia diet per RD
2 L $D_5W$

**U<sub>H</sub>** *UNIVERSITY HOSPITAL*

NAME: R. Leaming                          DOB: 4/5
AGE: 77                                   SEX: Female
PHYSICIAN: S. Goldman

\*\*\*\*\*\*\*\*\*\*\*\*\*\*\*\*\*\*\*\*\*\*\*\*\*\*\*\*\*\*\*\*\*\*\*\*\*\*\*\*\*\*\*\*\*\*CHEMISTRY\*\*\*\*\*\*\*\*\*\*\*\*\*\*\*\*\*\*\*\*\*\*\*\*\*\*\*\*\*\*\*\*\*\*\*\*\*\*\*\*\*\*\*\*

DAY:                                      Admit              2
DATE:
TIME:
LOCATION:

| | NORMAL | | | UNITS |
|---|---|---|---|---|
| Albumin | 3.6–5 | 5.1 | 2.8 | g/dL |
| Total protein | 6–8 | 7.9 | 6.4 | g/dL |
| Prealbumin | 19–43 | 35 | 11 | mg/dL |
| Transferrin | 200–400 | 250 | 148 | mg/dL |
| Sodium | 135–155 | 150 | 135 | mEq/L |
| Potassium | 3.5–5.5 | 4.5 | 4.5 | mEq/L |
| Chloride | 98–108 | 100 | 102 | mEq/L |
| $PO_4$ | 2.5–4.5 | 3.5 | 3.6 | mEq/L |
| Magnesium | 1.6–2.6 | 2.6 | 1.5 | mEq/L |
| Osmolality | 275–295 | 350 | 297 | mmol/kg $H_2O$ |
| Total $CO_2$ | 24–30 | | | mmol/L |
| Glucose | 70–120 | 135 | 118 | mg/dL |
| BUN | 8–26 | | | mg/dL |
| Creatinine | 0.6–1.3 | | | mg/dL |
| Uric acid | 2.6–6 (women) | | | mg/dL |
| | 3.5–7.2 (men) | | | |
| Calcium | 8.7–10.2 | | | mg/dL |
| Bilirubin | 0.2–1.3 | | | mg/dL |
| Ammonia ($NH_3$) | 9–33 | | | $\mu$mol/L |
| SGPT (ALT) | 10–60 | | | U/L |
| SGOT (AST) | 5–40 | | | U/L |
| Alk phos | 98–251 | | | U/L |
| CPK | 26–140 (women) | | | U/L |
| | 38–174 (men) | | | |
| LDH | 313–618 | | | U/L |
| CHOL | 140–199 | 180 | 138 | mg/dL |
| HDL-C | 40–85 (women) | 45 | 43 | mg/dL |
| | 37–70 (men) | | | |
| VLDL | | | | mg/dL |
| LDL | < 130 | 128 | 120 | mg/dL |
| LDL/HDL ratio | < 3.22 (women) | | | |
| | < 3.55 (men) | | | |
| Apo A | 101–199 (women) | | | mg/dL |
| | 94–178 (men) | | | |
| Apo B | 60–126 (women) | | | mg/dL |
| | 63–133 (men) | | | |
| TG | 35–160 | 100 | 30 | mg/dL |
| $T_4$ | 5.4–11.5 | | | $\mu$g/dL |
| $T_3$ | 80–200 | | | ng/dL |
| $HbA_{1c}$ | 4.8–7.8 | | | % |

**U<sub>H</sub> _UNIVERSITY HOSPITAL_**

NAME: R. Leaming                          DOB: 4/5
AGE: 77                                    SEX: Female
PHYSICIAN: S. Goldman

\*\*\*\*\*\*\*\*\*\*\*\*\*\*\*\*\*\*\*\*\*\*\*\*\*\*\*\*\*\*\*\*\*\*\*\*\*\*\*\*\*\*\*\*HEMATOLOGY\*\*\*\*\*\*\*\*\*\*\*\*\*\*\*\*\*\*\*\*\*\*\*\*\*\*\*\*\*\*\*\*\*\*\*\*\*\*\*\*\*\*\*\*

DAY:                                      Admit            2
DATE:
TIME:
LOCATION:

|  | NORMAL | Admit | 2 | UNITS |
|---|---|---|---|---|
| WBC | 4.3–10 | 10 | 4.0 | $\times\ 10^3/mm^3$ |
| RBC | 4–5  (women) | 5.5 | 4.3 | $\times\ 10^6/mm^3$ |
|  | 4.5–5.5 (men) |  |  |  |
| HGB | 12–16 (women) | 17 | 13 | g/dL |
|  | 13.5–17.5 (men) |  |  |  |
| HCT | 37–47 (women) | 48 | 38 | % |
|  | 40–54 (men) |  |  |  |
| MCV | 84–96 | 82 | 81 | fL |
| MCH | 27–34 | 28 | 28 | pg |
| MCHC | 31.5–36 | 33 | 34 | % |
| RDW | 11.6–16.5 | 11.5 | 11.8 | % |
| Plt ct | 140–440 |  |  | $\times\ 10^3$ |
| Diff TYPE |  |  |  |  |
| % GRANS | 34.6–79.2 |  |  | % |
| % LYM | 19.6–52.7 | 51.9 | 18.7 | % |
| SEGS | 50–62 |  |  | % |
| BANDS | 3–6 |  |  | % |
| LYMPHS | 25–40 |  |  | % |
| MONOS | 3–7 |  |  | % |
| EOS | 0–3 |  |  | % |
| TIBC | 65–165 (women) | 80 | 85 | μg/dL |
|  | 75–175 (men) |  |  |  |
| Ferritin | 18–160 (women) | 150 | 155 | μg/dL |
|  | 18–270 (men) |  |  |  |
| Vitamin B$_{12}$ | 100–700 | 450 | 450 | pg/mL |
| Folate | 2–20 | 18 | 17 | ng/mL |
| Total T cells | 812–2318 |  |  | $mm^3$ |
| T-helper cells | 589–1505 |  |  | $mm^3$ |
| T-suppressor cells | 325–997 |  |  | $mm^3$ |
| PT | 11–13 | 11.5 | 11.5 | sec |

## Case Questions

1. Describe Parkinson's disease.

2. What symptoms are common in Parkinson's patients that place those individuals at risk?

3. Define *dysphagia.*

4. Could Mrs. Leaming's dysphagia be linked to Parkinson's disease? If so, how?

5. What problems/complications are associated with dysphagia?

6. Describe the four phases of swallowing:

| Phase | Description |
|---|---|
| Oral preparation | |
| Oral transit | |
| Pharyngeal | |
| Esophageal | |

7. Give an example of one problem or condition related to each phase.

| Phase | Problem | Signs and Symptoms | Dietary Considerations |
|---|---|---|---|
| Oral preparation | | | |
| Oral transit | | | |
| Pharyngeal | | | |
| Esophageal | | | |

8.  Describe a bedside swallowing assessment.

9.  Describe a modified barium swallow or fiberoptic endoscopic evaluation of swallowing.

10. For this patient, compare the values of labs that are significantly different from admission and day 2 of hospitalization.

| Lab | Admit Value | Day 2 Value |
|-----|-------------|-------------|
|     |             |             |
|     |             |             |
|     |             |             |
|     |             |             |
|     |             |             |
|     |             |             |
|     |             |             |
|     |             |             |
|     |             |             |
|     |             |             |
|     |             |             |
|     |             |             |
|     |             |             |

11. Which values are more accurate? Why?

12. Mrs. Leaming's usual body weight is approximately 100 lb. On admission, she was found to weigh 90 lb. Calculate her percent usual body weight (%UBW).

13. How would you interpret her %UBW?

14. She has reportedly lost 7 lb. in the past 6 months. Calculate her percent weight change.

**15.**   How would you interpret her percent weight loss?

**16.**   Are any signs and symptoms documented in the physical examination that may indicate a poor nutritional status? If so, which ones? What type of nutritional deficiency may they indicate?

**17.**   Which of Mrs. Leaming's lab values may reflect her nutritional status?

**18.**   Albumin, prealbumin, and transferrin lab values are often used to evaluate visceral protein stores. Compare Mrs. Leaming's values to the norms, and indicate whether these reflect mild, moderate, or severe deficits of her visceral protein stores.

| Lab | Norms | Pt Values | Mild/Moderate/Severe Deficit |
|---|---|---|---|
| Albumin | | | |
| Prealbumin | | | |
| Transferrin | | | |

**19.**   By evaluating Mrs. Leaming's anthropometric data, biochemical data, clinical data, and what you know about her recent dietary intake, what would you conclude regarding her nutritional status?

**20.**   Estimate Mrs. Leaming's energy needs using the Harris-Benedict equation.

**21.**   Should current weight or usual weight be used to estimate energy needs? Is there much difference in the estimates for this patient?

**22.**   Calculate her protein needs. What standards would you use? Why?

**23.**   Estimate Mrs. Leaming's fluid needs using the following methods: weight, age and weight, energy needs.

**24.** Which of the preceding methods for estimating fluid needs is the easiest to calculate? Which method appears most reasonable for this patient? Explain.


**25.** To maintain or attain normal nutritional status while reducing danger of aspiration and choking, texture (of foods) and/or viscosity (of fluids) are personalized for a patient with dysphagia. In the following table, define each term used to describe characteristics of foods and give an example.

| Term | Definition | Example |
|------|------------|---------|
| Consistency | | |
| Texture | | |
| Viscosity | | |


**26.** Diets for dysphagia are described in four stages. In the following table, describe each diet stage and give examples of five foods that could be included in each diet.

| Diet | Description | Examples of Foods |
|------|-------------|-------------------|
| Phase 1: Puréed diet | | |
| Phase 2: Ground/minced diet | | |
| Phase 3: Soft/easy-to-chew diet | | |
| Phase 4: Modified general diet | | |


**27.** It is determined that Mrs. Leaming's dysphagia is centered in the esophageal transit phase and she has reduced esophageal peristalsis. Which diet is appropriate to try with Mrs. Leaming?

28. Using her usual dietary intake, make suggestions for food substitutions to Mrs. Leaming and her family.

Orange juice _____

Cream of Wheat _____

Raisin bran _____

2% milk _____

Banana _____

Coffee _____

Sweetener _____

Chicken tortellini soup _____

Saltine crackers _____

Canned pears _____

Iced tea _____

Baked chicken _____

Baked potato _____

Steamed broccoli _____

Margarine _____

Canned peaches _____

Popcorn _____

Coca-Cola _____

Ice cream _____

29. Because her foods will be ground or chopped and few raw fruits and vegetables are tolerated/allowed, how can Mrs. Leaming get adequate amounts of fiber in her diet? (What are some high-fiber foods that could be included?)

## Bibliography

American Dietetic Association. *Manual of Clinical Dietetics,* 6th ed. Chicago: American Dietetic Association, 2000.

Beck AM, Oversen L. At which body mass index and degree of weight loss should hospitalized elderly patients be considered at nutritional risk? *Clin Nutr.* 1998;17:195–198.

Brody RA, Tougher-Decker R, VonHagen S, Maillet JO. Role of registered dietitians in dysphagia screening. *J Am Diet Assoc.* 2000;100(9):1029–1037.

Commission on Accreditation for Dietetics Education. Knowledge, skills, and competencies for dietitians. *Accreditation Manual for Dietetics Education Programs,* rev. 4th ed. Chicago: American Dietetic Association, 2000.

Franklin VL. Parkinson's disease. Available at http://www.medical-library.org. Accessed September 22, 2000.

Grant A, DeHoog S. *Nutrition: Assessment, Support, and Management,* 5th ed., rev. and expanded. Seattle, WA: Ruth Grant and Susan DeHoog, 1999.

Holden K, Remig VM. Parkinson's disease: assessing and managing unique nutrition needs. In *Manual of Clinical Dietetics,* 6th ed. Chicago: American Dietetic Association, 2000.

*Living with Parkinson's Disease: Disease Progression.* Available at http://www.parkinson.org. Accessed October 31, 2000.

MedlinePlus Health information. Available at http://www.nlm.nih.gov/medlineplus/druginformation.html. Accessed November 1, 2000.

Parkinson's disease. *The Merck Manual.* Available at http://www.merck.com. Accessed October 31, 2000.

Shronts EF, Fish JA, Pesce-Hammond K. Assessment of nutritional status. In *The ASPEN Nutrition Support Practice Manual.* Silver Spring, MD: ASPEN, 1998.

Swallowing problems. *Signs and Symptoms of Parkinson's Dysphagia.* Available at http://www.parkinson.org. Accessed October 31, 2000.

# MEDICAL NUTRITION THERAPY FOR PULMONARY DISORDERS

## Introduction

The two cases in this section portray the interrelationship between nutrition and the respiratory system. In a healthy individual the respiratory system receives oxygen for cellular metabolism and expires waste products—primarily carbon dioxide. Fuels—carbohydrate, protein, and lipid—are metabolized, using oxygen and producing carbon dioxide. The type of fuel an individual receives can affect physiological conditions and interfere with normal respiratory function.

Nutritional status and pulmonary function are interdependent. Malnutrition can evolve from pulmonary disorders and can contribute to declining pulmonary status. The incidence of malnutrition is common for people with COPD, ranging anywhere from 25 to 50%. In respiratory disease, maintaining nutritional status improves muscle strength needed for breathing, decreases risk of infection, facilitates weaning from mechanical ventilation, and improves ability for physical activity.

The American Thoracic Society defines chronic obstructive pulmonary disease (COPD) as a disease process of chronic airway obstruction caused by chronic bronchitis, emphysema, or both. These conditions place a significant burden on the health care systems in the United States, with an estimated cost of over $25 billion dollars each year. Prevalence is increasing as well, with approximately 15 million people affected in the United States alone. COPD is the fourth leading cause of death, resulting in approximately 100,000 deaths each year.

In Cases 24 and 25, nutritional assessment and evaluation demonstrate the effects of COPD on nutritional status. As patients are started on nutrition support, you will examine the role of nutrition on declining respiratory status.

# Chronic Obstructive Pulmonary Disease

*Introductory Level*

## Objectives

After completing this case, the student will be able to:

1. Identify the potential effects of pulmonary disease on nutritional status.
2. Identify the effects of malnutrition on pulmonary status.
3. Distinguish between emphysema and chronic bronchitis, which together constitute chronic obstructive pulmonary disease (COPD).
4. Differentiate the effects of nutrient metabolism for oxygen consumption and carbon dioxide production.

5. Interpret pertinent laboratory values for assessment of nutritional status.
6. Interpret body composition data for nutritional assessment.
7. Determine methods to increase an individual's intake of energy and protein.

Stella Bernhardt is initially diagnosed with stage 1 COPD (emphysema). She is now admitted with increasing shortness of breath and possible upper respiratory infection.

 **UNIVERSITY HOSPITAL**

## ADMISSION DATABASE

Name: Stella Bernhardt
DOB: 10/23   age 62
Physician: D. Feinstein, MD

| BED#<br>1 | DATE:<br>1/25 | TIME:<br>1340 | TRIAGE STATUS (ER ONLY):<br>☐ Red  ☐ Yellow  ☐ Green  ☐ White |
|---|---|---|---|

### Initial Vital Signs

| TEMP:<br>98.8 | RESP:<br>22 | | SAO2: |
|---|---|---|---|

| HT:<br>5'3" | WT (lb):<br>119 | B/P:<br>130/88 | PULSE:<br>92 |
|---|---|---|---|

| LAST TETANUS<br>unknown | LAST ATE<br>today noon | LAST DRANK<br>today noon |
|---|---|---|

**PRIMARY PERSON TO CONTACT:**
Name: Pete Bernhardt
Home #: 404-339-6543
Work #: n/a

**ORIENTATION TO UNIT:** ☒ Call light  ☒ Television/telephone
☒ Bathroom  ☒ Visiting  ☒ Smoking  ☒ Meals
☒ Patient rights/responsibilities

### CHIEF COMPLAINT/HX OF PRESENT ILLNESS

I'm hardly able to do anything for myself right now. Even taking a
bath or getting dressed makes me so short of breath. My husband had
to help me out of the shower this morning.

**PERSONAL ARTICLES:** (Check if retained/describe)
☐ Contacts  ☐ R  ☐ L          ☒ Dentures  ☒ Upper  ☒ Lower
☒ Jewelry: wedding band
☒ Other: glasses

### ALLERGIES: Meds, Food, IVP Dye, Seafood: Type of Reaction

NKA

**VALUABLES ENVELOPE:**
☐ Valuables instructions

### PREVIOUS HOSPITALIZATIONS/SURGERIES

4 live births

last hospitalization—January 1 year ago for pneumonia

**INFORMATION OBTAINED FROM:**
☒ Patient          ☐ Previous record
☒ Family           ☐ Responsible party

Signature  *Stella Bernhardt*

### Home Medications (including OTC)   Codes: A=Sent home   B=Sent to pharmacy   C=Not brought in

| Medication | Dose | Frequency | Time of Last Dose | Code | Patient Understanding of Drug |
|---|---|---|---|---|---|
| Combivent | 2 puffs | qid | today noon | c | yes |
| | | | | | |
| | | | | | |
| | | | | | |
| | | | | | |
| | | | | | |
| | | | | | |
| | | | | | |
| | | | | | |
| | | | | | |

Do you take all medications as prescribed?   ☒ Yes   ☐ No   If no, why?

### PATIENT/FAMILY HISTORY

| | | |
|---|---|---|
| ☒ Cold in past two weeks | ☐ High blood pressure | ☐ Kidney/urinary problems |
| ☐ Hay fever | ☐ Arthritis | ☐ Gastric/abdominal pain/heartburn |
| ☒ Emphysema/lung problems Patient | ☐ Claustrophobia | ☐ Hearing problems |
| ☐ TB disease/positive TB skin test | ☐ Circulation problems | ☐ Glaucoma/eye problems |
| ☒ Cancer Mother | ☐ Easy bleeding/bruising/anemia | ☐ Back pain |
| ☐ Stroke/past paralysis | ☐ Sickle cell disease | ☐ Seizures |
| ☐ Heart attack | ☐ Liver disease/jaundice | ☐ Other |
| ☐ Angina/chest pain | ☐ Thyroid disease | |
| ☐ Heart problems | ☐ Diabetes | |

### RISK SCREENING

Have you had a blood transfusion?   ☒ Yes   ☐ No
Do you smoke?   ☒ Yes   ☐ No   Quit last year
If yes, how many pack(s)   1/day for   46 years
Does anyone in your household smoke?   ☒ Yes   ☐ No
Do you drink alcohol?   ☐ Yes   ☒ No
If yes, how often?   How much?
When was your last drink?   /   /
Do you take any recreational drugs?   ☐ Yes   ☒ No
If yes, type:_____   Route
Frequency:_____   Date last used:_____/_____/

**FOR WOMEN Ages 12–52**

Is there any chance you could be pregnant?   ☐ Yes   ☐ No
If yes, expected date (EDC):
Gravida/Para:

**ALL WOMEN**

Date of last Pap smear:   /10   /this year
Do you perform regular breast self-exams?   ☒ Yes   ☐ No

**ALL MEN**

Do you perform regular testicular exams?   ☐ Yes   ☐ No

Additional comments:

✗ *Betty Larson, RN*
Signature/Title

**Client name:** Stella Bernhardt
**DOB:** 10/23
**Age:** 62
**Sex:** Female
**Education:** Some college. *What grade/level?* Client completed two years of college.
**Occupation:** Retired office manager for independent insurance agency
**Hours of work:** N/A
**Household members:** Husband, age 68. PMH of CAD
**Ethnic background:** Caucasian
**Religious affiliation:** Methodist
**Referring physician:** Debra Feinstein, MD (pulmonology)

## Chief complaint:

"I'm hardly able to do anything for myself right now. Even taking a bath or getting dressed makes me short of breath. My husband had to help me out of the shower this morning. I feel that I am gasping for air. I am coughing up a lot of stuff that is a dark brownish-green. I always am short of breath, but I can tell when things change. I was at a church meeting with a lot of people—I might have caught something there. My husband says that I am confused in the morning. I know it is hard for me to get going in the morning. Do you think my confusion is related to my COPD?"

## Patient history:

*Onset of disease:* Initially diagnosed with stage 1 COPD (emphysema) five years ago. Medical records at last admission indicate pulmonary function tests: baseline $FEV^1 = 0.7L$, $FVC = 1.5L$, $FEV^1/FVC$ 46%.
*Type of Tx:* Combivent (metered-dose inhaler)—2 inhalations qid (each inhalation delivers 18 mcg ipratropium bromide; 130 mcg albuterol sulfate)
*PMH:* No occupational exposures; bronchitis and upper respiratory infections during winter months for most of adult life. Four live births; 2 miscarriages
*Meds:* Combivent (see *Type of Tx*)
*Smoker:* Yes. 46 year, 1 ppd history—has quit for past 1 year
*Family Hx: What?* CA    *Who?* mother, 2 aunts died from lung cancer

## Physical exam:

*General appearance:* 62-year-old female in no acute distress
*Vitals:* Temp: 98.8°F, BP 130/88, HR 92 bpm, RR 22 bpm
*Heart:* Regular rate and rhythm; mild jugular distension noted
*HEENT*
    *Eyes:* PERRLA, no hemorrhages
    *Ears:* Slight redness
    *Nose:* Clear
    *Throat:* Clear
*Genitalia:* Deferred
*Neurologic:* Alert, oriented; cranial nerves intact
*Extremities:* 1+ bilateral pitting edema. No cyanosis or clubbing
*Skin:* Warm, dry
*Chest/lungs:* Decreased breath sounds, percussion hyperresonant; prolonged expiration with wheezing; rhonchi throughout; using accessory muscles at rest
*Abdomen:* Liver, spleen palpable; nondistended, nontender, normal bowel sounds

**Nutrition Hx:**

*General:* Patient states that her appetite is poor. "I fill up so quickly—after just a few bites." Relates that meal preparation is difficult—"By the time I fix a meal, I am too tired to eat it." In the previous two days, she states that she has eaten very little. Increased coughing has made it very hard to eat. "I don't think food tastes as good, either. Everything has a bitter taste." Highest adult weight was 145–150# (5 years ago). States that her family constantly tells her how thin she has gotten—"I haven't weighed myself for a while but I know my clothes are bigger." Dentures are present but fit loosely.

*Usual dietary intake:*

| | |
|---|---|
| AM: | Coffee, juice or fruit, dry cereal with small amount of milk |
| Lunch: | Large meal of the day—meat, vegetables, rice, potato, or pasta but patient admits she eats only very small amounts. |
| Dinner/evening meal: | Eats very light in evening—usually soup, scrambled eggs, or sandwich Drinks Pepsi throughout day (usually 3 12-oz cans). |

*24-hr recall:*

| | |
|---|---|
| AM: | .5 c coffee with nondairy creamer, few sips of orange juice, .5 c oatmeal with 1 tsp sugar, .75 c chicken noodle soup, 2 saltine crackers, .5 c coffee with nondairy creamer; sips of Pepsi throughout day and evening—estimated amount 32 oz |

*Food allergies/intolerances/aversions (specify):* Avoids milk—"People say it will increase mucus production."

*Previous MNT?* No

*Food purchase/preparation:* Self; "My daughters come and help sometimes."

*Vit/min intake:* None

**Admitting diagnosis:**

Acute exacerbation of COPD, increasing dyspnea, hypercapnia, r/o pneumonia

**Tx plan:**

$O^2$ 1 L/minute via nasal cannula with humidity—keep $O^2$ saturation 90–91%

IVF $D_5$ 1/2 NS with 20 mEq KCL @ 75 cc/hr

Solumedrol 10 mg/kg q 6 hr

Ancef 500 mg q 6 hr

Ipratropium bromide via nebulizer 2.5 mg q 30 minutes × 3 treatments then q 2 hr

Albuterol sulfate via nebulizer 4 mg q 30 minutes × 3 doses then 2.5 mg q 4 hr

ABGs q 6 hours. CXR — EPA/LAT. Sputum cultures and Gram stain

**Hospital course:**

Mrs. Bernhardt was diagnosed with acute exacerbation of COPD secondary to bacterial pneumonia. This was confirmed by CXR and sputum culture. She responded well to aggressive medical treatment for her emphysema, although her physician does feel her underlying condition has progressed. She will be discharged on home $O^2$ therapy for the first time and referred to an outpatient pulmonary rehabilitation program. Her discharge medications will be the same (Combivent), but she will complete an oral course of corticosteroids and an additional 10-day course of Keflex. Dr. Feinstein ordered a nutrition consult in-house with recommendations for nutritional follow-up through the pulmonary rehabilitation program.

# U H *UNIVERSITY HOSPITAL*

NAME: Stella Bernhardt                    DOB: 10/23
AGE: 62                                   SEX: F
PHYSICIAN: Dr. Feinstein

\*\*\*\*\*\*\*\*\*\*\*\*\*\*\*\*\*\*\*\*\*\*\*\*\*\*\*\*\*\*\*\*\*\*\*\*\*\*\*\*\*CHEMISTRY\*\*\*\*\*\*\*\*\*\*\*\*\*\*\*\*\*\*\*\*\*\*\*\*\*\*\*\*\*\*\*\*\*\*\*\*\*\*\*

DAY:                                          1
DATE:
TIME:
LOCATION:

| | NORMAL | | UNITS |
|---|---|---|---|
| Albumin | 3.6–5 | 3.4 | g/dL |
| Total protein | 6–8 | 5.9 | g/dL |
| Prealbumin | 19–43 | 19 | mg/dL |
| Transferrin | 200–400 | 219 | mg/dL |
| Sodium | 135–155 | 136 | mmol/L |
| Potassium | 3.5–5.5 | 3.7 | mmol/L |
| Chloride | 98–108 | 101 | mmol/L |
| $PO_4$ | 2.5–4.5 | 3.1 | mmol/L |
| Magnesium | 1.6–2.6 | 2.1 | mmol/L |
| Osmolality | 275–295 | 280 | mmol/kg $H_2O$ |
| Total $CO_2$ | 24–30 | 31 | mmol/L |
| Glucose | 70–120 | 92 | mg/dL |
| BUN | 8–26 | 9 | mg/dL |
| Creatinine | 0.6–1.3 | 0.9 | mg/dL |
| Uric acid | 2.6–6 (women) | 3.4 | mg/dL |
| | 3.5–7.2 (men) | | |
| Calcium | 8.7–10.2 | 9.1 | mg/dL |
| Bilirubin | 0.2–1.3 | 0.4 | mg/dL |
| Ammonia ($NH_3$) | 9–33 | | $\mu$mol/L |
| SGPT (ALT) | 10–60 | | U/L |
| SGOT (AST) | 5–40 | | U/L |
| Alk phos | 98–251 | | U/L |
| CPK | 26–140 (women) | | U/L |
| | 38–174 (men) | | |
| LDH | 313–618 | | U/L |
| CHOL | 140–199 | | mg/dL |
| HDL–C | 40–85 (women) | | mg/dL |
| VLDL | 37–70 (men) | | mg/dL |
| LDL | <130 | | mg/dL |
| LDL/HDL ratio | <3.22 (women) | | |
| | <3.55 (men) | | |
| Apo A | 101–199 (women) | | mg/dL |
| | 94–178 (men) | | |
| Apo B | 60–126 (women) | | mg/dL |
| | 63–133 (men) | | |
| TG | 35–160 | | mg/dL |
| $T_4$ | 5.4–11.5 | | $\mu$g/dL |
| $T_3$ | 80–200 | | ng/dL |
| $HbA_{1c}$ | 4.8–7.8 | | % |

# U<sub>H</sub> *UNIVERSITY HOSPITAL*

NAME: Stella Bernhardt                    DOB: 10/23
AGE: 62                                   SEX: F
PHYSICIAN: D. Feinstein, MD

\*\*\*\*\*\*\*\*\*\*\*\*\*\*\*\*\*\*\*\*\*\*\*\*\*\*\*\*\*\*\*\*\*\*\*\*\*HEMATOLOGY\*\*\*\*\*\*\*\*\*\*\*\*\*\*\*\*\*\*\*\*\*\*\*\*\*\*\*\*\*\*\*\*\*\*\*\*\*\*\*

DAY:                                      1
DATE:
TIME:
LOCATION:

|  | NORMAL |  | UNITS |
|---|---|---|---|
| WBC | 4.3–10 | 15.0 | $\times\ 10^3/mm^3$ |
| RBC | 4–5 (women) | 4 | $\times\ 10^6/mm^3$ |
|  | 4.5–5.5 (men) |  |  |
| HGB | 12–16 (women) | 11.5 | g/dL |
|  | 13.5–17.5 (men) |  |  |
| HCT | 37–47 (women) | 35 | % |
|  | 40–54 (men) |  |  |
| MCV | 84–96 |  | fL |
| MCH | 27–34 |  | pg |
| MCHC | 31.5–36 |  | % |
| RDW | 11.6–16.5 |  | % |
| Plt ct | 140–440 |  | $\times\ 10^3$ |
| Diff TYPE |  |  |  |
| % GRANS | 34.6–79.2 |  | % |
| % LYM | 19.6–52.7 |  | % |
| SEGS | 50–62 | 83 | % |
| BANDS | 3–6 | 5 | % |
| LYMPHS | 25–40 | 10 | % |
| MONOS | 3–7 | 3 | % |
| EOS | 0–3 | 1 | % |
| TIBC | 65–165 (women) |  | µg/dL |
|  | 75–175 (men) |  |  |
| Ferritin | 18–160 (women) |  | µg/dL |
|  | 18–270 (men) |  |  |
| Vitamin B$_{12}$ | 100–700 |  | pg/mL |
| Folate | 2–20 |  | ng/mL |
| Total T cells | 812–2318 |  | mm$^3$ |
| T-helper cells | 589–1505 |  | mm$^3$ |
| T-suppressor cells | 325–997 |  | mm$^3$ |
| PT | 11–13 |  | sec |

# U<sub>H</sub> *UNIVERSITY HOSPITAL*

NAME: Stella Bernhardt                    DOB: 10/23
AGE: 62                                   SEX: F
PHYSICIAN: Dr. Feinstein

**********************************ARTERIAL BLOOD GASES (ABGs)**********************************

DAY:                                  1        2        3
DATE:
TIME:
LOCATION:

|  | NORMAL |  |  |  | UNITS |
|---|---|---|---|---|---|
| pH | 7.35–7.45 | 7.29 |  | 7.4 |  |
| $PaCO_2$ | 35–45 | 50.9 |  | 40.1 | mmHg |
| $PaO_2$ | ≥80 | 77.7 |  | 90.2 | mmHg |
| $SaO_2$ | 95% or higher | 92 |  |  | % |
| $CO_2$ content | 25–30 | 31 |  | 30.8 | mmol/L |
| $O_2$ content | 15–22 |  |  |  | vol % |
| Carbonic acid |  | 2.4 |  | 1.2 | mmol/L |
| Base excess | > 3 |  |  | 6.0 | mEq/L |
| Base deficit | < 3 | 3.6 |  |  | mEq/L |
| $HCO_3^-$ | 24–28 | 24.7 |  | 29.6 | mEq/L |
| HGB | 12–16 (women) | 11.5 |  |  | g/dL |
|  | 13.5–17.5 (men) |  |  |  |  |
| HCT | 37–47 (women) | 35 |  |  | % |
|  | 40–54 (men) |  |  |  |  |
| COHb | < 2 |  |  |  | % |
| $[NA^+]$ | 135–148 | 136 |  |  | mmol/L |
| $[K^+]$ | 3.5–5 | 3.7 |  |  | mEq/L |

## Case Questions

1. Mrs. Bernhardt was diagnosed with emphysema/COPD five years ago. What risk factors does Mrs. Bernhardt have for this disease?

2. **a.** Identify the symptoms described in the MD's history and physical that are consistent with Mrs. Bernhardt's diagnosis. Then describe the pathophysiology that may be responsible for each symptom.

| Symptom | Etiology |
|---|---|
| Shortness of breath (dyspnea) | |
| Early morning confusion (hypercapnia) | |
| Increased production of brown-green sputum | |
| Fatigue | |
| Early satiety | |
| Anorexia | |
| Dysgeusia | |

**b.** Now identify features of the physician's physical examination that are consistent with her admitting diagnosis. Describe the pathophysiology that might be responsible for each physical finding.

| Physical Finding | Etiology |
| --- | --- |
|  |  |
|  |  |
|  |  |
|  |  |
|  |  |
|  |  |
|  |  |

**3.** Outline the similarities and differences between emphysema and chronic bronchitis.

**4.** Mrs. Bernhardt has quit smoking. Shouldn't her condition now improve? Explain.

**5.** Mrs. Bernhardt's medical record indicates previous pulmonary function tests as follows: baseline $FEV^1$ = 0.7L, FVC=1.5L, $FEV^1$/FVC 46%. Define FEV, FVC, and FEV/FVC, and indicate how they are used in the diagnosis of COPD. How can these measurements be used in treating COPD?

**6.** Look at Mrs. Bernhardt's arterial blood gas report when she was admitted.

**a.** Why would arterial blood gases (ABGs) be drawn for this patient?

**b.** Define each of the following and interpret Mrs. Bernhardt's values:

pH:

$PaCO_2$:

$SaO_2$:

$HCO_3^{2-}$:

**c.** Once Mrs. Bernhardt was placed on oxygen therapy, what lab values tell you that the therapy is working?

7. Calculate Mrs. Bernhardt's UBW, %UBW, IBW, %IBW, and BMI. Do any of these values indicate that she is at nutritional risk?

8. Calculate Mrs. Bernhardt's energy and protein requirements. What activity and stress factors would you use? What is your rationale?

9. Using Mrs. Bernhardt's nutrition history and 24-hour recall as reference, do you feel she is consuming an adequate oral intake? Explain.

10. What factors can you identify from her nutrition interview that probably contribute to her difficulty in eating?

11. What is a respiratory quotient? How is this figure related to nutritional intake and respiratory status?

**12.** What is the current recommendation on the appropriate mix of calories from carbohydrate, protein, and fat?

**13.** For each nutrition problem you identified in Question 10, determine one or two interventions.

**14.** What goals might you set for Mrs. Bernhardt as she is discharged and beginning pulmonary rehabilitation?

**15.** You are now seeing Mrs. Bernhardt at her second visit to pulmonary rehabilitation. She provides you with the following information from her food record. Her weight is now 116 lbs. She explains that adjustment to her medications and oxygen at home has been difficult, so she hasn't felt like eating very much. When you talk with her, you find she is hungriest in the morning and often by evening she is too tired to eat. She is having no specific intolerances. She does tell you that she hasn't consumed any milk products because she thought they would cause more sputum to be produced.

*Food Diary*
*Monday*
*Breakfast:* Coffee, 1 c with 2 T nondairy creamer; orange juice ½ c; 1 poached egg; ½ slice toast
*Lunch:* ½ tuna salad sandwich (3 T tuna salad on 1 slice wheat bread); coffee, 1 c with 2 T nondairy creamer
*Supper:* Cream of tomato soup, 1 c; ½ slice toast; ½ banana; Pepsi—approx 36 oz

*Tuesday*
*Breakfast:* Coffee, 1 c with 2 T nondairy creamer; orange juice, ½ c; ½ c oatmeal with 2 T brown sugar
*Lunch:* 1 chicken leg from Kentucky Fried Chicken; ½ c mashed potatoes; 2 T gravy; coffee, 1 c with 2 T nondairy creamer
*Supper:* Cheese, 2 oz; 8 saltine crackers; 1 can V8 juice (6 oz); Pepsi, approx 36 oz

  **a.** Is she meeting her calorie and protein goals?

  **b.** What would you tell her regarding the use of supplements and/or milk and sputum production?

  **c.** Using her food diary, identify three interventions that you would propose for Mrs. Bernhardt to increase her caloric and protein intake.

# Bibliography

American Dietetic Association. Respiratory disease. In *Manual of Clinical Dietetics*, 6th ed. Chicago: American Dietetic Association, 2000:579–586.

American Thoracic Society. Standards for the diagnosis and care of patients with chronic obstructive pulmonary disease. *Am J Respir Crit Care Med.* 1995; 152(suppl):S77–S120.

American Thoracic Society. Standards for the diagnosis and care of patients with chronic obstructive pulmonary disease (COPD) and asthma, Chapter One. *Am Rev Respir Dis.* 1996;154:225–228.

American Thoracic Society. Statement: standardization of spirometry, update. *Am Rev Respir Dis.* 1987; 136:1285–1298.

Anthonisen NR, Connect JE, Kiley JP, et al. for the Lung Health Study Research Group. Effects of smoking intervention and the use of an anticholinergic bronchodilator on the rate of decline in FEV. *JAMA.* 1994;272:1497–1503.

Brashers VL. Chronic obstructive pulmonary disease. In *Clinical Applications of Pathophysiology.* St. Louis: Mosby-Times Mirror, 1998.

Burdet L, de Muralt B, Schultz Y, Pichard C, Fitting JW. Administration of growth hormone to underweight patients with chronic obstructive pulmonary disease: a prospective, randomized, controlled study. *Am J Respir Crit Care Med.* 1997;156:1800–1806.

Callahan CM, Dittus RS, Kate BP. Oral corticosteroid therapy for patients with stable chronic obstructive pulmonary disease: a meta-analysis. *Ann Intern Med.* 1991;114:216–223.

Commission on Accreditation for Dietetics Education. Knowledge, skills, and competencies for dietitians. *Accreditation Manual for Dietetics Education Programs,* rev. 4th ed. Chicago: American Dietetic Association, 2000.

Dewan NA, Bell W, Moore J, Anderson B, Kirchain W, O'Donohue WJ. Smell and taste function in subjects with chronic obstructive pulmonary disease. *Chest.* 1990;97:595–599.

Dompeling E, van Schayck CP, van Grunsven YM, et al. Slowing the deterioration of asthma and chronic obstructive pulmonary disease observed during bronchodilator therapy by adding inhaled corticosteroids: a 4-year prospective study. *Ann Intern Med.* 1993; 118:770–778.

Edelman NH, Rucker RB, Peavy HH. Nutrition and the respiratory system. *Am Rev Respir Dis.* 1999;34: 347–352.

Farzan S. Chronic obstructive pulmonary disease. In Farzan S (ed.), *A Concise Handbook of Respiratory Diseases,* 3rd ed. Norwalk, CT: Appleton and Lange, 1992:127–139.

Fletcher GM, Peto R. The natural history of chronic airflow obstruction. *Br Med J.* 1978;1:1645–1648.

Gardner P, Rosenberg HM, Wilson RW. *Leading Causes of Death by Age, Sex, Race, and Hispanic Origin: United States, 1992.* Vital and Health Statistics No. 20(29). Hyattsville, MD: National Center for Health Statistics, 1996.

Goldstein S, Askanazi J, Weissman C, Thomashow B, Kinney JM. Energy expenditure in patients with chronic obstructive pulmonary disease. *Chest.* 1987; 91:222–224.

Gray-Donald K, Carrey Z, Martin JIG. Postprandial dyspnea and malnutrition in patients with chronic obstructive pulmonary disease. *Clin Invest Med.* 1998;21:135–141.

Green JH, Muers MI. The thermic effect of food in underweight patients with emphasematous chronic obstructive lung disease. *Eur Respir J.* 1991;4:813–819.

Johannsen JM. Chronic obstructive pulmonary disease: current comprehensive care for emphysema and bronchitis. *Nurs Pract.* 1994;19:59–67.

Jones PW, Quirk FH, Baveystock CM, Littlejohns P. A self-complete measure of health status for chronic airflow limitation. The St George's respiratory questionnaire. *Am Rev Respir Dis.* 1992;145:1321–1327.

Mahler DA. Chronic obstructive pulmonary disease. In: Mahler DA (ed.), *Pulmonary Disease in the Elderly Patient.* New York: Dekker, 1993:1159–1184.

Mueller DH. Medical nutrition therapy for upper gastrointestinal tract disorders. In Mahan LK, Escott-Stump S. (eds.), *Krause's Food, Nutrition, and Diet Therapy,* 10th ed. Philadelphia: Saunders, 2000:816–832.

Muers MF, Green JH. Weight loss in chronic obstructive pulmonary disease. *Eur Respir J.* 1993;6:729–734.

Murray CJL, Lopez AD. Alternative projections of mortality and disability by cause 1990–2020: global burden of disease study. *Lancet.* 1997;349:1498–1504.

Murray GJ, Lopez AD. Mortality by cause for eight regions of the world: global burden of disease study. *Lancet.* 1997;349:1269–1276.

Osman LM, Godden DJ, Friend JAR, Legge JS, Douglas JG. Quality of life and hospital re-admission in patients with chronic obstructive pulmonary disease. *Thorax.* 1997;52:67–71.

Paggiaro PL, Dahle K, Bakran I, Frith L, Hollingworth K, Efthimiou J. Multicentre randomised placebo-controlled trial of inhaled fluticasone propionate in patients with chronic obstructive pulmonary disease. *Lancet.* 1998;351:773–780.

Pauwels RA, Lofdahl CG, Laitinen LA, Schouten JP, Postma DS, Pride NB, et al. Long-term treatment with inhaled budesonide in persons with mild chronic obstructive pulmonary disease who continue smoking. *N Engl J Med.* 1999;340:1948–1953.

Petty TL. Spirometry made simple. *Adv Manage Respir Care.* 1999;8:3741. Available at http://www.nlhep.org/SpirometryMadeSimple.htm. Accessed October 26, 2000.

Ryan CF, Buckley PA, Whittaker JS. Energy balance in stable malnourished patients with COPD. *Chest.* 1993;103:1038–1044.

Schols AM, Fredrix WHIM, Soeters PB, Wetererp KA, Wouters EFM. Resting energy expenditure in patients with chronic obstructive pulmonary disease. *Am J Clin Nutr.* 1991;54:983–987.

Schols AM, Schoffelen PFM, Ceulemans H, Wouters EFM, Saris WHM. Measurement of resting energy expenditure in patients with chronic obstructive pulmonary

disease in a clinical setting. *J Parenter Enteral Nutr.* 1992;16:364–368.

Thom TJ. International comparisons in COPD mortality. *Am Rev Respir Dis.* 1989;140:27–34.

Vestbo J, Sorensen T, Lange P, Brix A, Torre P, Viskum K. Long-term effect of inhaled budesonide in mild and moderate chronic obstructive pulmonary disease: a randomised controlled trial. *Lancet.* 1999; 353: 1819–1823.

Wilson D, Donahoe M, Rogers RM, Pennock BE. Metabolic rate and weight loss in chronic obstructive pulmonary disease. *J Parenter Enteral Nutr.* 1990; 14:7–11.

Zeman FJ. Nutrition in pulmonary diseases. In Zeman FJ (ed.), *Clinical Nutrition and Dietetics*, 2nd ed. New York: Macmillan, 1991:599–614.

# Chronic Obstructive Pulmonary Disease with Respiratory Failure

*Advanced Practice*

## Objectives

After completing this case, the student will be able to:

1. Define the pathophysiology of chronic obstructive pulmonary disease and the relationship to acute respiratory failure.

2. Identify the role of nutrition in mechanical ventilation.

3. Determine the metabolic implications of acute respiratory failure.

4. Interpret biochemical indices for assessment of respiratory function.

5. Interpret biochemical indices for assessment of nutritional status.

6. Plan, interpret, and evaluate nutrition support for mechanically ventilated patients.

Daishi Hayoto, a 65-year-old male, is brought to the University Hospital emergency room by his wife when he experiences severe shortness of breath. The patient has a long-standing history of COPD secondary to tobacco use.

# UNIVERSITY HOSPITAL

## ADMISSION DATABASE

Name: Daishi Hayato
DOB: 7/14  age 65
Physician: M. McFarland, MD

| BED# 1 | DATE: 3/26 | TIME: 1700 | TRIAGE STATUS (ER ONLY): ☐ Red ☐ Yellow ☐ Green ☐ White |
|---|---|---|---|

**PRIMARY PERSON TO CONTACT:**
Name: M. Hayato
Home #: 312-456-3422
Work #: N/A

### Initial Vital Signs

| TEMP: 98 | RESP: 36 | SAO2: |
|---|---|---|

| HT: 5'4" | WT (lb): 122 | B/P: 110/80 | PULSE: 118 |
|---|---|---|---|

**ORIENTATION TO UNIT:** ☒ Call light ☒ Television/telephone ☒ Bathroom ☒ Visiting ☐ Smoking ☒ Meals ☐ Patient rights/responsibilities

| LAST TETANUS unknown | LAST ATE this am | LAST DRANK about 3 hours ago |
|---|---|---|

### CHIEF COMPLAINT/HX OF PRESENT ILLNESS

"My husband has had emphysema for many years. He was working in the yard today and got really short of breath. I called our doctor and she said to go straight to the emergency room."

**PERSONAL ARTICLES:** (Check if retained/describe)
☐ Contacts ☐ R ☐ L      ☒ Dentures ☒ Upper ☒ Lower
☐ Jewelry:
☒ Other: eyeglasses

### ALLERGIES: Meds, Food, IVP Dye, Seafood: Type of Reaction

penicillin

**VALUABLES ENVELOPE:**
☐ Valuables instructions

### PREVIOUS HOSPITALIZATIONS/SURGERIES

cholecystectomy–20 years ago

dental extraction–5 years ago

**INFORMATION OBTAINED FROM:**
☐ Patient          ☐ Previous record
☒ Family           ☐ Responsible party

Signature  *M. Hayato*

| Home Medications (including OTC) | Codes: A=Sent home | | B=Sent to pharmacy | | C=Not brought in |
|---|---|---|---|---|---|
| Medication | Dose | Frequency | Time of Last Dose | Code | Patient Understanding of Drug |
| Combivent | 2 inhalations | qid | 1200 | A | yes |
| Lasix | 40 mg | qd | 0700 | A | yes |
| oxygen via nasal cannula | | during sleep | | | |
| | | | | | |
| | | | | | |
| | | | | | |
| | | | | | |
| | | | | | |
| | | | | | |
| | | | | | |

Do you take all medications as prescribed? ☒ Yes  ☐ No    If no, why?

### PATIENT/FAMILY HISTORY

☐ Cold in past two weeks
☐ Hay fever
☒ Emphysema/lung problems  Patient
☐ TB disease/positive TB skin test
☒ Cancer  Father
☐ Stroke/past paralysis
☐ Heart attack
☐ Angina/chest pain
☐ Heart problems

☐ High blood pressure
☐ Arthritis
☐ Claustrophobia
☒ Circulation problems  Patient
☐ Easy bleeding/bruising/anemia
☐ Sickle cell disease
☐ Liver disease/jaundice
☐ Thyroid disease
☐ Diabetes

☐ Kidney/urinary problems
☐ Gastric/abdominal pain/heartburn
☐ Hearing problems
☐ Glaucoma/eye problems
☐ Back pain
☐ Seizures
☐ Other

### RISK SCREENING

Have you had a blood transfusion? ☐ Yes ☒ No
Do you smoke? ☒ Yes ☐ No
If yes, how many pack(s)  2/day for  50 years
Does anyone in your household smoke? ☐ Yes ☒ No
Do you drink alcohol? ☒ Yes ☐ No
If yes, how often?  1-2 × week  How much? 1-2 drinks
When was your last drink?  last week /  /
Do you take any recreational drugs? ☐ Yes ☒ No
If yes, type:  /  / Route
Frequency:  /  / Date last used:  /  /

**FOR WOMEN Ages 12–52**

Is there any chance you could be pregnant? ☐ Yes ☐ No
If yes, expected date (EDC):
Gravida/Para:

**ALL WOMEN**

Date of last Pap smear:
Do you perform regular breast self-exams? ☐ Yes ☐ No

**ALL MEN**

Do you perform regular testicular exams? ☐ Yes ☒ No

Additional comments:

**✗** *Carolyn Masterson, RN*
Signature/Title

**Client name:** Daishi Hayato
**DOB:** 7/14
**Age:** 65
**Sex:** Male
**Education:** Bachelor's degree
**Occupation:** Retired manager of local grocery chain
**Hours of work:** N/A
**Household members:** Wife, age 62, well; four adult children not living in the area
**Ethnic background:** Asian American
**Religious affiliation:** Methodist
**Referring physician:** Marie McFarland, MD (pulmonary)

## Chief complaint:

"My husband has had emphysema for many years. He was working in the yard today and got really short of breath. I called our doctor and she said to go straight to the emergency room."

## Patient history:

*Onset of disease:* The patient has a long-standing history of COPD secondary to chronic tobacco use, 2 PPD for 50 years. He was in his usual state of health today with marked limitation on his exercise capacity due to dyspnea on exertion. He also notes two-pillow orthopnea, swelling in both lower extremities. Today while performing some yard work he noted the sudden onset of marked dyspnea. His wife brought him to the emergency room right away. There a chest radiograph showed a tension pneumothorax involving the left lung. Patient also states that he gets cramping in his right calf when he walks.

*PMH:* Had cholecystectomy twenty years ago. Total dental extraction five years ago. Patient describes intermittent claudication. Claims to be allergic to penicillin. Diagnosed with emphysema more than 10 years ago. Has been treated successfully with Combivent (metered dose inhaler)—2 inhalations qid (each inhalation delivers 18 mcg ipratropium bromide; 130 mcg albuterol sulfate).

*Meds:* Combivent, Lasix, $O_2$ 2 L/hour via nasal cannula at night

*Smoker:* Yes, 2 PPD for 50 years

*Family Hx: What?* Lung cancer   *Who?* Father

## Physical exam:

*General appearance:* Acutely dyspneic Asian American male in acute respiratory distress
*Vitals:* Temp 97.6° F, BP 110/80, HR 118 bpm, RR 36 bpm
*Heart:* Normal heart sounds; no murmurs or gallops
*HEENT:* Within normal limits; funduscopic exam reveals AV nicking
    *Eyes:* Pupil reflex normal
    *Ears:* Slight neurosensory deficit acoustically
    *Nose:* Unremarkable
    *Throat:* Jugular veins appear distended. Trachea is shifted to the right. Carotids are full, symmetrical and without bruits.
*Genitalia:* Unremarkable
*Rectal:* Prostate normal; stool hematest negative
*Neurologic:* DTR full and symmetric; alert and oriented × 3
*Extremities:* Cyanosis, $1^+$ pitting edema

*Skin:* Warm, dry to touch

*Chest/lungs:* Hyperresonance to percussion over the left chest anteriorly and posteriorly. Harsh inspiratory breath sounds are noted over the right chest with absent sounds on the left. Using accessory muscles at rest.

*Abdomen:* Old surgical scar RUQ. No organomegaly or masses. BS reduced.

*Circulation:* R femoral bruit present. Right PT and DP pulses were absent.

**Nutrition Hx:**

*General:* Wife relates general appetite is only fair. Usually breakfast is the largest meal. Has been decreased for past several weeks. She states that his highest weight was 135# but feels he weighs much less than that now.

*Usual dietary intake:*

| | |
|---|---|
| AM: | Egg, hot cereal, bread or muffin, hot tea (with milk and sugar) |
| Lunch: | Soup, sandwich, hot tea (with milk and sugar) |
| Dinner: | Small amount of meat, rice, 2–3 kinds of vegetables, hot tea (with milk and sugar) |
| 24-hr recall: | 2 scrambled eggs, few bites of Cream of Wheat, sips of hot tea, bite of toast; ate nothing rest of day—sips of hot tea |

*Food allergies/intolerances/aversions:* NKA

*Previous MNT?* No

*Food purchase/preparation:* Wife

*Vit/min intake:* None

*Anthropometric data:* Ht 5′4″, Wt 122#, UBW 135#

**Tx plan:**

ABG, pulse oximetry, CBC, chemistry panel, UA

Chest X-ray, ECG

Proventil 0.15 in 1.5 cc NS q 30 min × 3 followed by Proventil 0.3 cc in 3 cc normal saline q 2 hr per HHN (hand-held nebulizer)

Spirogram post nebulizer Tx

IVF D5 1/2 NS at TKO

Solumedrol 10–40 mg q 4–6 hr—high dose = 30 mg/kg q 4–6 hr (2 days max)

NPO

**Dx:**

Acute respiratory distress, COPD, peripheral vascular disease with intermittent claudication

**Hospital course:**

In the emergency room, a chest tube was inserted to the left thorax with drainage under suction. Subsequently the oropharynx was cleared. A resuscitation bag and mask was used to ventilate the patient with high-flow oxygen. Endotracheal intubation was then carried out, using the laryngoscope so that the trachea could be directly visualized. The patient was then ventilated with the help of a volume-cycled ventilator. Ventilation is 15 breaths/min with an $FiO_2$ of 100%, a positive end expiratory pressure of 6, and a tidal volume of 700 mL. Daily chest radiographs and ABGs were used each AM to guide settings on the ventilator. A nutrition consult was completed on day 2 of admission, and enteral feedings were initiated. Due to high residuals, the patient was started on ProcalAmine. Enteral feedings were restarted on day 4. Respiratory status actually became worse on day 5 but improved thereafter. ProcalAmine was discontinued on day 5, and enteral feeding continued until day 8. The patient was weaned from the ventilator on day 8 and discharged to home on day 11.

# UH UNIVERSITY HOSPITAL

NAME: Daishi Hayato                    DOB: 7/14
AGE: 65                                SEX: M
PHYSICIAN: M. McFarland, MD

\*\*\*\*\*\*\*\*\*\*\*\*\*\*\*\*\*\*\*\*\*\*\*\*\*\*\*\*\*\*\*\*\*\*\*\*\*\*\*\*\*\*CHEMISTRY\*\*\*\*\*\*\*\*\*\*\*\*\*\*\*\*\*\*\*\*\*\*\*\*\*\*\*\*\*\*\*\*\*\*\*\*\*\*\*\*\*\*

DAY:                                          1          4
DATE:
TIME:
LOCATION:

| | NORMAL | 1 | 4 | UNITS |
|---|---|---|---|---|
| Albumin | 3.6–5 | 3.6 | 3.9 | g/dL |
| Total protein | 6–8 | 6.1 | 6.5 | g/dL |
| Prealbumin | 19–43 | 26 | 17 | mg/dL |
| Transferrin | 200–400 | 250 | 200 | mg/dL |
| Sodium | 135–155 | 138 | 137 | mmol/L |
| Potassium | 3.5–5.5 | 3.9 | 3.5 | mmol/L |
| Chloride | 98–108 | 101 | 104 | mmol/L |
| $PO_4$ | 2.5–4.5 | 3.2 | 2.6 | mmol/L |
| Magnesium | 1.6–2.6 | 1.8 | 1.6 | mmol/L |
| Osmolality | 275–295 | 285 | 285 | mmol/kg $H_2O$ |
| Total $CO_2$ | 24–30 | 29 | 30 | mmol/L |
| Glucose | 70–120 | 108 | 110 | mg/dL |
| BUN | 8–26 | 11 | 15 | mg/dL |
| Creatinine | 0.6–1.3 | 0.7 | 0.9 | mg/dL |
| Uric acid | 2.6–6 (women) | 3.9 | | mg/dL |
| | 3.5–7.2 (men) | | | |
| Calcium | 8.7–10.2 | 9.1 | | mg/dL |
| Bilirubin | 0.2–1.3 | 0.8 | | mg/dL |
| Ammonia ($NH_3$) | 9–33 | 9 | | μmol/L |
| SGPT (ALT) | 10–60 | 15 | | U/L |
| SGOT (AST) | 5–40 | 22 | | U/L |
| Alk phos | 98–251 | 114 | | U/L |
| CPK | 26–140 (women) | 152 | | U/L |
| | 38–174 (men) | | | |
| LDH | 313–618 | 412 | | U/L |
| CHOL | 140–199 | 155 | | mg/dL |
| HDL-C | 40–85 (women) | 32 | | mg/dL |
| | 37–70 (men) | | | |
| VLDL | | | | mg/dL |
| LDL | < 130 | 142 | | mg/dL |
| LDL/HDL ratio | < 3.22 (women) | 4.4 | | |
| | < 3.55 (men) | | | |
| Apo A | 101–199 (women) | | | mg/dL |
| | 94–178 (men) | | | |
| Apo B | 60–126 (women) | | | mg/dL |
| | 63–133 (men) | | | |
| TG | 35–160 | 155 | | mg/dL |
| $T_4$ | 5.4–11.5 | | | μg/dL |
| $T_3$ | 80–200 | | | ng/dL |
| $HbA_{1C}$ | 4.8–7.8 | | | % |

# UH *UNIVERSITY HOSPITAL*

NAME: Daishi Hayato                    DOB: 7/14
AGE: 65                                 SEX: M
PHYSICIAN: M. McFarland, MD

\*\*\*\*\*\*\*\*\*\*\*\*\*\*\*\*\*\*\*\*\*\*\*\*\*\*\*\*\*\*\*\*\*\*\*\*\*\*\*\*\*HEMATOLOGY\*\*\*\*\*\*\*\*\*\*\*\*\*\*\*\*\*\*\*\*\*\*\*\*\*\*\*\*\*\*\*\*\*\*\*\*\*\*\*

DAY:                                                      1
DATE:
TIME:
LOCATION:

| | NORMAL | | UNITS |
|---|---|---|---|
| WBC | 4.3–10 | 5.6 | $\times 10^3/mm^3$ |
| RBC | 4–5 (women) | 4.7 | $\times 10^6/mm^3$ |
| | 4.5–5.5 (men) | | |
| HGB | 12–16 (women) | 13.2 | g/dL |
| | 13.5–17.5 (men) | | |
| HCT | 37–47 (women) | 39 | % |
| | 40–54 (men) | | |
| MCV | 84–96 | | fL |
| MCH | 27–34 | | pg |
| MCHC | 31.5–36 | | % |
| RDW | 11.6–16.5 | | % |
| Plt ct | 140–440 | | $\times 10^3$ |
| Diff TYPE | | | |
| % GRANS | 34.6–79.2 | 52.3 | % |
| % LYM | 19.6–52.7 | 48.5 | % |
| SEGS | 50–62 | 83 | % |
| BANDS | 3–6 | 5 | % |
| LYMPHS | 25–40 | 10 | % |
| MONOS | 3–7 | 3 | % |
| EOS | 0–3 | | % |
| TIBC | 65–165 (women) | | μg/dL |
| | 75–175 (men) | | |
| Ferritin | 18–160 (women) | | μg/dL |
| | 18–270 (men) | | |
| Vitamin $B_{12}$ | 100–700 | | pg/mL |
| Folate | 2–20 | | ng/mL |
| Total T cells | 812–2318 | | $mm^3$ |
| T-helper cells | 589–1505 | | $mm^3$ |
| T-suppressor cells | 325–997 | | $mm^3$ |
| PT | 11–13 | | sec |

# UH *UNIVERSITY HOSPITAL*

NAME: Daishi Hayato        DOB: 8/2
AGE: 65        SEX: M
PHYSICIAN: M. McFarland, MD

```
*********************************ARTERIAL BLOOD GASES (ABGs)*********************************
```

| | NORMAL | 1 | 2 | 3 | 5 | UNITS |
|---|---|---|---|---|---|---|
| DAY: | | | | | | |
| DATE: | | | | | | |
| TIME: | | | | | | |
| LOCATION: | | | | | | |
| pH | 7.35–7.45 | 7.2 | 7.30 | 7.36 | 7.22 | |
| $pCO_2$ | 35–45 | 65 | 59 | 50 | 66 | mm Hg |
| $SO_2$ | ≥95 | | | | | % |
| $CO_2$ content | 23–30 | 35 | 30 | 29 | 36 | mmol/L |
| $O_2$ content | 15–22 | | | | | vol % |
| $pO_2$ | ≥80 | 56 | 58 | 60 | 57 | mm Hg |
| Base excess | > 3 | | | | | mEq/L |
| Base deficit | < 3 | | | | | mEq/L |
| $HCO_3^-$ | 24–28 | 38 | 33 | 32 | 37 | mEq/L |
| HGB | 12–16 (women) | | | | | g/dl |
| | 13.5–17.5 (men) | | | | | |
| HCT | 37–47 (women) | | | | | % |
| | 40–54 (men) | | | | | |
| COHb | < 2 | | | | | % |
| $[NA^+]$ | 135–148 | | | | | mmol/L |
| $[K^+]$ | 3.5–5 | | | | | mEq/L |

# UH UNIVERSITY HOSPITAL

Name: Daishi Hayato
Physician: M. McFarland, MD

## PATIENT CARE SUMMARY SHEET

Date: 3   Room:   Wt Yesterday 121 lb   Today: 122 lb   Postdialysis   lb

| Temp °F | NIGHTS | | | | | | | | DAYS | | | | | | | | EVENINGS | | | | | | | |
|---|---|---|---|---|---|---|---|---|---|---|---|---|---|---|---|---|---|---|---|---|---|---|---|---|
| | 00 | 01 | 02 | 03 | 04 | 05 | 06 | 07 | 08 | 09 | 10 | 11 | 12 | 13 | 14 | 15 | 16 | 17 | 18 | 19 | 20 | 21 | 22 | 23 |
| 105 | | | | | | | | | | | | | | | | | | | | | | | | |
| 104 | | | | | | | | | | | | | | | | | | | | | | | | |
| 103 | | | | | | | | | | | | | | | | | | | | | | | | |
| 102 | | | | | | | | | | | | | | | | | | | | | | | | |
| 101 | | | | | | | | | | | | | | | | | | | | | | | | |
| 100 | | | | | | | | | | | | | | | | | | | | | | | | |
| 99 | | | | | | | | | | | | | | | | | | | | | | | | |
| 98 | | | | | | | | | | | | | | | | | | | | | | | | |
| 97 | | | | | | | | | | | | | | | | | | | | | | | | |
| 96 | | | | | | | | | | | | | | | | | | | | | | | | |
| Pulse | 80 | | | | | | | | 75 | | | | | | | | | | | | | | | |
| Respiration | | | | | | | | | | | | | | | | | | | | | | | | |
| BP | 110/80 | | | | | | | | 112/72 | | | | | | | | | | | | | | | |
| Blood Glucose | | | | | | | | | | | | | | | | | | | | | | | | |
| Appetite/Assist | | | | | | | | | | | | | | | | | | | | | | | | |
| INTAKE | NPO | | | | | | | | NPO | | | | | | | | NPO | | | | | | | |
| Oral | | | | | | | | | | | | | | | | | | | | | | | | |
| IV | 50→ | → | → | → | → | → | → | → | 50 | → | → | → | → | → | → | → | 100 | 100 | 100 | 100 | 100 | 100 | | |
| TF Formula/Flush | 25 | 25 | 25 | 25 | 25/50 | 25 | 25 | 25 | 25 | 25 | 25 | 25/50 | 25 | 25 | 25 | 25 | * | | | | | | | |
| Shift Total | 650 | | | | | | | | 650 | | | | | | | | 800 | | | | | | | |
| OUTPUT | | | | | | | | | | | | | | | | | | | | | | | | |
| Void | 125 | | | 200 | | | | | 275 | | | | 300 | | | | 200 | 175 | 150 | | 240 | | | |
| Cath. | | | | | | | | | | | | | | | | | | | | | | | | |
| Emesis | | | | | | | | | | | | | | | | | | | | | | | | |
| BM | 200 | | | | | | | | | | | | 100 | | | | | | | | | | | |
| Drains | | | | | | | | | | | | | | | | | | | | | | | | |
| Shift Total | 525 | | | | | | | | 675 | | | | | | | | 765 | | | | | | | |
| Gain | +125 | | | | | | | | | | | | | | | | +35 | | | | | | | |
| Loss | | | | | | | | | 50 | | | | | | | | | | | | | | | |
| Signatures | Mary Rogers, RN | | | | | | | | Patricia Elkins, RN | | | | | | | | Frannie Lowe, RN | | | | | | | |

*0tube held due to high residuals.

## Case Questions

1. Mr. Hayato was diagnosed with emphysema more than 10 years ago. Define *emphysema* and its underlying pathophysiology.

2. Define the following terms found in the history and physical for Mr. Hayato.

   **a.** *Dyspnea:*

   **b.** *Orthopnea:*

   **c.** *Pneumothorax:*

   **d.** *Endotracheal intubation:*

   **e.** *Cyanosis:*

3. Identify features of the physician's physical examination that are consistent with her admitting diagnosis. Describe the pathophysiology that might be responsible for each physical finding.

| Physical Finding | Etiology |
|---|---|
|  |  |
|  |  |
|  |  |
|  |  |
|  |  |
|  |  |

4.  Evaluate Mr. Hayato's biochemical indices for nutritional assessment on day 1.

5.  Evaluate Mr. Hayato's admitting anthropometric data for nutritional assessment.

6.  Determine Mr. Hayato's energy and protein requirements using the Harris-Benedict equation, the Ireton-Jones equation, and the COPD predictive equations. Compare them. As Mr. Hayato's clinician, what would you set as your goal for meeting his energy needs?

7.  Mr. Hayato was started on enteral feedings on day 2 of his admission. How does nutrition affect ventilation?

8.  Mr. Hayato was started on Isosource @ 25 cc/hr continuously over 24 hours. What does this provide? What would be his goal rate to meet his calculated nutritional requirements?

9.  Should the patient have been started on a special pulmonary feeding to prevent the overproduction of carbon dioxide? Explain. What is the rationale for pulmonary formulas?

10.  Examine the patient care summary sheet. How much enteral feeding did the patient receive?

11.  You read in the physician's orders that the patient experienced high gastric residuals and the enteral feeding was discontinued. What does this mean, and what is the potential cause for the problem?

12.  Dr. McFarland elected to begin peripheral parenteral nutrition using a formula called Procal-Amine. She began the PPN @ 100 cc/hr and discontinued Mr. Hayato's regular IV of D5 1/2 NS at TKO. What is ProcalAmine, and how much nutrition does this provide?

13.  Was this adequate to meet the patient's nutritional needs? Explain.

**14.**   Do you feel it was a good idea to begin peripheral parenteral nutrition (PPN)? Are there pros and cons? What are the limitations of using this form of nutrition support? Were other nutrition support options available for the health care team?

**15.**   On day 4, the enteral feeding was restarted at 25 cc/hr and then increased to 50 cc/hr after 12 hours. You notice that the ProcalAmine @ 100 cc/hr was also continued. What would have been the total energy intake for Mr. Hayato?

**16.**   Examine the values documented for arterial blood gases (ABGs).

   **a.**   On the day Mr. Hayato was intubated, his ABGs were as follows: pH 7.2, $pCO_2$ 65, $CO_2$ 35, $pO_2$ 56, $HCO_3^-$ 38. What can you determine from each of these values?

   **b.**   On day 3 while Mr. Hayato was on the ventilator, his ABGs were as follows: pH 7.36, $pCO_2$ 50, $CO_2$ 29, $pO_2$ 60, $HCO_3^-$ 32. What can you determine from each of these values?

   **c.**   On day 5, after restarting enteral feeding and continuing on ProcalAmine, his ABGs were as follows: pH 7.22, $pCO_2$ 66, $pO_2$ 57, $CO_2$ 36, $HCO_3^-$ 37. In addition, indirect calorimetry indicated a RQ of .95 and measured energy intake to be 1350 kcal. How does the patient's measured energy intake compare to your previous calculations? What does the RQ indicate?

   **d.**   What contribution do you think his nutrition support might have made to his change in respiratory status?

**17.**   Mr. Hayato was weaned successfully from the ventilator. What role if any does nutrition play in this process?

**18.**   As Mr. Hayato is prepared for discharge, what nutritional goals might you set with him and his wife to improve his overall nutritional status?

# Bibliography

Akrabawi SS, Mobarhan S, Stoltz RR, Ferguson PW. Gastric emptying, pulmonary function, gas exchange, and respiratory quotient after feeding a moderate versus high fat enteral formula meal in chronic obstructive pulmonary disease patients. *Nutrition.* 1996; 12:260–265.

Charney P. Enteral nutrition: indications, options, and formulations. In Gottschlich M (ed.), *The Science and Practice of Nutrition Support.* Dubuque, IA: Kendall/Hunt, 2001:141–166.

Commission on Accreditation for Dietetics Education. Knowledge, skills, and competencies for dietitians. *Accreditation Manual for Dietetics Education Programs,* rev. 4th ed. Chicago: American Dietetic Association, 2000.

Confalonieri M, Rossi A. Burden of chronic obstructive pulmonary disease. *Lancet.* 2000;356:S56–65.

Hogg JH, Klapholz A, Reid-Hector J. Pulmonary disease. In Gottschlich M (ed.), *The Science and Practice of Nutrition Support.* Dubuque, IA: Kendall/Hunt, 2001:491–516.

Ireton-Jones CS, Boprman KR, Turner WW. Nutrition considerations in the management of ventilator-dependent patients. *Nutr Clin Prac.* 1993;8:60–64.

# Unit Seven

# MEDICAL NUTRITION THERAPY FOR ENDOCRINE DISORDERS

## Introduction

The cases in this section focus primarily on diabetes mellitus, because of the large number of patients with diabetes treated by dietitians. Diabetes mellitus is a chronic disease that has no cure and is the seventh leading cause of death in the United States. Nearly 6% of the population has diabetes, and approximately one-third of that number is undiagnosed.

Diabetes affects men and women equally, but minorities (especially American Indians and Alaska Natives) are almost twice as likely as non-Hispanic whites to develop diabetes in their lifetime. In addition, diabetes is one of the most costly health problems in the United States. Health care and other direct medical costs, as well as indirect costs (such as loss of productivity), are approximately $98 billion annually. Each year at least 190,000 people die as a result of diabetes and its complications. For example, diabetes is the leading cause of new blindness in the United States. Diabetes is the leading cause of nephropathy, which leads to end-stage renal disease requiring dialysis or organ transplant for survival.

Diabetes mellitus is actually a group of diseases characterized by hyperglycemia resulting from cessation of insulin production or impairment in insulin secretion and/or insulin action. There are four major categories of diabetes mellitus including type 1, type 2, gestational, and other secondary diseases resulting in diabetes.

Medical nutrition therapy is integral to total diabetes care and management. The Diabetes Control and Complications Trial (DCCT) corroborates the significance of integrating nutrition and blood glucose self-management education in achieving and maintaining target blood glucose levels. Nutrition and meal planning are among the most challenging aspects of diabetes care for the person with diabetes and the health care team. The major components of successful nutrition management are learning about nutrition, altering eating habits, implementing new behaviors, participating in exercise, evaluating changes, and integrating this information into diabetes care. Observance of meal-planning principles requires people with diabetes to make demanding lifestyle changes.

Modern diabetes nutrition therapy no longer "puts people on diets," but instead develops nutrition treatment plans around an individual patient. To be effective, the registered dietitian must be able to customize his or her approach to the personal lifestyle and diabetes management goals of the individual with diabetes. The four cases included here allow you to put this guideline into practice.

The final case in this section spotlights an example of a inherited metabolic disorder—phenylketonuria (PKU). Most patients with PKU are managed by a metabolic nutritionist as part of a special health care team. It does serve, though, as an excellent example of the role of medical nutrition therapy. The nutritional care can prevent the dire consequences of this condition.

Throughout this section, you will put to use nutrition assessment skills for all age groups. These cases let you build appropriate experience in developing care plans and planning appropriate nutrition education.

# Type 1 Diabetes Mellitus

*Introductory Level*

## Objectives

After completing this case, the student will be able to:

1. Apply working knowledge of pathophysiology of type 1 diabetes mellitus to demonstrate understanding of medical care.
2. Interpret anthropometric, biochemical, clinical, and dietary data to complete comprehensive nutritional assessment.
3. Apply appropriate MNT recommendations for type 1 diabetes mellitus.

4. Determine appropriate nutrient requirements.
5. Identify appropriate MNT goals.
6. Use nutrition and medical data to communicate in the medical record.

Susan Cheng, a 15-year-old high school student, is admitted to the hospital complaining of thirst, hunger, problems with urination, and fatigue. There is a history of diabetes in the family.

 **U H** *UNIVERSITY HOSPITAL*

Name: S. Cheng
DOB: 9/25  age 15
Physician: P. Green

# ADMISSION DATABASE

| BED#<br>1 | DATE:<br>9/28 | TIME:<br>1100 | TRIAGE STATUS (ER ONLY):<br>☐ Red ☐ Yellow ☐ Green ☐ White |
|---|---|---|---|

**PRIMARY PERSON TO CONTACT:**
Name: Mai or David Cheng
Home #: 390-8217
Work #: 390-2234

## Initial Vital Signs

| TEMP:<br>98.6 | RESP:<br>18 | SAO2: |
|---|---|---|

**ORIENTATION TO UNIT:** ☒ Call light ☒ Television/telephone
☒ Bathroom ☒ Visiting ☒ Smoking ☒ Meals
☒ Patient rights/responsibilities

| HT:<br>5'2" | WT (lb):<br>100 | B/P:<br>124/70 | PULSE:<br>85 |
|---|---|---|---|

| LAST TETANUS<br>4 years ago | LAST ATE<br>this AM | LAST DRANK<br>1 hour ago |
|---|---|---|

**PERSONAL ARTICLES:** (Check if retained/describe)
☒ Contacts ☒ R ☒ L  ☐ Dentures ☐ Upper ☐ Lower
☐ Jewelry:
☐ Other:

### CHIEF COMPLAINT/HX OF PRESENT ILLNESS

excessive thirst, urination, hunger, & fatigue

**VALUABLES ENVELOPE:**
☐ Valuables instructions

### ALLERGIES: Meds, Food, IVP Dye, Seafood: Type of Reaction

**INFORMATION OBTAINED FROM:**
☒ Patient  ☐ Previous record
☒ Family  ☐ Responsible party

### PREVIOUS HOSPITALIZATIONS/SURGERIES

Signature  *Susan Cheng*

### Home Medications (including OTC)    Codes: A = Sent home    B = Sent to pharmacy    C = Not brought in

| Medication | Dose | Frequency | Time of Last Dose | Code | Patient Understanding of Drug |
|---|---|---|---|---|---|
|  |  |  |  |  |  |
|  |  |  |  |  |  |
|  |  |  |  |  |  |
|  |  |  |  |  |  |
|  |  |  |  |  |  |
|  |  |  |  |  |  |
|  |  |  |  |  |  |
|  |  |  |  |  |  |
|  |  |  |  |  |  |
|  |  |  |  |  |  |

Do you take all medications as prescribed?  ☐ Yes  ☐ No  If no, why?

### PATIENT/FAMILY HISTORY

| | | |
|---|---|---|
| ☐ Cold in past two weeks | ☐ High blood pressure | ☐ Kidney/urinary problems |
| ☐ Hay fever | ☐ Arthritis | ☐ Gastric/abdominal pain/heartburn |
| ☐ Emphysema/lung problems | ☐ Claustrophobia | ☐ Hearing problems |
| ☐ TB disease/positive TB skin test | ☐ Circulation problems | ☐ Glaucoma/eye problems |
| ☐ Cancer | ☐ Easy bleeding/bruising/anemia | ☐ Back pain |
| ☐ Stroke/past paralysis | ☐ Sickle cell disease | ☐ Seizures |
| ☐ Heart attack | ☐ Liver disease/jaundice | ☐ Other |
| ☐ Angina/chest pain | ☐ Thyroid disease | |
| ☐ Heart problems | ☒ Diabetes Maternal grandmother | |

### RISK SCREENING

Have you had a blood transfusion?  ☐ Yes  ☒ No
Do you smoke?  ☐ Yes  ☒ No
If yes, how many pack(s) ___ /day for ___ years
Does anyone in your household smoke?  ☐ Yes  ☒ No
Do you drink alcohol?  ☐ Yes  ☐ No
If yes, how often? _____  How much? ___
When was your last drink? ___ / ___ / ___
Do you take any recreational drugs?  ☐ Yes  ☐ No
If yes, type: _____  Route ___
Frequency: _____  Date last used: ___ / ___ / ___

**FOR WOMEN Ages 12–52**

Is there any chance you could be pregnant?  ☐ Yes  ☒ No
If yes, expected date (EDC): ___ / ___ / ___
Gravida/Para:

**ALL WOMEN**

Date of last Pap smear: ___ / ___ / ___
Do you perform regular breast self-exams?  ☐ Yes  ☐ No

**ALL MEN**

Do you perform regular testicular exams?  ☐ Yes  ☐ No

Additional comments:

✗ *Francis Miller, RN*
Signature/Title

**Client name:** Susan Cheng
**DOB:** 9/25
**Age:** 15
**Sex:** Female
**Education:** Less than high school. *What grade/level?* 9th grade, HS student
**Occupation:** Student
**Hours of work:** N/A
**Household members:** Mother age 40, father age 42, sister age 16, brother age 9—all in excellent health
**Ethnic background:** Asian American
**Religious affiliation:** Protestant
**Referring physician:** Pryce Green, MD (endocrinology)

## Chief complaint:

"I've been so thirsty and hungry. I haven't slept through the night for two weeks. I have to get up several times a night to go to the bathroom. It's a real pain. I've also noticed that my clothes are getting loose. My Mom and Dad think I must be losing weight."

## Patient history:

*Onset of disease:* Susan is a 15-yo female who lives with her parents, brother, and sister. She is in the 9th grade and a member of the girls' volleyball team. She has had an uneventful medical history with no significant illness until the past several weeks. Her parents brought her to the office with c/o polydipsia, polyuria, polyphagia, weight loss, and fatigue. Blood was drawn in the ER to measure blood glucose and glycated hemoglobin levels.
*Type of Tx:* Sliding scale insulin, diabetes education for Pt and family, 2400 kcal ADA
*PMH:* Normal adolescence
*Meds:* None PTA
*Smoker:* No
*Family Hx: What?* DM *Who?* Maternal grandmother

## Physical exam:

*General appearance:* Tired-appearing adolescent female
*Vitals:* Temp 98.6°F, BP 124/70 mm Hg, HR 85 bpm, RR 18 bpm
*Heart:* Regular rate and rhythm, heart sounds normal
*HEENT:* Noncontributory
*Genitalia:* Normal adolescent female
*Neurologic:* Alert and oriented
*Extremities:* Noncontributory
*Skin:* Smooth, warm, and dry; excellent turgor; no edema
*Chest/lungs:* Lungs are clear
*Peripheral vascular:* Pulse 4+ bilaterally, warm, no edema
*Abdomen:* Nontender, no guarding

## Nutrition Hx:

*General:* Mother describes appetite as good. Meals are somewhat irregular due to Susan's volleyball practice/game schedule. Susan eats lunch in the school cafeteria.

*Usual dietary intake:*

| | |
|---|---|
| AM: | ~ 1½ c dry cereal, usually sugar-coated, 1 c 2% milk, 1 c orange juice (unsweetened), hot chocolate in the winter (made from mix) |
| Lunch: | 6-in pepperoni pizza, mixed salad w/ 1000 island dressing (~ ¼ c), 1 can Coke (regular), dessert—usually cake or a candy bar |
| Snack: | Peanut butter and jelly sandwich (2 slices white bread, 1 tbsp grape jelly, 2–3 tbsp crunchy peanut butter), 1 12-oz can Coke |
| PM: | Spaghetti w/meat sauce (about 2 c noodles and ½ c sauce w/ 1 oz ground beef), steamed broccoli—3 stalks (will eat with salt and butter, but prefers cheese sauce), 16 oz 2% milk |
| HS snack: | 2 c ice cream (different flavors ) or popcorn with melted butter and salt (about 6 c popcorn with ~ ¼ c melted butter) with 12-oz Coke |
| 24-hr recall: | N/A |

*Food allergies/intolerances/aversions:* NKA
*Previous MNT?* No
*Food purchase/preparation:* Parents
*Vit/min intake:* None
*Current diet order:* 2400 kcal ADA diet

**Tx plan:**
Achieve glycemic control.
Evaluate serum lipid levels.
Monitor blood glucose levels.
Initiate self-management training for Pt and parents on insulin administration, nutrition prescription, meal planning, signs/symptoms and Tx of hypo-/hyperglycemia, monitoring instructions (SBGM, urine ketones, and use of record system), exercise.
Baseline visual examination
Contraception education

# UH UNIVERSITY HOSPITAL

NAME: S. Cheng                          DOB: 9/25
AGE: 15                                 SEX: Female
PHYSICIAN: P. Green

\*\*\*\*\*\*\*\*\*\*\*\*\*\*\*\*\*\*\*\*\*\*\*\*\*\*\*\*\*\*\*\*\*\*\*\*\*\*\*\*\*CHEMISTRY\*\*\*\*\*\*\*\*\*\*\*\*\*\*\*\*\*\*\*\*\*\*\*\*\*\*\*\*\*\*\*\*\*\*\*\*\*\*

DAY:                                    Admit          d/c
DATE:
TIME:
LOCATION:

|  | NORMAL | | | UNITS |
|---|---|---|---|---|
| Albumin | 3.6–5 | 4.2 | 4.5 | g/dL |
| Total protein | 6–8 | 7.5 | 7.6 | g/dL |
| Prealbumin | 19–43 | 40 | 39 | mg/dL |
| Transferrin | 200–400 | | | mg/dL |
| Sodium | 135–155 | 142 | 145 | mEq/L |
| Potassium | 3.5–5.5 | 4.5 | 4.7 | mEq/L |
| Chloride | 98–108 | 98 | 99 | mEq/L |
| $PO_4$ | 2.5–4.5 | 3.7 | 3.8 | mEq/L |
| Magnesium | 1.6–2.6 | 2.1 | 1.9 | mEq/L |
| Osmolality | 275–295 | 305 | | mmol/kg $H_2O$ |
| Total $CO_2$ | 24–30 | | | mmol/L |
| Glucose | 70–120 | 250 | 120 | mg/dL |
| BUN | 8–26 | 20 | 18 | mg/dL |
| Creatinine | 0.6–1.3 | 0.9 | 0.8 | mg/dL |
| Uric acid | 2.6–6 (women) | | | mg/dL |
|  | 3.5–7.2 (men) | | | |
| Calcium | 8.7–10.2 | 9.5 | 9.7 | mg/dL |
| Bilirubin | 0.2–1.3 | | | mg/dL |
| Ammonia ($NH_3$) | 9–33 | | | $\mu$mol/L |
| SGPT (ALT) | 10–60 | | | U/L |
| SGOT (AST) | 5–40 | | | U/L |
| Alk phos | 98–251 | | | U/L |
| CPK | 26–140 (women) | | | U/L |
|  | 38–174 (men) | | | |
| LDH | 313–618 | | | mg/dL |
| CHOL | 140–199 | 169 | 170 | mg/dL |
| HDL-C | 40–85 (women) | | | mg/dL |
|  | 37–70 (men) | | | |
| VLDL | | | | mg/dL |
| LDL | < 130 | 109 | | mg/dL |
| LDL/HDL ratio | < 3.22 (women) | | | |
|  | < 3.55 (men) | | | |
| Apo A | 101–199 (women) | | | mg/dL |
|  | 94–178 (men) | | | |
| Apo B | 60–126 (women) | | | mg/dL |
|  | 63–133 (men) | | | |
| TG | 35–160 | | | mg/dL |
| $T_4$ | 5.4–11.5 | | | $\mu$g/dL |
| $T_3$ | 80–200 | | | ng/dL |
| $HbA_{1C}$ | 4.8–7.8 | 7.95 | | % |

## Case Questions

1. Describe the etiology of type 1 diabetes mellitus.

2. How long will Susan need to take exogenous insulin?

3. Why did Dr. Green order a lipid profile?

4. How often should lipid profiles be assessed?

5. Describe the metabolic events that led to Susan's symptoms (polyuria, polydipsia, polyphagia, weight loss, fatigue) and integrate with the pathophysiology of the disease.

6. Compare the pharmacological differences in Humulin (human insulin).

| | Regular Insulin | Insulin Lispro (Humalog) | Lente Insulin | NPH Insulin | Ultralente Insulin | Glargine Insulin (Lantus) |
|---|---|---|---|---|---|---|
| Classification | | | | | | |
| Onset of action | | | | | | |
| Peak (hr) | | | | | | |
| Duration (hr) | | | | | | |

7. Explain the difference between a mixed dose of insulin and a split dose of insulin.

8. When Susan's blood glucose level is tested at 2 AM, she is hypoglycemic. In addition, her plasma ketones are elevated. When she is tested early in the morning before breakfast, she is hyperglycemic. Describe the dawn phenomenon and the Somogyi phenomenon. Which one is Susan likely to be experiencing? How might this be prevented?

9. Once Susan's blood glucose levels were under control, Dr. Green prescribed the following insulin regimen. AM—10 units NPH, 5 units regular; PM—4 units NPH, 4 units regular. How did Dr. Green arrive at this dosage?

10. Dietitians must obtain and use information from all components of a nutrition assessment to develop appropriate interventions and goals that are achievable for the patient. This assessment is ongoing, and continuously modified and updated throughout the MNT process. For each of the following components of an initial nutrition assessment, list at least three assessments you would perform for each component:

| Component | Assessments |
|---|---|
| Clinical data | |
| Nutrition history | |
| Weight history | |
| Physical activity history | |
| Monitoring | |
| Psychosocial/economic | |
| Knowledge and skills level | |
| Expectations and readiness to change | |

11. Determine Susan's stature for age and weight for age.

12. How would you interpret these values?

13. Estimate Susan's daily energy needs.

   *Method 1:*   1000 kcal for 1st year
   Add 100 kcal per year up to age 11.
   Girls 11–15 yr, add 100 kcal or less per year after age 10.
   Girls > 15 yr, calculate as adults.
   Boys 11–15 yr, add 200 kcal per year after age 10.
   Boys > 15 yr,   add 23 kcal/lb if very active.
   18 kcal/lb if moderately active.
   16 kcal/lb if sedentary.

   *Method 2:*   1000 kcal for 1st year
   Add 125 kcal × age for boys.
   Add 100 kcal × age for girls.
   Add up to 20% more kcal for activity.

14. What is the best method of determining whether or not the energy level is adequate?

15. Using a computer dietary analysis program or food composition table, calculate the kcal, protein, fat (saturated, polyunsaturated, and monounsaturated), CHO, fiber, and cholesterol content of Susan's typical diet.

16. List at least 10 more questions you would need to ask Susan and/or her family to begin developing an appropriate meal plan approach for her individual lifestyle.

17. Does Susan's diet meet the Food Guide Pyramid guidelines? If not, which groups are under- or overrepresented?

18. What goals would you discuss with Susan to help her achieve a healthy, balanced diet?

19. Does the current diet order meet the patient's overall nutritional needs? If yes, explain why it is appropriate. If no, what would you recommend? Justify your answer.

20. Using Susan's "usual" dietary intake as a guide, calculate a meal plan for 2800 kcal (60% CHO, 10% protein, 30% fat). Show Susan and her mother how to incorporate her favorite foods and/or her usual eating habits into the meal plan.

| Food Group | No. of Exchanges | CHO g | Protein g | Fat g | kcal |
|---|---|---|---|---|---|
| Starch/bread | | | | | |
| Meat, lean | | | | | |
| Meat, medium-fat | | | | | |
| Meat, high-fat | | | | | |
| Vegetables | | | | | |
| Fruit | | | | | |
| Milk, skim | | | | | |
| Milk, low-fat | | | | | |
| Milk, whole | | | | | |
| Fat | | | | | |
| | | TOTAL × 4 = _____ | TOTAL × 4 = _____ | TOTAL × 9 = _____ | |
| | | % kcal = _____ | % kcal = _____ | % kcal = _____ | |
| | | TOTAL KCAL = _____ | | | |

**Meal Plan**

| Number of Exchanges/Choices | Sample Menu | Sample Menu |
|---|---|---|
| _____ Carbohydrate group<br>   _____ Starch<br>   _____ Fruit<br>   _____ Milk _____<br>_____ Meat group _____<br>_____ Fat group | | |
| _____ _____<br>_____ _____ | | |
| _____ Carbohydrate group<br>   _____ Starch<br>   _____ Fruit<br>   _____ Milk _____<br>   _____ Vegetables<br>_____ Meat group _____<br>_____ Fat group | | |
| _____ _____<br>_____ _____ | | |
| _____ Carbohydrate group<br>   _____ Starch<br>   _____ Fruit<br>   _____ Milk _____<br>   _____ Vegetables<br>_____ Meat group _____<br>_____ Fat group | | |

© 1993, American Dietetic Association. "A Healthy Food Guide: Kidney Disease." Used with permission.

**21.** List four nutrition behaviors that will help Susan improve blood glucose control and prevent complications.

**22.** What tools can Susan use to coordinate her eating patterns with her insulin and physical activity?

**23.** Susan is discharged Friday morning. She and her family have received information on insulin administration, SMBG, urine ketones, recordkeeping, exercise, signs, symptoms, and Tx of hypo-/hyperglycemia, meal planning (CHO counting), and contraception. Susan and her parents verbalize understanding of the instructions and have no further questions at this time. Appointments to see the outpatient dietitian and CDE are made for them to return in two weeks. When you come in to work Monday morning, you see that Susan was admitted through the ER Saturday night with a BG of 50 mg/dL. You see her when you make rounds and review her chart. During an interview, Susan tells you she was invited to a party Saturday night after her discharge on Friday. She tested her blood glucose before going to the party, and it measured 95 mg/dL. She took 2 units of insulin and knew she needed to have a snack that contained ~15 grams of CHO, so she drank one beer when she arrived at the party. She remembers getting lightheaded and then woke up in the ER. What happened to Susan? What kind of educational information will you give her before this discharge?*

---

*Author's note:* This question may seem out of place considering the patient's age, but this exact situation was experienced by the author when she was practicing as a clinical dietitian. The diabetes education team had not considered asking the patient or educating her about alcohol because she was a minor. But we learned (the hard way) that underage teens do consume alcohol.

# Bibliography

American Diabetes Association. Clinical practice recommendations: hospital admission guidelines for diabetes mellitus. *Diabetes Care* (suppl). Available at http://www.diabetes.org/diabetescare/supplement/s37.htm. Accessed August 16, 2000.

American Diabetes Association. Clinical practice recommendations: position statement: standards of medical care for patients with diabetes mellitus. *Diabetes Care* (suppl). Available at http://www.diabetes.org/clinicalrecommendations/CareSup1Jan01.htm. Accessed August 25, 2001.

American Diabetes Association. *Maximizing the Role of Nutrition in Diabetes Management.* Alexandria, VA: American Diabetes Association, 1994.

American Dietetic Association. *Manual of Clinical Dietetics,* 6th ed. Chicago: American Dietetic Association, 2000.

American Dietetic Association and Morrison Health Care. *Medical Nutrition Therapy Across the Continuum of Care.* Chicago: American Dietetic Association, 1998.

Commission on Accreditation for Dietetics Education. Knowledge, skills, and competencies for dietitians. *Accreditation Manual for Dietetics Education Programs,* rev. 4th ed. Chicago: American Dietetic Association, 2000.

Fischbach F. *Manual of Laboratory and Diagnostic Tests,* 6th ed. Philadelphia: Lippincott, 2000.

Franz MJ. Medical nutrition therapy for diabetes mellitus and hypoglycemia of nondiabetic origin. In Mahan LK, Escott-Stump S (eds.), *Krause's Food, Nutrition, and Diet Therapy,* 10th ed. Philadelphia: Saunders, 2000.

Goldstein D, Rife D, Derrick K, Kirchoff K. The test with a memory. *Diabetes Forecast,* April 96. Available at http://www.diabetes.org/diabetesforecast/96apr/memory.htm. Accessed August 16, 2000.

Holler HJ, Pastors JG. *Diabetes Medical Nutrition Therapy: A Professional Guide to Management and Nutrition Education Resources.* Chicago: American Dietetic Association/American Diabetes Association, 1997.

Karlsen M, Thomson LL. Efficacy of medical nutrition therapy: are your patients getting what they need? *Clin Diabetes.* 1996;14(3). Available at http://www.diabetes.org/clinicaldiabetes/v14n3m-j96/pg54.htm. Accessed September 11, 2000.

MedlinePlus Health information. Available at http://www.nlm.nih.gov/medlineplus/druginformation.html. Accessed September 8, 2000.

Monk A, Barry B, McClain K, Weaver T, Cooper N, Franz MJ. Practice guidelines for medical nutrition therapy provided by dietitians for persons with non-insulin dependent diabetes mellitus. *J Am Diet Assoc.* 1995;95:999–1006.

Nonketotic hyperglycemic-hyperosmolar coma. *The Merck Manual.* Available at http://www.merck.com. Accessed August 16, 2000.

Pastors JG. Nutrition assessment for diabetes medical nutrition therapy. *Diabetes Spectrum.* 9(2):99–103, 1996. Available at http://www.diabetes.org/diabetesspectrum/96v9n02/nutas.htm. Accessed September 11, 2000.

Prochaska JO, DiClemente CC, Norcross JC. In search of how people change: applications to addictive behaviors. *Am Psychol.* 1992;47:1102–1114.

Prochaska JO, Ruggiero L (eds.). From research to practice: readiness for change. *Diabetes Spectrum.* 1993; 6:22–60.

Setter SM. New drug therapies for the treatment of diabetes. *Diabetes Care and Education on the Cutting Edge.* 1998,19:3–7.

Swearingen PL, Ross DG. Endocrine disorders. *Manual of Medical-Surgical Nursing Care,* 4th ed. St. Louis: Mosby, 1999.

# Type 2 Diabetes Mellitus

*Introductory Level*

## Objectives

After completing this case, the student will be able to:

1. Collect pertinent information and use nutrition assessment techniques to determine baseline nutritional status.
2. Evaluate laboratory indices for nutritional implications and significance.
3. Integrate working knowledge of pathophysiology to nutrition care of type 2 diabetes mellitus.
4. Develop appropriate behavior outcomes for the patient.
5. Determine appropriate nutrient requirements.
6. Identify appropriate MNT goals.
7. Complete appropriate documentation in the medical record.

Eileen Douglas is a 71-year-old woman who was admitted for surgical debridement of a nonhealing foot wound. On admission, Mrs. Douglas was found to be hyperglycemic, and a diagnosis of type 2 diabetes mellitus was determined.

# UNIVERSITY HOSPITAL

## ADMISSION DATABASE

Name: E. Douglas
DOB: 7/27   age 71
Physician: R. Case, D. Shyne

| BED# 1 | DATE: 6/8 | TIME: 1523 | TRIAGE STATUS (ER ONLY): ☐ Red ☐ Yellow ☐ Green ☐ White |
|---|---|---|---|

**Initial Vital Signs**

PRIMARY PERSON TO CONTACT:
Name: Connie Locher
Home #: 555-8217
Work #: 555-7512

| TEMP: 99.2 | RESP: 12 | SAO2: |
|---|---|---|

| HT: 5'0" | WT (lb): 155 | B/P: 150/97 | PULSE: 75 |
|---|---|---|---|

ORIENTATION TO UNIT: ☒ Call light ☒ Television/telephone
☒ Bathroom ☒ Visiting ☒ Smoking ☒ Meals
☒ Patient rights/responsibilities

| LAST TETANUS 10 years ago | LAST ATE noon | LAST DRANK noon |
|---|---|---|

### CHIEF COMPLAINT/HX OF PRESENT ILLNESS

unhealed ulcer on L foot

PERSONAL ARTICLES: (Check if retained/describe)
☐ Contacts ☐ R ☐ L      ☒ Dentures ☒ Upper ☒ Lower
☐ Jewelry:
☐ Other:

### ALLERGIES: Meds, Food, IVP Dye, Seafood: Type of Reaction

VALUABLES ENVELOPE:
☐ Valuables instructions

### PREVIOUS HOSPITALIZATIONS/SURGERIES

INFORMATION OBTAINED FROM:
☒ Patient      ☐ Previous record
☐ Family       ☐ Responsible party

Signature   *Eileen Douglas*

| Home Medications (including OTC) | | Codes: A = Sent home | | B = Sent to pharmacy | C = Not brought in |
|---|---|---|---|---|---|
| Medication | Dose | Frequency | Time of Last Dose | Code | Patient Understanding of Drug |
| sliding scale Humulin | | | | | |
| Cipro | | | | | |
| Capoten | | | | | |
| | | | | | |
| | | | | | |
| | | | | | |
| | | | | | |
| | | | | | |
| | | | | | |
| | | | | | |
| | | | | | |

Do you take all medications as prescribed?   ☒ Yes   ☐ No   If no, why?

### PATIENT/FAMILY HISTORY

| | | |
|---|---|---|
| ☐ Cold in past two weeks | ☒ High blood pressure Patient | ☐ Kidney/urinary problems |
| ☐ Hay fever | ☐ Arthritis | ☐ Gastric/abdominal pain/heartburn |
| ☐ Emphysema/lung problems | ☐ Claustrophobia | ☐ Hearing problems |
| ☐ TB disease/positive TB skin test | ☐ Circulation problems | ☐ Glaucoma/eye problems |
| ☐ Cancer | ☐ Easy bleeding/bruising/anemia | ☐ Back pain |
| ☐ Stroke/past paralysis | ☐ Sickle cell disease | ☐ Seizures |
| ☐ Heart attack | ☐ Liver disease/jaundice | ☐ Other |
| ☐ Angina/chest pain | ☐ Thyroid disease | |
| ☐ Heart problems | ☒ Diabetes Sibling | |

### RISK SCREENING

Have you had a blood transfusion?   ☐ Yes   ☒ No
Do you smoke?   ☐ Yes   ☒ No
If yes, how many pack(s) _____ /day for _____ years
Does anyone in your household smoke?   ☐ Yes   ☒ No
Do you drink alcohol?   ☐ Yes   ☒ No
If yes, how often?_____   How much?
When was your last drink?   /   /
Do you take any recreational drugs?   ☐ Yes   ☒ No
If yes, type:_____   Route
Frequency:_____   Date last used:_____/_____/_____

**FOR WOMEN Ages 12–52**

Is there any chance you could be pregnant?   ☐ Yes   ☐ No
If yes, expected date (EDC):_____/_____/
Gravida/Para: 2 / 2

**ALL WOMEN**

Date of last Pap smear: 12 / 12 /last year
Do you perform regular breast self-exams?   ☐ Yes   ☒ No

**ALL MEN**

Do you perform regular testicular exams?   ☐ Yes   ☐ No

Additional comments:

✗ *Ruth Long, RN*
Signature/Title

**Client name:** Eileen Douglas
**DOB:** 7/27
**Age:** 71
**Sex:** Female
**Education:** Less than high school    *What grade/level?* 10th grade
**Occupation:** Homemaker
**Hours of work:** N/A
**Household members:** Sister, age 80, Dx with type 2 DM 10 years ago. Mrs. Douglas cares for her sister.
**Ethnic background:** African American
**Religious affiliation:** Protestant
**Referring physician:** Richard Case, MD (internal medicine); Dennis Shyne, MD (general surgery)

## Chief complaint:

"This cut on my foot happened over two months ago and has not healed. And I don't think I see as well. Maybe I need my eyes checked again. I have been having trouble reading the newspaper for the past few months."

## Patient history:

*Onset of disease:* Mrs. Douglas is a 71-year-old widow who lives with her 80-year-old sister whom she cares for. They live in low-income housing in a third-floor walk-up apartment. In addition to the un-healed wound and blurry vision, Mrs. Douglas complains of frequent bladder infections, which are documented in her clinic chart, and a slight tingling and numbness in her feet. On admission to the hospital, her blood glucose measured 325 mg/dL. Surgical debridement of wound is indicated, along with normalization of blood glucose and alleviation of blurred vision.

*Type of Tx:* Surgical debridement of wound, sliding scale insulin, 1200-kcal diet, DM self-management education
*PMH:* HTN
*Meds:* Humulin, sliding scale
Cipro (ciprofloxacin), 100 mg PO q 12 hr
Capoten (captopril), 50 mg PO bid
*Smoker:* No
*Family Hx: What?* DM    *Who?* sister, for 10 years

## Physical exam:

*General appearance:* Overweight elderly African American female
*Vitals:* Temp 99.2°F, BP 150/97 mm Hg, HR 75 bpm, RR 12 bpm
*Heart:* Regular rate and rhythm, no gallops or rubs, point of maximal impulse at the fifth intercostal space in the midclavicular line
*HEENT:*
  *Head:* Normocephalic
  *Eyes:* Wears glasses for myopia, mild retinopathy
  *Ears:* Tympanic membranes normal
  *Nose:* Dry mucous membranes w/out lesions
  *Throat:* Slightly dry mucous membranes w/out exudates or lesions
*Genitalia:* Normal w/out lesions

*Neurologic:* Alert and oriented. Cranial nerves II–XII grossly intact, strength 5/5 throughout, sensation to light touch intact in hands, mildly diminished in feet, normal gait, normal reflexes

*Extremities:* Normal muscular tone for age, normal ROM, nontender

*Skin:* Warm and dry, 2 × 3 cm ulcer on lateral left foot

*Chest/lungs:* Respirations normal; no crackles, rhonchi, wheezes, or rubs noted

*Peripheral vascular:* Pulse 2+ bilaterally, cool, mild edema

*Abdomen:* Audible bowel sounds, soft and nontender, w/out masses or organomegaly

## Nutrition Hx:

*General:* Because her sister "has sugar," Mrs. Douglas does not purchase cakes, candy, and other desserts. In fact, Mrs. Douglas reports that she and her sister try to avoid "all starchy foods" because that's what they were told to do when her sister received a printed diet sheet from her MD (10 years ago). Once a month, though, she and her sister have cake and ice cream at the Senior Center birthday party.

*Usual dietary intake:*

| | |
|---|---|
| AM: | One egg, fried in bacon fat, 2 strips of bacon or sausage, 1 cup coffee, black, ½ cup orange juice (unsweetened) |
| Lunch: | Lunchmeat sandwich: 2 slices enriched white bread, 1 slice (1 oz) bologna, 1 slice (1 oz) American cheese, mustard, 1 glass (8 oz) iced tea unsweetened |
| PM: | 1 cup turnip greens seasoned with fatback, salt, and pepper (simmered on stove top for at least 3 hours), 2 small new potatoes, boiled, seasoned with salt and pepper, 2-inch square of cornbread with 1 tsp butter, 1 cup beans and ham (Great Northern beans cooked with ham)—approximately ¾ cup beans and ¼ cup or 1 oz ham), 1 cup coffee, black |
| Snack: | 2 vanilla wafers |
| 24-hr recall: | N/A |

*Food allergies/intolerances/aversions:* N/A

*Previous MNT?* No

*Food purchase/preparation:* Self

*Vit/min intake:* None

*Current diet order:* 1200 kcal ADA exchange diet

## Tx plan:

Debride wound.

Normalize blood glucose levels.

Provide adequate kcalories and nutrients to meet Pt's needs.

Begin self-management training on nutrition prescription, meal planning, signs/symptoms, and Tx of hypo-/hyperglycemia, SMBG, appropriate exercise, potential food–drug interaction.

# U<sub>H</sub> *UNIVERSITY HOSPITAL*

NAME: E. Douglas                    DOB: 7/27
AGE: 71                             SEX: Female
PHYSICIAN: R. Case

\*\*\*\*\*\*\*\*\*\*\*\*\*\*\*\*\*\*\*\*\*\*\*\*\*\*\*\*\*\*\*\*\*\*\*\*\*\*\*\*\*\*\*\*CHEMISTRY\*\*\*\*\*\*\*\*\*\*\*\*\*\*\*\*\*\*\*\*\*\*\*\*\*\*\*\*\*\*\*\*\*\*\*\*\*\*\*\*\*\*\*\*

DAY:                                Admit          d/c
DATE:
TIME:
LOCATION:

|  | NORMAL | | | UNITS |
|---|---|---|---|---|
| Albumin | 3.6–5 | 4.0 | 4.1 | g/dL |
| Total protein | 6–8 | 7 | 7.2 | g/dL |
| Prealbumin | 19–43 | 23 | 24.5 | mg/dL |
| Transferrin | 200–400 | 310 | 305 | mg/dL |
| Sodium | 135–155 | 140 | 145 | mmol/L |
| Potassium | 3.5–5.5 | 4.2 | 4.5 | mmol/L |
| Chloride | 98–108 | 103 | 100 | mmol/L |
| $PO_4$ | 2.5–4.5 | 3.6 | 3.2 | mmol/L |
| Magnesium | 1.6–2.6 | 2.1 | 1.8 | mmol/L |
| Osmolality | 275–295 | 285 | 286 | mmol/kg $H_2O$ |
| Total $CO_2$ | 24–30 | 25 | 26 | mmol/L |
| Glucose | 70–120 | 325 | 121 | mg/dL |
| BUN | 8–26 | 26 | 26 | mg/dL |
| Creatinine | 0.6–1.3 | 1.2 | 1.2 | mg/dL |
| Uric acid | 2.6–6 (women) | | | mg/dL |
|  | 3.5–7.2 (men) | | | |
| Calcium | 8.7–10.2 | | | mg/dL |
| Bilirubin | 0.2–1.3 | | | mg/dL |
| Ammonia ($NH_3$) | 9–33 | | | μmol/L |
| SGPT (ALT) | 10–60 | | | U/L |
| SGOT (AST) | 5–40 | | | U/L |
| Alk phos | 98–251 | | | U/L |
| CPK | 26–140 (women) | | | U/L |
| LDH | 313–618 | | | U/L |
| CHOL | 140–199 | 300 | 250 | mg/dL |
| HDL–C | 40–85 (women) | 35 | 37 | mg/dL |
|  | 37–70 (men) | | | |
| VLDL | | | | mg/dL |
| LDL | < 130 | 140 | 138 | mg/dL |
| LDL/HDL ratio | < 3.22 (women) | | | |
|  | < 3.55 (men) | | | |
| Apo A | 101–199 (women) | | | mg/dL |
|  | 94–178 (men) | | | |
| Apo B | 60–126 (women) | | | mg/dL |
|  | 63–133 (men) | | | |
| TG | 35–160 | 400 | 300 | mg/dL |
| $T_4$ | 5.4–11.5 | | | μg/dL |
| $T_3$ | 80–200 | | | ng/dL |
| $HbA_{1C}$ | 4.8–7.8 | 8.5 | | % |

**U**<sub>**H**</sub> *UNIVERSITY HOSPITAL*

NAME: E. Douglas                    DOB: 7/27
AGE: 71                             SEX: Female
PHYSICIAN: R. Case

∗∗∗∗∗∗∗∗∗∗∗∗∗∗∗∗∗∗∗∗∗∗∗∗∗∗∗∗∗∗∗∗∗∗∗∗∗∗∗HEMATOLOGY∗∗∗∗∗∗∗∗∗∗∗∗∗∗∗∗∗∗∗∗∗∗∗∗∗∗∗∗∗∗∗∗∗∗∗∗∗∗∗

DAY:                                                    1          5
DATE:
TIME:
LOCATION:

| | NORMAL | | | UNITS |
|---|---|---|---|---|
| WBC | 4.3–10 | | | $\times\ 10^3/mm^3$ |
| RBC | 4–5 (women) | | | $\times\ 10^6/mm^3$ |
| | 4.5–5.5 (men) | | | |
| HGB | 12–16 (women) | 9.9 | 10.1 | g/dL |
| | 13.5–17.5 (men) | | | |
| HCT | 37–47 (women) | 30.4 | 29.7 | % |
| | 40–54 (men) | | | |
| MCV | 84–96 | | | fL |
| MCH | 27–34 | | | pg |
| MCHC | 31.5–36 | | | % |
| RDW | 11.6–16.5 | | | % |
| Plt ct | 140–440 | | | $\times\ 10^3$ |
| Diff TYPE | | | | |
| % GRANS | 34.6–79.2 | | | % |
| % LYM | 19.6–52.7 | | | % |
| SEGS | 50–62 | | | % |
| BANDS | 3–6 | | | % |
| LYMPHS | 25–40 | | | % |
| MONOS | 3–7 | | | % |
| EOS | 0–3 | | | % |
| TIBC | 65–165 (women) | | | μg/dL |
| | 75–175 (men) | | | |
| Ferritin | 18–160 (women) | | | μg/dL |
| | 18–270 (men) | | | |
| Vitamin $B_{12}$ | 100–700 | | | pg/mL |
| Folate | 2–20 | | | ng/mL |
| Total T cells | 812–2318 | | | $mm^3$ |
| T-helper cells | 589–1505 | | | $mm^3$ |
| T-suppressor cells | 325–997 | | | $mm^3$ |
| PT | 11–13 | | | sec |

## Case Questions

1. What is the difference between type 1 diabetes mellitus and type 2 diabetes mellitus?

2. Why are overweight individuals more susceptible to developing diabetes?

3. List the chronic complications of diabetes mellitus.

4. What risk factors does Mrs. Douglas have for developing diabetes mellitus?

5. **a.** Does Mrs. Douglas present with any complications of diabetes mellitus? If yes, which ones?

   **b.** How does the pathophysiology of the disease relate to her presenting signs and symptoms?

6. Prior to admission, Mrs. Douglas had not been diagnosed with diabetes mellitus. How could she present with complications?

7. Using the chart on the following page, compare the patient's laboratory values that were out of range on admission with normal values. How would you interpret this patient's labs? Make sure explanations are pertinent to *this* situation.

| Parameter | Normal Value | Pt's Value | Interpretation |
|---|---|---|---|
| Glucose | | | |
| HbA$_{1c}$ | | | |
| Cholesterol | | | |
| LDH-cholesterol | | | |
| HDL-cholesterol | | | |
| Triglycerides | | | |

8.   Calculate Mrs. Douglas's body mass index (BMI). What does this mean?

9.   What are the health implications for a BMI in this range?

10.   Calculate Mrs. Douglas's energy needs using the Harris-Benedict equation. Which weight (IBW, RBW, ABW, or current body weight) should you use to accurately calculate the patient's kcalorie needs? Should Mrs. Douglas's weight be adjusted for obesity?

11.   Calculate Mrs. Douglas's protein needs.

12.   Does Mrs. Douglas's "usual" dietary intake meet the Food Guide Pyramid guidelines? Is she deficient in any food groups? If so, which ones?

13.   Using a computer dietary analysis program or food composition table, calculate the kcal, protein, fat, CHO, fiber, cholesterol, and Na content of Mrs. Douglas's diet.

**14.** How would you compare Mrs. Douglas's "usual" dietary intake to her current nutritional needs?

**15.** What other observations can you make regarding Mrs. Douglas's "usual" dietary intake (in reference to her DM)?

**16.** What is the most important nutritional concern at this time?

**17.** What is the MNT goal at this time?

**18.** Is the diet order of 1200 kcal appropriate? If yes, explain why it is appropriate. If no, what would you recommend? Justify your answer.

**19.** Review the following initial nutrition note written for this patient. Is this progress note complete? Do you note any errors? Any omissions? Explain.

6/8, 1300 hr    Nutrition Note
S:   Pt lives with 80-yo sister whom she cares for. Sister has had "sugar" for 10 years, and Pt is responsible for shopping & meal preparation. Usual dietary intake reflects an intake of ~1700 kcal, 59 g protein, 142% fat rec, 151% Na rec, 126% cholesterol rec, and a more than adequate fiber intake.
O:   71 yo f, ht 5′0″, wt 155#, IBW 100#, ABW 106#
     *Dx:* Unhealed foot ulcer, type 2 DM
     *PMH:* HTN
     *Meds:* SS insulin, Cipro, Cardizem
A:   Glucose and HbA$_{1c}$ high due to DM. SS insulin for coverage. Pt will need education appropriate for her understanding level prior to and follow-up after discharge.
P:   (1) Recommend Pt ed on appropriate food choices or diet for glucose control.
     (2) Dietitian to follow daily. Monitor for adequate oral intake.

**20.** What additional information does the dietitian need to collect before he or she can mutually develop clinical and behavioral outcomes with the patient and health care team?

**21.** What type of follow-up would be optimal for Mrs. Douglas?

# Bibliography

American Dietetic Association. *Manual of Clinical Dietetics,* 6th ed. Chicago: American Dietetic Association, 2000.

American Dietetic Association and Morrison Health Care. *Medical Nutrition Therapy Across the Continuum of Care.* Chicago: American Dietetic Association, 1998.

American Diabetes Association position statement: standards of medical care for patients with diabetes mellitus, clinical practice recommendations 2000. *Diabetes Care,* 23 (suppl 1). Available at http://journal.diabetes.org. Accessed August 16, 2000.

Diabetes facts and figures. Available at http://www.diabetes.org/ada/facts.asp. Accessed August 16, 2000.

Diabetes mellitus. *The Merck Manual.* Available at http://www.merck.com. Accessed August 16, 2000.

Fischbach F. *A Manual of Laboratory and Diagnostic Tests,* 6th ed. Philadelphia: Lippincott, 2000.

Holler HJ, Pastors JG. *Diabetes Medical Nutrition Therapy. A Professional Guide to Management and Nutrition Education Resources.* Chicago: American Dietetic Association/American Diabetes Association, 1997.

MedlinePlus Health information. Available at http://www.nlm.nih.gov/medlineplus/druginformation.html. Accessed August 4, 2000.

Morrison G, Hark L. *Medical Nutrition and Disease,* 2nd ed. Malden, MA: Blackwell Science, 1999.

National Heart, Lung and Blood Institute, Obesity Education Initiative professional education materials. Available at http:// www.nhlbi.nih.gov/health/public/heart/obesity/lose_wt/profmats.htm. Accessed August 23, 2000.

National Institute of Diabetes and Digestive and Kidney Diseases (NIDDK), National Diabetes Information Clearinghouse. Diabetes. Available at http://intelihealth7161/203032.html?d=dmtContent. Accessed August 16, 2000.

Pronsky ZM. *Powers and Moore's Food–Medication Interactions,* 11th ed. Birchrunville, PA: Food–Medication Interactions, 2000.

Wallace, JI. Management of diabetes in the elderly. *Clin Diabetes.* 1999;17:19. Available at http://www.diabetes.org/clinicaldiabetes/v17n111999/Pg19.htm. Accessed August 16, 2000.

# Type 1 Diabetes Mellitus with Diabetic Ketoacidosis

*Advanced Practice*

## Objectives

After completing this case, the student will be able to:

1. Integrate pathophysiology of diabetic ketoacidosis (DKA) into MNT recommendations.
2. Assess the nutritional status of a patient with DKA.
3. Interpret laboratory parameters for nutritional implications and significance.
4. Interpret laboratory parameters to analyze fluid and electrolyte status and acid–base balance.

5. Manage monitoring of patient's food and nutrient intake.
6. Apply dietary history information to development of nutrition care plan.

Susan Cheng, introduced in Case 26, is admitted through the emergency room with confusion, nausea, vomiting, and difficulty breathing. A diagnosis of diabetic ketoacidosis is confirmed.

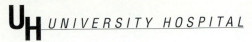
## ADMISSION DATABASE

Name: S. Cheng
DOB: 9/25 age 16
Physician: P. Green

| BED# 2 | DATE: 10/2 | TIME: 0900 | TRIAGE STATUS (ER ONLY): ☐ Red ☐ Yellow ☐ Green ☐ White |
|---|---|---|---|

**PRIMARY PERSON TO CONTACT:**
Name: Mai or David Cheng
Home #: 555-8217
Work #: 555-1234

### Initial Vital Signs

| TEMP: 98.6 | RESP: 20 | SAO2: |
|---|---|---|

| HT: 5'3" | WT (lb): 110 | B/P: 70/100 | PULSE: 105 |
|---|---|---|---|

**ORIENTATION TO UNIT:** ☒ Call light ☒ Television/telephone ☒ Bathroom ☒ Visiting ☒ Smoking ☒ Meals ☒ Patient rights/responsibilities

| LAST TETANUS 5 years ago | LAST ATE last night | LAST DRANK 2 hrs ago |
|---|---|---|

### CHIEF COMPLAINT/HX OF PRESENT ILLNESS

nausea, vomiting, fatigue

**PERSONAL ARTICLES:** (Check if retained/describe)
☒ Contacts ☒ R ☒ L   ☐ Dentures ☐ Upper ☐ Lower
☐ Jewelry:
☐ Other:

### ALLERGIES: Meds, Food, IVP Dye, Seafood: Type of Reaction

**VALUABLES ENVELOPE:**
☐ Valuables instructions

### PREVIOUS HOSPITALIZATIONS/SURGERIES

one year ago–Type 1 DM diagnosed

**INFORMATION OBTAINED FROM:**
☒ Patient   ☒ Previous record
☒ Family   ☐ Responsible party

Signature   *Mai Cheng*

### Home Medications (including OTC)    Codes: A = Sent home    B = Sent to pharmacy    C = Not brought in

| Medication | Dose | Frequency | Time of Last Dose | Code | Patient Understanding of Drug |
|---|---|---|---|---|---|
| regular insulin | 5u | q AM | 0600 | c | yes |
|  | 4 | q HS | 0600 | c | yes |
| NPH insulin | 10 | q AM | 2200 10/1 | c | yes |
|  | 4 | q HS | 2200 10/1 | c | yes |
|  |  |  |  |  |  |
|  |  |  |  |  |  |
|  |  |  |  |  |  |
|  |  |  |  |  |  |
|  |  |  |  |  |  |
|  |  |  |  |  |  |

Do you take all medications as prescribed?   ☒ Yes   ☐ No   If no, why?

### PATIENT/FAMILY HISTORY

☐ Cold in past two weeks
☐ Hay fever
☐ Emphysema/lung problems
☐ TB disease/positive TB skin test
☐ Cancer
☐ Stroke/past paralysis
☐ Heart attack
☐ Angina/chest pain
☐ Heart problems

☐ High blood pressure
☐ Arthritis
☐ Claustrophobia
☐ Circulation problems
☐ Easy bleeding/bruising/anemia
☐ Sickle cell disease
☐ Liver disease/jaundice
☐ Thyroid disease
☒ Diabetes Maternal grandmother

☐ Kidney/urinary problems
☐ Gastric/abdominal pain/heartburn
☐ Hearing problems
☐ Glaucoma/eye problems
☐ Back pain
☐ Seizures
☐ Other

### RISK SCREENING

Have you had a blood transfusion?   ☐ Yes   ☒ No
Do you smoke?   ☐ Yes   ☒ No
If yes, how many pack(s)     /day for     years
Does anyone in your household smoke?   ☐ Yes   ☒ No
Do you drink alcohol?   ☐ Yes   ☐ No
If yes, how often?_____   How much?
When was your last drink?     /     /
Do you take any recreational drugs?   ☐ Yes   ☒ No
If yes, type:_____   Route
Frequency:_____   Date last used:_____/_____/_____

**FOR WOMEN Ages 12–52**

Is there any chance you could be pregnant?   ☐ Yes   ☒ No
If yes, expected date (EDC):_____/_____/_____
Gravida/Para:

**ALL WOMEN**

Date of last Pap smear:_____/_____/_____
Do you perform regular breast self-exams?   ☐ Yes   ☒ No

**ALL MEN**

Do you perform regular testicular exams?   ☐ Yes   ☐ No

Additional comments:

✗ *Tina McDonald, RD*
Signature/Title

**Client name:** Susan Cheng
**DOB:** 9/25
**Age:** 16
**Sex:** Female
**Education:** Less than high school *What grade/level?* 10th grade, HS student
**Occupation:** Student
**Hours of work:** 8:30 AM–3:30 PM, volleyball practice 3:45–6 PM
**Household members:** Mother age 41, father age 43, sister age 17, brother age 10—all in excellent health
**Ethnic background:** Asian American
**Religious affiliation:** Protestant
**Referring physician:** Pryce Green, MD (endocrinology)

## Chief complaint:

"I don't understand why my blood glucose levels are so high. I check my blood glucose levels every day. I've been following my meal plan. I feel just like I did when I was first diagnosed with diabetes."

## Patient history:

*Onset of disease:* Susan is a 16-yo female who was diagnosed with type 1 diabetes mellitus one year ago. She lives with her parents, brother, and sister and is currently in the 10th grade. She is a starter on the girls' volleyball team, practices four evenings per week, and participates in approximately two games per week, some of which are away games. Susan was admitted through the ER complaining of fatigue, nausea, vomiting, and intense thirst. The ER physician observed that she was confused and breathing with difficulty and further noted the smell of acetone on her breath. Urine tests were positive for glycosuria and ketonuria, and her blood glucose was 400 mg/dL.

*PMH:* Type 1 DM diagnosed one year ago. Was initially treated with a 2800-kcal diet and a split, mixed dose of insulin. Pt and family received self-management training including insulin administration, exchange lists and meal pattern, signs/symptoms of hypo-/hyperglycemia, SBGM, urine ketones, recordkeeping, exercise management, and contraception education. Shortly after her initial admission and discharge, she was readmitted one day later with hypoglycemia resulting from alcohol consumption. Pt received education regarding appropriate use and planning of alcohol consumption.

*Meds:* 10 u NPH, 5 u regular in AM, 4 u NPH, 4 u regular in PM
*Smoker:* No
*Family Hx: What?* DM    *Who?* Maternal grandmother

## Physical exam:

*General appearance:* Tired-looking young woman, c/o fatigue, nausea, and vomiting
*Vitals:* Temp 98.6°F, BP 70/100 mm Hg, HR 105 bpm, RR 20 bpm
*HEENT:*
   *Heart:* Tachycardia
   *Eyes:* Sunken, contact lenses in both eyes
   *Ears:* Membranes dry
   *Nose:* Dry mucous membranes
   *Throat:* Noncontributory
*Genitalia:* Noncontributory

*Neurologic:* Irritable, lethargic
*Extremities:* Noncontributory
*Skin:* Dry, flushed skin, poor turgor
*Chest/lungs:* Deep, rapid Kussmaul's respirations
*Abdomen:* Tender with guarding, decreased bowel sounds

## Nutrition Hx:

*General:* Susan describes her appetite as good, but she has lost 5 pounds in the last 2 weeks. Her usual weight is 115#. She has a snack before volleyball practice.
*Usual dietary intake:* N/A
*24-hr recall:* N/A

*Food allergies/intolerances/aversion:* NKA
*Previous MNT?* Yes    *If yes, when:* 1 year ago
*Where?* University Medical Center
*Food purchase/preparation:* Parents and school cafeteria
*Vit/min intake:* Iron supplement
*Current diet order:* 3000 kcal

## Tx plan:

1 L 0.9% NS w/ 10 u regular insulin IV over first 30 minutes
Reduce rate to 1 L/hr 0.9% NS with 10 u regular insulin (per L)
Measure glucose, acetone, electrolytes, and ABGs every hour
At blood glucose = 250 mg/dL, decrease insulin to 5 u in 5% dextrose and 0.4% NS, continue infusion until plasma cleared of ketones
Maintain blood glucose at 250 mg/dL and continue dextrose/NS infusion until acidosis corrected
When Pt stable, begin sliding scale regular insulin SQ q 4 hr
        Sliding scale for glucometer:
        BG     > 400     Regular insulin SQ— 20 units
               > 300                          15 units
               > 200                          10 units
300 mg Tagamet IV piggyback q 8 hr
Liquids when PO fluids tolerated
When stable, evaluate for insulin pump
Diabetes education consult

# U<sub>H</sub> UNIVERSITY HOSPITAL

NAME: S. Cheng
AGE: 16
PHYSICIAN: P. Green

DOB: 9/25
SEX: Female

\*\*\*\*\*\*\*\*\*\*\*\*\*\*\*\*\*\*\*\*\*\*\*\*\*\*\*\*\*\*\*\*\*\*\*\*\*\*\*\*\*\*\*\*\*\*CHEMISTRY\*\*\*\*\*\*\*\*\*\*\*\*\*\*\*\*\*\*\*\*\*\*\*\*\*\*\*\*\*\*\*\*\*\*\*\*\*\*\*\*\*

| DAY:<br>DATE:<br>TIME:<br>LOCATION: | | Admit | day 2 | d/c | |
|---|---|---|---|---|---|
| | NORMAL | | | | UNITS |
| Albumin | 3.6–5 | 4.3 | 4.1 | 4.0 | g/dL |
| Total protein | 6–8 | 7 | 6.9 | 6.8 | g/dL |
| Prealbumin | 19–43 | 40 | 39 | 40 | mg/dL |
| Transferrin | 200–400 | 300 | 302 | 304 | mg/dL |
| Sodium | 135–155 | 150 | 140 | 137 | mEq/L |
| Potassium | 3.5–5.5 | 5.8 | 5.1 | 5.0 | mEq/L |
| Chloride | 98–108 | 110 | 102 | 100 | mEq/L |
| $PO_4$ | 2.5–4.5 | 4.9 | 4.0 | 3.8 | mEq/L |
| Magnesium | 1.6–2.6 | | | | mEq/L |
| Osmolality | 275–295 | 336 | 298 | 287 | mmol/kg $H_2O$ |
| Total $CO_2$ | 24–30 | 22 | 24 | 26 | mmol/L |
| Glucose | 70–120 | 475 | 200 | 120 | mg/dL |
| BUN | 8–26 | 29 | 21 | 20 | mg/dL |
| Creatinine | 0.6–1.3 | 1.8 | 1.2 | 1.0 | mg/dL |
| Uric acid | 2.6–6 (women)<br>3.5–7.2 (men) | | | | mg/dL |
| Calcium | 8.7–10.2 | 10 | 9.5 | 9.0 | mg/dL |
| Bilirubin | 0.2–1.3 | | | | mg/dL |
| Ammonia ($NH_3$) | 9–33 | | | | $\mu$mol/L |
| SGPT (ALT) | 10–60 | | | | U/L |
| SGOT (AST) | 5–40 | | | | U/L |
| Alk phos | 98–251 | | | | U/L |
| CPK | 26–140 (women)<br>38–174 (men) | | | | U/L |
| LDH | 313–618 | | | | U/L |
| CHOL | 140–199 | 201 | 200 | 198 | mg/dL |
| HDL-C | 40–85 (women)<br>37–70 (men) | | | | mg/dL |
| VLDL | | | | | mg/dL |
| LDL | < 130 | | | | mg/dL |
| LDL/HDL ratio | < 3.22 (women)<br>< 3.55 (men) | | | | |
| Apo A | 101–199 (women)<br>94–178 (men) | | | | mg/dL |
| Apo B | 60–126 (women)<br>63–133 (men) | | | | mg/dL |
| TG | 35–160 | | | | mg/dL |
| $T_4$ | 5.4–11.5 | | | | $\mu$g/dL |
| $T_3$ | 80–200 | | | | ng/dL |
| $HbA_{1C}$ | 4.8–7.8 | 12.0 | | | % |

# U H UNIVERSITY HOSPITAL

NAME: S. Cheng                          DOB: 9/25
AGE: 16                                 SEX: Female
PHYSICIAN: P. Green

**************************************HEMATOLOGY**************************************

| | NORMAL | Admit | d/c | UNITS |
|---|---|---|---|---|
| DAY: | | | | |
| DATE: | | | | |
| TIME: | | | | |
| LOCATION: | | | | |
| WBC | 4.3-10 | 12 | 9.0 | $\times 10^3/mm^3$ |
| RBC | 4-5 (women) | 4.4 | 4.5 | $\times 10^6/mm^3$ |
| | 4.5-5.5 (men) | | | |
| HGB | 12-16 (women) | 18 | 15 | g/dL |
| | 13.5-17.5 (men) | | | |
| HCT | 37-47 (women) | 50 | 45 | % |
| | 40-54 (men) | | | |
| MCV | 84-96 | | | fL |
| MCH | 27-34 | | | pg |
| MCHC | 31.5-36 | | | % |
| RDW | 11.6-16.5 | | | % |
| Plt ct | 140-440 | | | $\times 10^3$ |
| Diff TYPE | | | | |
| % GRANS | 34.6-79.2 | | | % |
| % LYM | 19.6-52.7 | 50.1 | 50.3 | % |
| SEGS | 50-62 | | | % |
| BANDS | 3-6 | | | % |
| LYMPHS | 25-40 | 35 | 39 | % |
| MONOS | 3-7 | | | % |
| EOS | 0-3 | | | % |
| TIBC | 65-165 (women) | | | $\mu$g/dL |
| | 75-175 (men) | | | |
| Ferritin | 18-160 (women) | | | $\mu$g/dL |
| | 18-270 (men) | | | |
| Vitamin B$_{12}$ | 100-700 | | | pg/mL |
| Folate | 2-20 | | | ng/mL |
| Total T cells | 812-2318 | | | $mm^3$ |
| T-helper cells | 589-1505 | | | $mm^3$ |
| T-suppressor cells | 325-997 | | | $mm^3$ |
| PT | 11-13 | | | sec |

**U**<sub>H</sub> *UNIVERSITY HOSPITAL*

NAME: S. Cheng                    DOB: 9/25
AGE: 16                           SEX: Female
PHYSICIAN: P. Green

\*\*\*\*\*\*\*\*\*\*\*\*\*\*\*\*\*\*\*\*\*\*\*\*\*\*\*\*\*\*\*\*\*\*\*\*\*\*\*\*\*\*\*URINALYSIS\*\*\*\*\*\*\*\*\*\*\*\*\*\*\*\*\*\*\*\*\*\*\*\*\*\*\*\*\*\*\*\*\*\*\*\*\*\*\*\*\*\*

| DAY: | | Admit | 2 | d/c | |
|------|--------|-------|---|-----|------|
| DATE: | | | | | |
| TIME: | | | | | |
| LOCATION: | | | | | |
| | NORMAL | | | | UNITS |
| Coll meth | | First morning | First morning | First morning | |
| Color | | Amber | Pale yellow | Pale yellow | |
| Appear | | Hazy | Clear | Clear | |
| Sp grv | 1.003–1.030 | | | | |
| pH | 5–7 | | | | |
| Prot | NEG | | | | mg/dL |
| Glu | NEG | pos | neg | neg | mg/dL |
| Ket | NEG | pos | tr | neg | |
| Occ bld | NEG | | | | |
| Ubil | NEG | | | | |
| Nit | NEG | | | | |
| Urobil | < 1.1 | | | | EU/dL |
| Leu bst | NEG | | | | |
| Prot chk | NEG | | | | |
| WBCs | 0–5 | | | | /HPF |
| RBCs | 0–5 | | | | /HPF |
| EPIs | 0 | | | | /LPF |
| Bact | 0 | | | | |
| Mucus | 0 | | | | |
| Crys | 0 | | | | |
| Casts | 0 | | | | /LPF |
| Yeast | 0 | | | | |

# U**H** *UNIVERSITY HOSPITAL*

NAME: S. Cheng                          DOB: 9/25
AGE: 16                                 SEX: Female
PHYSICIAN: P. Green

**********************************ARTERIAL BLOOD GASES (ABGs)**********************************

| | NORMAL | Admit | 2 | UNITS |
|---|---|---|---|---|
| DAY: | | Admit | 2 | |
| DATE: | | | | |
| TIME: | | | | |
| LOCATION: | | | | |
| pH | 7.35–7.45 | 7.31 | 7.35 | |
| pCO$_2$ | 35–45 | | | mmHg |
| SO$_2$ | ≥95 | | | % |
| CO$_2$ content | 23–30 | 22 | 23 | mEq/L |
| O$_2$ content | 15–22 | | | vol % |
| pO$_2$ | ≥80 | | | mmHg |
| Base excess | >3 | | | mEq/L |
| Base deficit | <3 | | | mEq/L |
| HCO$_3$– | 24–28 | 21 | 24 | mEq/L |
| HGB | 12–16 (women) | | | g/dL |
| | 13.5–17.5 (men) | | | |
| HCT | 37–47 (women) | | | % |
| | 40–54 (men) | | | |
| COHb | <2 | | | % |
| [NA$^+$] | 135–148 | | | mmol/L |
| [K$^+$] | 3.5–5 | | | mEq/L |

 UNIVERSITY HOSPITAL

Name: Susan Cheng
AM dose: 10u NPH/5u reg
HS dose: 4u NPH/4u reg
BG Target Range:  80 mg/dL to 120 mg/dL

## BLOOD GLUCOSE MONITORING RECORD

| Day/Date | Breakfast | | Lunch | | Dinner | | Bedtime | | Comments |
|---|---|---|---|---|---|---|---|---|---|
| | BG | I | BG | I | BG | I | BG | I | |
| 9/1 | | 10/5 | 100 | | | | 110 | 4/4 | |
| 9/2 | 120 | 10/5 | | | 150 | | | 4/4 | Ate lunch late |
| 9/3 | | 10/5 | 115 | | | | 120 | 4/4 | |
| 9/4 | 90 | 10/5 | | | 117 | | | 4/4 | |
| 9/5 | | 10/5 | 120 | | | | | 4/4 | |
| 9/6 | | 10/5 | | | 118 | | | 4/4 | |
| 9/7 | | 10/5 | | | | | 125 | 4/4 | |
| 9/8 | 80 | 10/5 | 118 | | | | 120 | 4/4 | |
| 9/9 | | 10/5 | | | 115 | | | 4/4 | |
| 9/10 | 95 | 10/5 | | | | | 120 | 4/4 | |
| 9/11 | | 10/5 | 122 | | 120 | | | 4/4 | |
| 9/12 | 70 | 10/5 | 100 | | | | 120 | 4/4 | Forgot to eat last evening's snack |
| 9/13 | 90 | 10/5 | | | 130 | | | 4/4 | |
| 9/14 | | 10/5 | 120 | | | | 122 | 4/4 | |
| 9/15 | 115 | 10/5 | 120 | | | | 95 | 4/4 | |
| 9/16 | 80 | 10/5 | 110 | | 120 | | 120 | 4/4 | |
| 9/17 | | 10/5 | 118 | | | | 125 | 4/4 | |
| 9/18 | 100 | 10/5 | | | 125 | | | 4/4 | |
| 9/19 | 110 | 10/5 | | | 125 | | 120 | 4/4 | |
| 9/20 | 115 | 10/5 | | | | | 130 | 4/4 | Period started |
| 9/21 | 140 | 10/5 | 135 | | 120 | | 130 | 4/4 | Period |
| 9/22 | 135 | 10/5 | 150 | | 170 | | 185 | 4/4 | Period |
| 9/23 | 200 | 10/5 | 170 | | 150 | | 200 | 4/4 | Period |
| 9/24 | 200 | 10/5 | 220 | | 230 | | 300 | 4/4 | Volleyball tourney |
| 9/25 | 250 | 10/5 | 250 | | 275 | | 280 | 4/4 | My birthday! |
| | | | | | | | | | 10/5 Volleyball tourney |
| 9/26 | 275 | 10/5 | 300 | | 305 | | 320 | 4/4 | |
| 9/27 | 300 | 10/5 | 305 | | 310 | | 295 | 4/4 | |
| 9/28 | 280 | 10/5 | 275 | | 300 | | 280 | 4/4 | |
| 9/29 | 250 | 10/5 | 260 | | 305 | | 325 | 4/4 | |
| 9/30 | 350 | 10/5 | 350 | | 375 | | 350 | 4/4 | |
| 10/1 | | | | | | | | | |
| 10/2 | | | | | | | | | |
| 10/3 | | | | | | | | | |
| 10/4 | | | | | | | | | |
| 10/5 | | | | | | | | | |

## Case Questions

1. There are precipitating factors for diabetic ketoacidosis. List at least seven possible factors.

2. Describe the metabolic events that led to the symptoms associated with DKA.

3. Assess Susan's physical examination. What is consistent with diabetic ketoacidosis? Give the physiological rationale for each that you identify.

4. Examine Susan's biochemical indices both in the chemistry section and in her ABG report. Which are consistent with DKA? Why?

5. If Susan's symptoms were left untreated, what would happen?

6. Assuming Susan's SMBG records are correct, what events seem to have precipitated the development of DKA?

7. What, if anything, could Susan have done to avoid DKA?

8. While Susan is being stabilized, Tagamet is being given IV piggyback. What does "IV piggyback" mean? What is Tagamet, and why has it been prescribed?

9. The Diabetes Control and Complications Trial was a landmark multicenter trial designed to test the proposition that complications of diabetes mellitus are related to elevation of plasma glucose. It is the longest and largest prospective study showing that lowering blood glucose concentration slows or prevents development of complications common to individuals with diabetes. The trial compared "intensive" insulin therapy ("tight control") with "conventional" insulin therapy. Define "intensive" insulin therapy. Define "conventional" insulin therapy.

10. List the microvascular and neurologic complications associated with type 1 diabetes.

**11.** What are the advantages of intensive insulin therapy?

**12.** What are the risks of intensive therapy (tight control)?

**13.** Dr. Green consults with you, and the two of you decide that Susan would benefit from insulin pump therapy combined with CHO counting for intensive insulin therapy. This will give Susan better glycemic control and more flexibility. What are some of the key characteristics of candidates for intensive insulin therapy?

**14.** Explain how an insulin pump works.

**15.** How would you describe CHO counting to Susan and her family?

**16.** How is CHO counting used with intensive insulin therapy?

**17.** Estimate Susan's daily energy needs using the Harris-Benedict equation.

**18.** Using the 1-week food diary from Susan (next page), calculate the average amount of CHO usually consumed each meal and snack.

___gm CHO breakfast

___gm CHO lunch

___gm CHO snack

___gm CHO dinner

___gm CHO HS

| Day | Meal | Food |  |
|-----|------|------|---|
| Monday | AM | 1½ c Rice Krispies, 1 c 2% milk, 1 c orange juice (calcium fortified), 1 med banana |  |
|  | Lunch | 6-in personal vegetarian pizza, 12 oz Diet Coke, 1 large apple |  |
|  | Snack | PBJ sandwich: 2 slices whole wheat bread, 2 tbsp crunch peanut butter, 1 tsp grape jelly (regular), 12 oz Diet Coke |  |
|  | Dinner | Spaghetti w/ meat sauce: 3 c cooked spaghetti, ½ c sauce with ~ 2 oz. cooked ground beef, large tossed salad w/ diet dressing, 12 oz 2% milk, 2 stalks cooked broccoli, 6 vanilla wafers |  |
|  | HS | ½ c vanilla ice cream (Ben & Jerry's) |  |
| Tuesday | AM | 1½ c Rice Krispies, 1 c 2% milk, 1 c orange juice (calcium fortified), 1 med banana |  |
|  | Lunch | Cheeseburger: 1 bun, 1 slice American cheese, 2 oz beef patty, 1 small bag potato chips (~1 oz), 12 oz Diet Mountain Dew, 1 med. orange |  |
|  | Snack | 6 saltines, 2 oz Colby cheese, 12 grapes, water |  |
|  | Dinner | 3 tacos, ¼ c refried beans, 1 c 2% milk |  |
|  | HS | ½ c orange sherbet |  |
| Wednesday | AM | 1½ c Cap'n Crunch cereal, 1 c 2% milk, 1 c orange juice (calcium fortified) |  |
|  | Lunch | 4 tacos, 12 oz Diet Coke |  |
|  | Snack | 6 saltines, 2 tbsp peanut butter, 12 oz Diet Coke |  |
|  | Dinner | 3 oz baked chicken, 1 large baked potato w/ 1 tsp butter & 1 tsp sour cream, 1 c green beans, 2 Fig Newtons |  |
|  | HS | 1½ oz pretzels, 2 tsp mustard, 12 oz caffeine-free Diet Coke |  |
| Thursday | AM | 1½ c Rice Krispies, 1 c 2% milk, 1 c orange juice (calcium fortified), 1 med. banana |  |
|  | Lunch | Meatloaf sandwich on 2 slices whole wheat bread (~3 oz ground beef), 1 c mashed potatoes, ½ c cooked carrots, 1 c 2% milk |  |
|  | Snack | ½ c cottage cheese, 1 c unsweetened canned peaches |  |
|  | Dinner | 3 oz baked pork chop, 1 large baked potato with 2 tsp butter, large tossed salad w/ diet dressing, 12 oz 2% milk, 1 small slice angel food cake (no icing) |  |
|  | HS | ½ c vanilla ice cream (Ben & Jerry's) |  |
| Friday | AM | 1½ c Rice Krispies, 1 c 2% milk, 1 c orange juice (calcium fortified), 1 med banana |  |
|  | Lunch | 6-in personal vegetarian pizza, 12 oz Diet Coke, 1 large apple |  |
|  | Snack | 2 tbs peanut butter, 1 English muffin, toasted, 1 c 2% milk |  |
|  | Dinner | Fried fish sandwich—3 oz fish, 1 bun, baked French fries, 1 c raw broccoli & cauliflower, low-fat ranch dressing, 12 oz 2% milk, 2 Fig Newtons |  |
|  | HS | 3 c popcorn (popped), 12 oz caffeine-free Diet Coke |  |
| Saturday | AM | 3 buttermilk pancakes w/ 2 tbsp maple syrup & 2 tsp butter, 3 strips crisp bacon, 1 c 2% milk, 1 c orange juice (calcium fortified) |  |
|  | Lunch | Chef's salad—~1 oz cubed ham, 1 oz cubed cheddar cheese, 1 oz cubed turkey, low-fat ranch dressing, 12 saltine crackers, 1 c 2% milk, 1 large apple |  |
|  | Snack | 1 c Dairy Queen ice cream (low-fat) in a cone |  |
|  | Dinner | 1 slice deep pan broccoli-and-Canadian-bacon pizza, 1 dinner salad, 1 tsp low-fat Italian dressing, 12 oz Diet Coke |  |
|  | HS | 3 c popped popcorn w/ salt & butter-flavored Pam spray, 6 oz Diet Coke (caffeine free) |  |
| Sunday | AM | 1 slice French toast w/ 2 tbsp maple syrup, 1 c 2% milk, 1 c orange juice (calcium fortified), ½ c sliced strawberries |  |
|  | Lunch | 3 oz fried chicken, 1 c mashed potatoes w/ 1 tbsp gravy, 1 c 2% milk, 1 c cooked carrots, 1 small slice angel food cake |  |
|  | Snack | 1 med banana |  |
|  | Dinner | Italian beef sandwich (2 oz sliced beef on 1 hoagie bun), 2 oz WOW potato chips, 1 dill pickle, 12 oz Diet Coke |  |
|  | HS | 3 c popped popcorn w/ salt & butter-flavored Pam spray, 6 oz Diet Coke (caffeine free) |  |

*Note:* Values are approximate.

**19.** After you have calculated Susan's usual CHO intake from her food record (Question 18), develop a CHO-counting meal plan that she could use. Include menu ideas.

Daily Total:   CHO _____ g
Protein _____ g
Fat _____ g
Kcalories _____

| Time | CHO Choice or Grams CHO | Menu Ideas |
|---|---|---|
| | _____ CHO choices or _____ grams CHO | |
| | _____ oz meat/meat substitutes | |
| | _____ serving fat | |
| | _____ CHO choices or grams | |
| | _____ CHO choices or _____ grams CHO | |
| | _____ oz meat/meat substitutes | |
| | _____ serving fat | |
| | _____ CHO choices or grams | |
| | _____ CHO choices or _____ grams CHO | |
| | _____ oz meat/meat substitutes | |
| | _____ serving fat | |
| | _____ CHO choices or grams | |

**20.** Just before Susan is discharged, her mother asks you, "My friend who owns a health food store told me that Susan should use stevia instead of artificial sweeteners or sugar. What do you think?" What will you tell Susan and her mother?

# Bibliography

American Diabetes Association. Clinical practice recommendations: position statement: standards of medical care for patients with diabetes mellitus. *Diabetes Care* (supp). Available at http://www.diabetes.org/clinicalrecommendations/CareSup1Jan01.htm. Accessed August 25, 2001.

American Diabetes Association Position Statement. Diabetes mellitus and exercise. Available at http://www.diabetes.org/diabetescare/supplement198/S40.htm. Accessed November 28, 1998.

American Diabetes Association Position Statement. Hospital admission guidelines for diabetes mellitus. Clinical practice recommendation. Available at http://www.diabetes.org/diabetescare/supplement/s37.htm. Accessed August 16, 2000.

American Dietetic Association. *Exchange Lists for Meal Planning.* Chicago: American Dietetic Association, 1995.

American Dietetic Association. *Manual of Clinical Dietetics,* 6th ed. Chicago: American Dietetic Association, 2000.

American Dietetic Association and Morrison Health Care. *Medical Nutrition Therapy Across the Continuum of Care.* Chicago: American Dietetic Association, 1998.

Brost B. The effect of menses on blood glucose control: How real? How common? How do you manage? *Diabetes Care and Education Newsflash.* 2000;21(1);6–8.

Commission on Accreditation for Dietetics Education. Knowledge, skills, and competencies for dietitians. *Accreditation Manual for Dietetics Education Programs,* rev. 4th ed. Chicago: American Dietetic Association, 2000.

Daly A, Barry B, Gillespie S, Kulkarni K, Richardson M. *Carbohydrate Counting: Moving On, Level 2.* Chicago: American Dietetic Association, 1995.

Diabetes Care and Education Practice Group. *Meal Planning Approaches for Diabetes Management,* 2nd ed. Chicago: American Dietetic Association, 1994.

Diabetes Control and Complications Trial Research Group. The effect of intensive treatment of diabetes on the development and progression of long-term complications in insulin-dependent diabetes mellitus. *N Engl J Med.* 1983;329(14):977–986.

Fischbach F. *Manual of Laboratory and Diagnostic Tests,* 6th ed. Philadelphia: Lippincott, 2000.

Franz MJ. Medical nutrition therapy for diabetes mellitus and hypoglycemia of nondiabetic origin. In Mahan LK, Escott-Stump S (eds.), *Krause's Food, Nutrition, and Diet Therapy,* 10th ed. Philadelphia: Saunders, 2000.

Goldstein D, Rife D, Derrick K, Kirchoff K. The test with a memory. Diabetes Forecast, April 96. Available at http://www.diabetes.org/diabetesforecast/96apr/memory.htm. Accessed August 16, 2000.

Holler HJ, Pastors JG. *Diabetes Medical Nutrition Therapy: A Professional Guide to Management and Nutrition Education Resources.* Chicago: American Dietetic Association/American Diabetes Association, 1997.

Mayeaux EJ. Diabetes ketoacidosis 2000. Louisiana State University Medical Center. Available at http://lib-sh.Isumc.edu/fammed/intern/dka.html. Accessed August 22, 2000.

Mayeaux EJ. Hemoglobin A1c and blood glucose 2000. Louisiana State University Medical Center. Available at http://lib-sh.Isumc.edu/fammed/intern/diabalc.html. Accessed September 19, 2000.

*Maximizing the Role of Nutrition in Diabetes Management.* Alexandria, VA: American Diabetes Association, 1994.

MedlinePlus health information. Available at http://www.nlm.nih.gov/medlineplus/druginformation.html. Accessed September 8, 2000.

Nonketotic hyperglycemic-hyperosmolar coma. *The Merck Manual.* Available at http://www.merck.com. Accessed August 16, 2000.

Poirier L. Honoring the woman who happens to have diabetes. *Diabetes Spectrum.* 1997;10(3):163–165.

Poirier L, Coburn K. *Women and Diabetes: Life Planning for Health and Wellness.* Alexandria, VA: American Diabetes Association, 1997.

Schardt D. Stevia: a bittersweet tale. *Nutrition Action Newsletter.* April 2000. Available at http://www.cspinet.org/nah/4_00/stevia.html. Accessed September 12, 2000.

Swearingen PL, Ross DG. Endocrine disorders. *Manual of Medical-Surgical Nursing Care,* 4th ed. St. Louis: Mosby, 1999.

# Type 2 Diabetes Mellitus with Hyperosmolar Hyperglycemic Nonketotic (HHNK) Syndrome

*Advanced Practice*

## Objectives

After completing this case, the student will be able to:

1. Integrate the pathophysiology of hyperglycemic hyperosmolar nonketotic (HHNK) syndrome into MNT recommendations.
2. Assess the nutritional status of a patient with HHNK syndrome.
3. Interpret laboratory parameters for nutritional implications and significance.
4. Demonstrate working knowledge of fluid and electrolyte requirements.
5. Manage monitoring of patient's food and nutrient intake.

Mrs. Eileen Douglas, an elderly woman with type 2 diabetes mellitus introduced in Case 27, is admitted to University Hospital after finding her blood glucose to be 905 mg/dL. She is subsequently diagnosed with pneumonia, dehydration, and hyperglycemic hyperosmolar nonketotic (HHNK) syndrome.

# UNIVERSITY HOSPITAL

## ADMISSION DATABASE

**Name:** E. Douglas
**DOB:** 7/27   age 72
**Physician:** R. Case

| BED# 1 | DATE: 6/15 | TIME: 1523 | TRIAGE STATUS (ER ONLY): ☐ Red ☐ Yellow ☐ Green ☐ White |
|---|---|---|---|

**PRIMARY PERSON TO CONTACT:**
Name: Connie Locher
Home #: 555-8217
Work #: 555-7512

### Initial Vital Signs

| TEMP: 99.0 | RESP: 28 | SAO2: |
|---|---|---|

| HT: 5'0" | WT (lb): 125 | B/P: 68/100 | PULSE: 102 |
|---|---|---|---|

| LAST TETANUS 11 years ago | LAST ATE this am | LAST DRANK this am |
|---|---|---|

**ORIENTATION TO UNIT:** ☒ Call light ☒ Television/telephone ☒ Bathroom ☒ Visiting ☒ Smoking ☒ Meals ☒ Patient rights/responsibilities

### CHIEF COMPLAINT/HX OF PRESENT ILLNESS

cough, increased thirst and urination

**PERSONAL ARTICLES:** (Check if retained/describe)
☐ Contacts ☐ R ☐ L      ☒ Dentures ☒ Upper ☒ Lower
☐ Jewelry:
☐ Other:

### ALLERGIES: Meds, Food, IVP Dye, Seafood: Type of Reaction

**VALUABLES ENVELOPE:**
☐ Valuables instructions

### PREVIOUS HOSPITALIZATIONS/SURGERIES

last year-Dx DM; surgical debridement of foot ulcer

**INFORMATION OBTAINED FROM:**
☒ Patient           ☐ Previous record
☐ Family           ☐ Responsible party

Signature  *Eileen Douglas*

### Home Medications (including OTC)   Codes: A = Sent home   B = Sent to pharmacy   C = Not brought in

| Medication | Dose | Frequency | Time of Last Dose | Code | Patient Understanding of Drug |
|---|---|---|---|---|---|
| Captopril | 5o mgl | bid | this am | A | yes |
| | | | | | |
| | | | | | |
| | | | | | |
| | | | | | |
| | | | | | |
| | | | | | |
| | | | | | |
| | | | | | |
| | | | | | |
| | | | | | |

Do you take all medications as prescribed?   ☒ Yes   ☐ No   If no, why?

### PATIENT/FAMILY HISTORY

| | | |
|---|---|---|
| ☒ Cold in past two weeks | ☒ High blood pressure Patient | ☐ Kidney/urinary problems |
| ☐ Hay fever | ☐ Arthritis | ☐ Gastric/abdominal pain/heartburn |
| ☐ Emphysema/lung problems | ☐ Claustrophobia | ☐ Hearing problems |
| ☐ TB disease/positive TB skin test | ☐ Circulation problems | ☐ Glaucoma/eye problems |
| ☐ Cancer | ☐ Easy bleeding/bruising/anemia | ☐ Back pain |
| ☐ Stroke/past paralysis | ☐ Sickle cell disease | ☐ Seizures |
| ☐ Heart attack | ☐ Liver disease/jaundice | ☐ Other |
| ☐ Angina/chest pain | ☐ Thyroid disease | |
| ☐ Heart problems | ☒ Diabetes Sibling, patient | |

### RISK SCREENING

Have you had a blood transfusion?   ☐ Yes   ☒ No
Do you smoke?   ☐ Yes   ☒ No
If yes, how many pack(s) _____ /day for _____ years
Does anyone in your household smoke?   ☐ Yes   ☒ No
Do you drink alcohol?   ☐ Yes   ☒ No
If yes, how often?_____   How much? _____
When was your last drink? ___ / ___
Do you take any recreational drugs?   ☐ Yes   ☒ No
If yes, type:_____   Route _____
Frequency:_____   Date last used:_____/_____/

**FOR WOMEN Ages 12–52**

Is there any chance you could be pregnant?   ☐ Yes   ☐ No
If yes, expected date (EDC):_____/_____/
Gravida/Para: 2/2

**ALL WOMEN**

Date of last Pap smear:  12 / 12 /two years ago
Do you perform regular breast self-exams?   ☐ Yes   ☒ No

**ALL MEN**

Do you perform regular testicular exams?   ☐ Yes   ☐ No

Additional comments:

**x** *Ruth Long, RN*
Signature/Title

**Client name:** Eileen Douglas
**DOB:** 7/27
**Age:** 72
**Sex:** Female
**Education:** Less than high school *What grade/level?* 10th grade
**Occupation:** Homemaker
**Hours of work:** N/A
**Household members:** sister age 81, Dx with type 2 DM 10 years ago. Mrs. Douglas cares for her sister.
**Ethnic background:** African American
**Religious affiliation:** Protestant
**Referring physician:** Richard Case, MD (internal medicine)

**Chief complaint:**
"I've been so thirsty since this cough started. But I hate to drink so much, because it makes me have to go to the bathroom."

**Patient history:**
*Onset of disease:* Mrs. Douglas is a 72-yo widow who lives with her 81-yo sister, whom she cares for. They live in low-income housing in a third-floor walk-up apartment. One year ago, Mrs. Douglas was hospitalized for surgical debridement of an ulcer on her lateral left foot. At that time she complained of frequent bladder infections, blurry vision, and also demonstrated mildly diminished sensation to light touch in her feet. Admitting blood glucose at that time was 325 mg/dL. Hyperglycemia was Tx with SS insulin. Captopril (Capoten) was prescribed to control her HTN. Ciprofloxacin (Cipro) was used to Tx bladder infection and prevent infection of debrided wound. She was d/c with instructions for diabetic education (nutrition and exercise) and a prescription for captopril, which has so far controlled her HTN. Glucose levels were well controlled for six months after d/c as demonstrated by home glucose monitoring records, monthly FBGs, and HgbA$_{1c}$ values. She has not kept her follow-up appointments at the clinic for the past six months. In addition, Mrs. Douglas states she has not been checking her BG or urine acetone levels at home because she can no longer afford to purchase the necessary supplies. On this admission, her blood glucose measures 905 mg/dL, and she is Dx with pneumonia, dehydration, and hyperglycemic hyperosmolar nonketotic (HHNK) syndrome.
*Type of Tx:* Rehydration, normalization of blood glucose, Tx of pneumonia
*PMH:* HTN, type 2 DM Dx 1 year ago
*Meds:*
    Regular insulin, 20 units IV until BG is < 600 mg/dL
    Captopril, 50 mg PO bid
    Isotonic (0.9%) normal saline solution via IV, 2 L over first 2 hours
    Penicillin G, 6 million units/day IV
*Smoker:* No
*Family Hx: What?* HTN, Pt; DM, sister, Pt

**Physical exam:**
*General appearance:* Elderly African American female in mild distress
*Vitals:* Temp 99.0°F, BP 68/100 mm Hg, HR 102 bpm, RR 28 bpm
*Heart:* Tachycardia

*HEENT:*
  *Head:* Normocephalic
  *Eyes:* Wears glasses for myopia, mild retinopathy
  *Ears:* Tympanic membranes
  *Nose:* Dry mucous membranes w/out lesions
  *Throat:* Dry mucous membranes w/out exudates or lesions
*Genitalia:* Normal w/out lesions
*Neurologic:* Mildly confused and irritable, bilateral weakness in hands
*Extremities:* Normal muscular tone for age, normal ROM, nontender
*Skin:* Dry, flushed skin, poor turgor
*Chest/lungs:* Shallow, tachypneic breathing
*Peripheral vascular:* Pulse 2+ bilaterally, cool
*Abdomen:* Audible bowel sounds, soft and nontender, w/out masses or organomegaly

**Nutrition Hx:**
*General:* When Mrs. Douglas's diabetes was Dx last year, a nonkcaloric-restricted, low-fat (< 30% total kcal), high CHO (> 50% total kcal) diet, in combination with a walking program, were prescribed.
*Usual dietary intake:* Unavailable on admission
*24-hr recall:* N/A

*Food allergies/intolerances/aversions* (specify): N/A
*Previous MNT?* Yes    *If yes, when:* 1 year ago
*Where?* University Medical Center—Outpt nutrition education
*Food purchase/preparation:* Self
*Vit/min intake:* None
*Current diet order:* NPO

**Tx plan:**
Infuse 2–3 L of normal (0.9%) saline over 1–2 hr. Measure electrolytes hourly, and assess BG q ½ hour. When BP stabilizes and urine flow is restored, change IV infusion to 0.45% normal saline. Adjust rate in accordance with frequent assessment of BP, cardiovascular status, and balance between I and O. Regular insulin, 20 units IV until BG is < 600 mg/dL; then regular insulin IV or IM until BG reaches 250 mg/dL. Add 5% glucose to IV fluids when BG reaches roughly 250 mg/dL. Adjust insulin dosage according to serial glucose levels and resolution of ketosis.

# UH UNIVERSITY HOSPITAL

NAME: E Douglas                    DOB: 7/27
AGE: 72                            SEX: Female
PHYSICIAN: R. Case

\*\*\*\*\*\*\*\*\*\*\*\*\*\*\*\*\*\*\*\*\*\*\*\*\*\*\*\*\*\*\*\*\*\*\*\*\*\*\*\*\*\*\*\*CHEMISTRY\*\*\*\*\*\*\*\*\*\*\*\*\*\*\*\*\*\*\*\*\*\*\*\*\*\*\*\*\*\*\*\*\*\*\*\*\*\*\*\*\*\*

DAY:                                          Admit          d/c
DATE:
TIME:
LOCATION:

| | NORMAL | Admit | d/c | UNITS |
|---|---|---|---|---|
| Albumin | 3.6–5 | 3.8 | 3.6 | g/dL |
| Total protein | 6–8 | | | g/dL |
| Prealbumin | 19–43 | | | mg/dL |
| Transferrin | 200–400 | | | mg/dL |
| Sodium | 135–155 | 160 | 145 | mEq/L |
| Potassium | 3.5–5.5 | 3.2 | 4.0 | mEq/L |
| Chloride | 98–108 | 94.5 | 99 | mEq/L |
| $PO_4$ | 2.5–4.5 | 1.6 | 2.7 | mEq/L |
| Magnesium | 1.6–2.6 | 1.48 | 1.8 | mEq/L |
| Osmolality | 275–295 | 386 | 280 | mmol/kg $H_2O$ |
| Total $CO_2$ | 24–30 | 32 | 25 | mmol/L |
| Glucose | 70–120 | 905 | 200 | mg/dL |
| BUN | 8–26 | 40 | 27 | mg/dL |
| Creatinine | 0.6–1.3 | 3.2 | 1.4 | mg/dL |
| Uric acid | 2.6–6 (women) | | | mg/dL |
| | 3.5–7.2 (men) | | | |
| Calcium | 8.7–10.2 | | | mg/dL |
| Bilirubin | 0.2–1.3 | | | mg/dL |
| Ammonia ($NH_3$) | 9–33 | | | $\mu$mol/L |
| SGPT (ALT) | 10–60 | | | U/L |
| SGOT (AST) | 5–40 | | | U/L |
| Alk phos | 98–251 | | | U/L |
| CPK | 26–140 (women) | | | U/L |
| | 38–174 (men) | | | |
| LDH | 313–618 | | | U/L |
| CHOL | 140–199 | 350 | 250 | mg/dL |
| HDL-C | 40–85 (women) | | | mg/dL |
| | 37–70 (men) | | | |
| VLDL | | | | mg/dL |
| LDL | < 130 | | | mg/dL |
| LDL/HDL ratio | < 3.22 (women) | | | |
| | < 3.55 (men) | | | |
| Apo A | 101–199 (women) | | | mg/dL |
| | 94–178 (men) | | | |
| Apo B | 60–126 (women) | | | mg/dL |
| | 63–133 (men) | | | |
| TG | 35–160 | | | mg/dL |
| $T_4$ | 5.4–11.5 | | | $\mu$g/dL |
| $T_3$ | 80–200 | | | ng/dL |
| $HbA_{1C}$ | 4.8–7.8 | 10.5 | | % |

# U<sub>H</sub> _UNIVERSITY HOSPITAL_

NAME: E. Douglas                    DOB: 7/27
AGE: 72                             SEX: Female
PHYSICIAN: R. Case

\*\*\*\*\*\*\*\*\*\*\*\*\*\*\*\*\*\*\*\*\*\*\*\*\*\*\*\*\*\*\*\*\*\*\*\*\*\*\*\*\*\*\*HEMATOLOGY\*\*\*\*\*\*\*\*\*\*\*\*\*\*\*\*\*\*\*\*\*\*\*\*\*\*\*\*\*\*\*\*\*\*\*\*\*\*

| | NORMAL | Admit | d/c | UNITS |
|---|---|---|---|---|
| DAY: | | | | |
| DATE: | | | | |
| TIME: | | | | |
| LOCATION: | | | | |
| WBC | 4.3–10 | | | $\times\ 10^3/mm^3$ |
| RBC | 4–5 (women) | | | $\times\ 10^6/mm^3$ |
| | 4.5–5.5 (men) | | | |
| HGB | 12–16 (women) | 20 | 14.5 | g/dL |
| | 13.5–17.5 (men) | | | |
| HCT | 37–47 (women) | 52 | 40 | % |
| | 40–54 (men) | | | |
| MCV | 84–96 | | | fL |
| MCH | 27–34 | | | pg |
| MCHC | 31.5–36 | | | % |
| RDW | 11.6–16.5 | | | % |
| Plt ct | 140–440 | | | $\times\ 10^3$ |
| Diff TYPE | | | | |
| % GRANS | 34.6–79.2 | | | % |
| % LYM | 19.6–52.7 | | | % |
| SEGS | 50–62 | | | % |
| BANDS | 3–6 | | | % |
| LYMPHS | 25–40 | | | % |
| MONOS | 3–7 | | | % |
| EOS | 0–3 | | | % |
| TIBC | 65–165 (women) | | | µg/dL |
| | 75–175 (men) | | | |
| Ferritin | 18–160 (women) | | | µg/dL |
| | 18–270 (men) | | | |
| Vitamin $B_{12}$ | 100–700 | | | pg/mL |
| Folate | 2–20 | | | ng/mL |
| Total T cells | 812–2318 | | | $mm^3$ |
| T-helper cells | 589–1505 | | | $mm^3$ |
| T-suppressor cells | 325–997 | | | $mm^3$ |
| PT | 11–13 | | | sec |

**U**<sub>H</sub> *UNIVERSITY HOSPITAL*

NAME: E. Douglas                          DOB: 7/27
AGE: 72                                    SEX: Female
PHYSICIAN: R. Case

\*\*\*\*\*\*\*\*\*\*\*\*\*\*\*\*\*\*\*\*\*\*\*\*\*\*\*\*\*\*\*\*\*\*\*\*\*\*\*\*URINALYSIS\*\*\*\*\*\*\*\*\*\*\*\*\*\*\*\*\*\*\*\*\*\*\*\*\*\*\*\*\*\*\*\*\*\*\*\*\*\*\*\*

| | NORMAL | Admit | 2 | d/c | UNITS |
|---|---|---|---|---|---|
| DAY: | | Admit | 2 | d/c | |
| DATE: | | | | | |
| TIME: | | | | | |
| LOCATION: | | | | | |
| Coll meth | | Random specimen | First morning | First morning | |
| Color | | Amber | Pale yellow | Straw | |
| Appear | | Slightly hazy | Clear | Clear | |
| Sp grv | 1.003–1.030 | 1.05 | | | |
| pH | 5–7 | | | | |
| Prot | NEG | | | | mg/dL |
| Glu | NEG | pos | | neg | mg/dL |
| Ket | NEG | neg | | neg | |
| Occ bld | NEG | | | | |
| Ubil | NEG | | | | |
| Nit | NEG | | | | |
| Urobil | < 1.1 | | | | EU/dL |
| Leu bst | NEG | | | | |
| Prot chk | NEG | | | | |
| WBCs | 0–5 | | | | /HPF |
| RBCs | 0–5 | | | | /HPF |
| EPIs | 0 | | | | /LPF |
| Bact | 0 | | | | |
| Mucus | 0 | | | | |
| Crys | 0 | | | | |
| Casts | 0 | | | | /LPF |
| Yeast | 0 | | | | |

## Case Questions

1.  Briefly describe hyperglycemic hyperosmolar nonketotic (HHNK) syndrome.

2.  Mrs. Douglas complains of increased thirst and frequent urination. How do these symptoms relate to the pathophysiology of HHNK and to her diabetes?

3.  **a.** How could Mrs. Douglas's blood glucose level become so high without producing ketosis?

    **b.** How could an infection precipitate HHNK?

4.  What is the immediate aim of treatment?

5.  If Mrs. Douglas's HHNK were not treated, how would you expect her disease to progress?

6.  Compare Mrs. Douglas's admitting laboratory values with normal values; provide explanation.

| Parameter | Normal Value | Pt's Value | Explanation |
|---|---|---|---|
| Sodium | 135–155 mEq/L | | |
| Potassium | 3.5–5.5 mEq/L | | |
| Chloride | 98–108 mEq/L | | |
| PO$_4$ | 2.5–4.5 mEq/L | | |
| Magnesium | 1.6–2.6 mEq/L | | |
| Osmolality | 275–295 mmol/kg H$_2$O | | |
| Total CO$_2$ | 24–30 mmol/k | | |
| Glucose | 70–120 mg/dL | | |
| BUN | 8–26 mg/dL | | |
| Creatinine | 0.6–1.3 mg/dL | | |
| Cholesterol | 140–199 mg/dL | | |
| HbA$_{1c}$ | 4.8–7.8% | | |
| Hgb | 12–16 g/dL | | |
| Hct | 37–47% | | |
| Urine glucose | NEG | | |

7. Why wasn't HbA$_{1c}$ measured at discharge?

8. Why does the treatment plan initially specify lowering Mrs. Douglas's blood glucose level to the 250 mg/dL range instead of a normal blood glucose level?

9. Why is regular insulin used to correct hyperglycemia?

10. The goal for healthy elderly patients with diabetes should be near-normal fasting plasma glucose levels without hypoglycemia. Although acceptable glucose control must be carefully individualized, the elderly tend to be predisposed to hypoglycemia. List five factors that predispose elderly patients to hypoglycemia.

11. Dr. Case decides to send Mrs. Douglas home on glipizide (Glucotrol) to help control her blood glucose levels. In what classification of oral hypoglycemics is glipizide?

12. Describe the (medication) action of glipizide.

**13.**    Compare the pharmacologic differences among the oral hypoglycemic agents.

| Characteristic | Biguanides | Sulfonylureas | α-Glucosidase Inhibitors | Meglitinides |
|---|---|---|---|---|
| Brand name(s) (and generic name) | | | | |
| Mechanism of action | | | | |
| Efficacy | | | | |
| Plasma insulin levels | | | | |
| Body weight | | | | |
| Plasma lipids | | | | |
| Side effects and contraindications | | | | |
| Adult daily maintenance dose (mg) | | | | |
| Number of daily doses | | | | |

**14.**    What will be the MNT goal(s) when Mrs. Douglas is discharged?

**15.**    List three behavioral outcomes that you could develop for Mrs. Douglas.

# Bibliography

American Dietetic Association. *Manual of Clinical Dietetics,* 6th ed. Chicago: American Dietetic Association, 2000.

American Dietetic Association and Morrison Health Care. *Medical Nutrition Therapy Across the Continuum of Care.* Chicago: American Dietetic Association, 1998.

Clark CM. Oral therapy in type 2 diabetes: pharmacological properties and clinical use of currently available agents. *Diabetes Spectrum.* 1998;11(4): 211–221. Available at http:// www.diabetes.org/ diabetesspectrum/98v11n4/pg211.htm. Accessed May 11, 1999.

Clinical Practice Recommendations 2000. American Diabetes Association position statement: implications of the United Kingdom prospective diabetes study. *Diabetes Care,* 23(1). Available at http:// journal.diabetes.org/FullText/Supplements/ DiabetesCare/Supplement100/s27.htm. Accessed August 16, 2000.

Commission on Accreditation for Dietetics Education. Knowledge, skills, and competencies for dietitians. *Accreditation Manual for Dietetics Education Programs,* rev. 4th ed. Chicago: American Dietetic Association, 2000.

Fischbach F. *Manual of Laboratory and Diagnostic Tests,* 6th ed. Philadelphia: Lippincott, 2000.

Franz MJ. Medical nutrition therapy for diabetes mellitus and hypoglycemia of nondiabetic origin. In Mahan LK, Escott-Stump S (eds.), *Krause's Food, Nutrition, and Diet Therapy,* 10th ed. Philadelphia: Saunders, 2000.

Goldstein D, Rife D, Derrick K, Kirchoff K. The test with a memory. *Diabetes Forecast,* April 96. Available at http://www.diabetes.org/diabetesforecast/96apr/ memory.htm. Accessed August 16, 2000.

Holler HJ, Pastors JG. *Diabetes Medical Nutrition Therapy: A Professional Guide to Management and Nutrition Education Resources.* Chicago: American Dietetic Association/American Diabetes Association, 1997.

*Maximizing the Role of Nutrition in Diabetes Management.* Alexandria, VA: American Diabetes Association, 1994.

MedlinePlus health information. Available at http:// www.nlm.nih.gov/medlineplus/druginformation .html. Accessed September 8, 2000.

Nonketotic hyperglycemic-hyperosmolar coma. *The Merck Manual.* Available at http://www.merck.com. Accessed August 16, 2000.

Setter SM. New drug therapies for the treatment of diabetes. *Diabetes Care and Education on the Cutting Edge.* 1998;19(2):3–7.

Swearingen PL, Ross DG. Endocrine disorders. *Manual of Medical-Surgical Nursing Care,* 4th ed. St. Louis: Mosby, 1999.

Wallace JI. Management of diabetes in the elderly. *Clin Diabetes.* 17(1);1999. Available at http:// www.diabetes.org/clinicaldiabetes/v17n111999/ Pg19.htm. Accessed August 16, 2000.

White JR. The pharmacological reduction of blood glucose in patients with type 2 diabetes mellitus. *Clin Diabetes.* 1998;16(2):58. Available at http:// www.diabetes.org/clinicaldiabetes/v16n21998/ pg58.htm. Accessed May 11, 1999.

## Case 30

# Metabolic Disorder: Phenylketonuria

*Laurie Bernstein, MS, RD, FADA*

*Introductory Level*

## Objectives

After completing this case, the student will be able to:

1. Integrate the principles of pathophysiology of the disease to support principles of medical nutrition therapy in phenylketonuria.
2. Evaluate principles and goals of nutrition management.
3. Develop appropriate nutrition prescription.
4. Interpret pertinent lab values.
5. Describe the genetics of autosomal recessive disorders.
6. Establish care plan to support medical nutrition therapy.

Laura Mayberry is a nine-day-old newborn admitted through the Metabolic Clinic at University Hospital. Laura had a positive newborn screen for PKU from a specimen collected at two days of age with a phenylalanine of 20 mg%. Her diagnostic follow-up levels collected on day 8 were as follows: phenylalanine 44.4 mg/dL and tyrosine 1.2 mg/dL. Based on these levels, a presumptive diagnosis of classic PKU was made.

 **UNIVERSITY HOSPITAL**

## ADMISSION DATABASE

Name: Laura Mayberry
DOB: 7/8 age 9 days
Physician: M. Bissell

| BED# 1 | DATE: 7/19 | TIME: 1300 | TRIAGE STATUS (ER ONLY): ☐ Red ☐ Yellow ☐ Green ☐ White |
|---|---|---|---|

### Initial Vital Signs

| TEMP: 98.6 | RESP: 16 | SAO2: | |
|---|---|---|---|
| HT: 18.5 | WT (lb): 5.8 | B/P: | PULSE: |
| LAST TETANUS | | LAST ATE | LAST DRANK |

**PRIMARY PERSON TO CONTACT:**
Name: Amber Mayberry
Home #: 314-224-7845
Work #:

ORIENTATION TO UNIT: ☒ Call light ☒ Television/telephone ☒ Bathroom ☒ Visiting ☒ Smoking ☒ Meals ☒ Patient rights/responsibilities

### CHIEF COMPLAINT/HX OF PRESENT ILLNESS

Mother states that her pediatrician asked her to bring in her newborn because there was a PKU test run on the baby in the hospital and it was positive

**PERSONAL ARTICLES:** (Check if retained/describe)
☐ Contacts ☐ R ☐ L ☐ Dentures ☐ Upper ☐ Lower
☐ Jewelry:
☐ Other:

**ALLERGIES: Meds, Food, IVP Dye, Seafood: Type of Reaction**

NKA

**VALUABLES ENVELOPE:**
☐ Valuables instructions

### PREVIOUS HOSPITALIZATIONS/SURGERIES

Live birth-9 days ago

**INFORMATION OBTAINED FROM:**
☐ Patient ☐ Previous record
☒ Family ☐ Responsible party

Signature *Amber Mayberry*

| Home Medications (including OTC) | | Codes: A = Sent home | B = Sent to pharmacy | | C = Not brought in |
|---|---|---|---|---|---|
| Medication | Dose | Frequency | Time of Last Dose | Code | Patient Understanding of Drug |
| | | | | | |
| | | | | | |
| | | | | | |
| | | | | | |
| | | | | | |
| | | | | | |
| | | | | | |
| | | | | | |
| | | | | | |
| | | | | | |
| | | | | | |

Do you take all medications as prescribed? ☐ Yes ☐ No   If no, why? N/A

### PATIENT/FAMILY HISTORY

☐ Cold in past two weeks
☐ Hay fever
☐ Emphysema/lung problems
☐ TB disease/positive TB skin test
☐ Cancer
☐ Stroke/past paralysis
☐ Heart attack
☐ Angina/chest pain
☐ Heart problems

☐ High blood pressure
☐ Arthritis
☐ Claustrophobia
☐ Circulation problems
☐ Easy bleeding/bruising/anemia
☐ Sickle cell disease
☐ Liver disease/jaundice
☐ Thyroid disease
☐ Diabetes Maternal grandmother

☐ Kidney/urinary problems
☐ Gastric/abdominal pain/heartburn
☐ Hearing problems
☐ Glaucoma/eye problems
☐ Back pain
☐ Seizures
☐ Other

### RISK SCREENING

Have you had a blood transfusion? ☐ Yes ☒ No
Do you smoke? ☐ Yes ☐ No
If yes, how many pack(s)   /day for   years
Does anyone in your household smoke? ☐ Yes ☐ No
Do you drink alcohol? ☐ Yes ☐ No
If yes, how often?   How much?
When was your last drink?   /   /
Do you take any recreational drugs? ☐ Yes ☐ No
If yes, type:   Route
Frequency:   Date last used:   /   /

**FOR WOMEN Ages 12–52**

Is there any chance you could be pregnant? ☐ Yes ☐ No
If yes, expected date (EDC):   /   /
Gravida/Para:

**ALL WOMEN**

Date of last Pap smear:   /   /
Do you perform regular breast self-exams? ☐ Yes ☐ No

**ALL MEN**

Do you perform regular testicular exams? ☐ Yes ☐ No

Additional comments:

✗ *Gary Jenkins, RN, BSN*
Signature/Title

**Client name:** Laura Mayberry
**DOB:** 7/8
**Age:** 9 days
**Sex:** Female
**Education:** N/A—newborn
**Occupation:** N/A
**Hours of work:** N/A
**Household members:** Mother age 31, father age 36, brother age 5, sister age 7
**Ethnic background:** Caucasian
**Religious affiliation:** Catholic
**Referring physician:** Michelle Bissel, MD (endocrinologist)

## Chief complaint:

Mother states that her pediatrician asked her to bring in her newborn because there was a PKU test run on the baby in the hospital and it was positive.

## Patient history:

*Onset of disease:* Laura is a 9-day-old female infant seen with her parents in the metabolic clinic. Laura has a diagnosis of phenylketonuria (PKU). Laura had a positive newborn screen for PKU from a specimen collected at 2 days of age with a phenylalanine of 20 mg%. Her diagnostic follow-up levels collected on day 8 were as follows: phenylalanine 44.4 mg/dL and tyrosine 1.2 mg/dL. Based on these levels, a presumptive diagnosis of classic PKU was made.
*Type of Tx:* None at present—routine pediatric care at birth
*PMH:* Laura is the result of her mother's unplanned pregnancy. Mother received routine prenatal care after approximately 1 month and describes the pregnancy as normal. Her only medications included prenatal vitamins and iron. By mother's report, Laura was delivered vaginally at 40 weeks gestation without difficulty. Laura was 5.5 lb at birth and was 18.5 inches long.
*Family Hx:* Laura's parents reported no family history of learning disabilities, mental retardation, or PKU. There is no consanguinity.

## Physical exam:

*General appearance:* Well 9-day-old female infant
*Vitals:* All WNL
*Heart:* Normal
*HEENT:* Unremarkable
*Neurologic:* Normal infant reflexes—positive Babinski, Moro, suck, rooting, plantar, and grasp
*Extremities:* Normal tone and strength
*Skin:* Warm, dry
*Chest/lungs:* Normal
*Peripheral vascular:* Normal
*Abdomen:* Positive bowel sounds in all quadrants

## Nutrition Hx:

*General:* Consuming soy formula—approximately 4 oz every 3–4 hours
*Food purchase/preparation:* Parents
*Vit/min intake:* None

Growth parameters obtained on the day of the visit were as follows: weight 5.8 lb, length 18.5 in, and HC 13 in

**Tx plan:**
Meet with metabolic nutritionist to establish metabolic prescription to reduce phenylalanine levels to treatment range of 2–6 mg/dL (120–360 μmol/L). The initial diet prescription for Laura was 52 grams of Phenex-1 and 15 grams of Similac powder with iron. This prescription yields 25mg/kg of phenylalanine, 316 mg/kg of tyrosine, 3.5 grams of protein/kg, and 122 kcal/kg. At a total of 16 ounces (480 mL), this prescription yields 20 kcal/oz.

## Case Questions

1. What is phenylketonuria?

2. It is common for a parent to ask if something happened during the pregnancy that could have caused the baby to be born with PKU. What would you tell a parent in this situation?

3. What is the screening procedure for phenylketonuria?

4. Why is it important to accomplish screening as soon after birth as possible?

5. What laboratory tests are used to screen for PKU?

6. Why does a baby with PKU have elevated phenylalanine levels?

7. What is the goal of the PKU diet?

8. During infancy, what is the major component of the diet to treat phenylketonuria?

9. What are "good" blood phenylalanine levels?

10. How long will this infant have to remain on a special diet?

11. Why do we measure tyrosine and phenylalanine?

12. Using the pediatric growth charts in Appendix D, evaluate Laura's growth at this visit.

13. Determine energy, protein, and fluid needs for Laura.

14.    Who should prescribe the medical nutrition therapy for the infant with phenylketonuria?

15.    Phenylalanine should be prescribed at 25–70 mg/kg, and tyrosine should be prescribed at 300–350 mg/kg for 0 to <3mo. Convert Laura's weight to kilograms, then calculate her requirements.

16.    What is the composition of the special metabolic formulas that are available to treat PKU? What are examples of these formulas?

17.    Is the special metabolic formula supplemented in any way? Explain.

18.    How will Laura be monitored after starting on the special diet?

19.    What should be monitored to evaluate the efficacy of the nutrition support? Identify appropriate outcomes for Laura.

20.    In general, how will Laura's diet change as she grows older?

# Bibliography

Acosta PB. The contribution of therapy of inherited amino acid disorders to knowledge of amino requirements. In Wapnir RA (ed.), *Congenital Metabolic Diseases: Diagnosis and Treatment.* New York: Dekker, 1985.

Acosta PB, Wenz E, Williamson M. Nutrient intakes of treated infants with phenylketonuria. *Am J Clin Nutr.* 1977;30:198–208.

Acosta PB, Yannicelli S. Protein intake affects phenylalanine requirements and growth of infants with phenylketonuria. *Acta Paediatr.* 1994; 407(suppl): 66–67.

Acosta PB, Yannicelli S. *The Ross Metabolic Formula System Nutrition Support Protocols,* 3rd ed. Columbus, OH: Ross Laboratories, 1993.

Allen JR, McCauley JC, Waters DL, et al: Fasting energy expenditure in children with phenylketonuria. *Am J Clin Nutr.* 1995;62:797–801.

Anderson VE, Siegel FS. Behavioral and biochemical correlates of diet change in phenylketonuria. *Pediatr Res.* 1976;10:10–17.

Behrman RE, Kleigman RM, Arvin M. *Nelson Textbook of Pediatrics,* 15th ed. Philadelphia: Saunders, 1996.

Bernstein LE, Freehauf CF. *Eat Right, Stay Bright Guide for Hyperphenylalaninemia.* Gaithersburg, MD: Scientific Hospital Supplies International, 2000.

Bohles H, Ullrich K, Endres W, et al: Inadequate iron availability as a possible cause of low serum carnitine concentrations in patients with phenylketonuria. *Eur J Pediatr.* 1991;150:425–428.

Doherty LB, Rohr R, Levy HL. Detection of phenylketonuria in the very early newborn blood specimen. *Pediatrics.* 1991;87:240–244.

Elsas LJ, Acosta PB. Nutrition support of inherited metabolic diseases. In Shils ME, et al. (eds.), *Modern Nutrition in Health and Disease,* 8th ed. Philadelphia: Lea & Febiger, 1994.

Food and Nutrition Board, Committee on Dietary Allowances. *Recommended Dietary Allowances,* 9th and 10th eds. Washington, DC: National Academy of Sciences, 1980 and 1989.

Gropper S, Acosta PB. Effect of simultaneous ingestion of L-amino acids and whole protein on plasma amino acid and urea nitrogen concentrations in humans. *JPEN.* 1991;15:48–53.

Gropper SS, Gropper DM, Acosta PB. Plasma amino acid response to ingestion of L-amino acids and whole protein. *J Pediatr Gastroent Nutr.* 1993;16:143–150.

Gropper SS, Trahms C, Cloud HA, et al. Iron deficiency without anemia in children with phenylketonuria. *Int Pediatr.* 1994;9:237–243.

Gropper SS, Yannicelli S. Plasma molybdenum concentrations in children with and without phenylketonuria. *Biol Trace Elem Res.* 1993;38:227–231.

Hanley WB, Linsao L, Davidson W, et al. Malnutrition with early treatment of phenylketonuria. *Pediatr Res.* 1970;4:318–327.

Kindt E, Motzfeldt K, Halvorsen S, Lie S. Protein requirements in infants and children treated for phenylketonuria. *Am J Clin Nutr.* 1983;37:778–785.

Kindt E, Motzfeldt K, Halvorsen S, Lie SO. Is phenylalanine requirement in infants and children related to protein intake? *Br J Nutr.* 1984;51:435–442.

Martin SB, Acosta PB. Osmotic behaviors of components of chemically-defined formulas. *J Pediatr Perinat Nutr.* 1987;1:1–17.

Michals K, Lopus M, Matalon R. Phenylalanine metabolites as indicators of dietary compliance in children with phenylketonuria. *Biochem Med.* 1988;39:18–23.

Nyerges H. (ed). *Newborn Screening Practitioner's Manual.* Denver, CO: Mountain States Regional Genetic Services Network, 1990.

Reilly C, Barrett JE, Patterson CM, et al. Trace element nutrition status and dietary intake of children with phenylketonuria. *Am J Clin Nutr.* 1990;52:159–165.

Schoeffer A, Hermann ME, Broesicke HG, Moench E. Effect of dosage and timing of amino acid mixtures on nitrogen retention in patients with phenylketonuria. *J Nutr Med.* 1994;4:415–418.

Scriver CR, Kaufman S, Eisensmith RC, Woo SLC. The hyperphenylalaninemias. In Scriver CR, et al. (eds.), *The Metabolic and Molecular Bases of Inherited Disease,* 7th ed. New York: McGraw-Hill, 1995.

Shortland D, Smith I, Francis DEM, et al. Amino acid and protein requirements in a preterm infant with classic phenylketonuria. *Arch Dis Child.* 1985;60:263–265.

Sibinga MS, Friedman CJ, Steisel IM, Baker EC. The depressing effect of diet on physical growth in phenylketonuria. *Develop Med Child Neurol.* 1971;13:63–70.

Smith I, Beasley MG, Ades AE. Effect on intelligence of relaxing the low phenylalanine diet in phenylketonuria. *Arch Dis Child.* 1990;65:311–316.

Spika JS, Shaffer N, Hargrett-Bean N, et al. Risk factors for infant botulism in the United States. *Am J Dis Child.* 1989;143:828–832.

Yannicelli S, Ernest A, Neifert MR, McCabe ERB. *Guide to Breastfeeding the Infant with PKU,* 2nd ed. Washington, DC: Department of Health and Human Services. Human Services Publication No. HRS-M-CH88-12, October 1988.

# Unit Eight

# MEDICAL NUTRITION THERAPY FOR RENAL DISEASE

## Introduction

There has been a noticeable growth in the field of medical nutrition therapy in patients with renal disease. The importance of nutrition in the care of patients with chronic renal failure is illustrated by the fact that indicators of nutritional status effectively predict morbidity and mortality in these patients.

Kidney and urologic diseases affect approximately 20 million Americans. Of this number, more than 50,000 Americans die each year because of kidney disease, and more than 260,000 Americans suffer from end-stage renal disease (ESRD) and need renal replacement therapy to stay alive. Kidney disease is one of the costliest illnesses in the United States today. Kidney and urologic diseases remain one of the foremost causes of work lost among men and women. Each year roughly 27 million physician visits are made, more than 6 million hospitalizations occur, and nearly 2.5 million procedures result from kidney and urologic problems.

The primary cause of ESRD is diabetes mellitus, accounting for about 35% of all new cases each year and 25% of all cases. Uncontrolled hypertension is the second leading cause of ESRD in the United States, accounting for ap-

proximately 30% of U.S. cases. Researchers from the National Institutes of Health have established that ESRD caused by diabetes mellitus is anywhere from 10 to 75 times more prevalent in Native Americans than in whites, and the prevalence differs among tribes. Among Pima Indians age 35 and over, 50% have type 2 diabetes mellitus—the highest rate in the world.

Each case presented in this section can stand alone or be used in tandem to illustrate progression of renal disease from impaired renal function to ESRD. Integrated into each case are aspects that predispose toward ESRD, such as diabetes mellitus and ethnicity. Fundamental principles such as modification of nutrient composition in impaired renal function, ESRD, and renal replacement therapy are included.

Maturity-onset diabetes of the young (MODY) is touched on in these cases, as well as results from the Modification of Diet in Renal Disease Study, the largest randomized, multicenter clinical trial designed to evaluate effects of dietary protein and phosphorus and of blood pressure control on the progression rate of chronic renal insufficiency.

# Impaired Renal Function

*Introductory Level*

*With assistance from Sandy Dunning, MS, RD, LD*

## Objectives

After completing this case, the student will be able to:

1. Integrate physiology of impaired renal function into MNT recommendations.
2. Interpret laboratory parameters for nutritional implications and significance.
3. Assess fluid and electrolyte requirements.
4. Manage monitoring of patient's food and nutrient intake.
5. Determine appropriate nutrient requirements.
6. Develop appropriate MNT goals.
7. Using medical nutritional data, make appropriate documentation in the medical record.
8. Integrate sociocultural and ethnic food consumption issues within nutrition care plan.

Enez Joaquin is a 24-year-old Pima Indian who has had type 2 diabetes mellitus since age 13. Mrs. Joaquin is admitted because of a declining glomerular filtration rate. Medical care for impaired renal function will now be determined.

**UNIVERSITY HOSPITAL**

# ADMISSION DATABASE

Name: E. Joaquin
DOB: 4/14  age 24
Physician: L. Nila

| BED#<br>2 | DATE:<br>4/8 | TIME:<br>1346 | TRIAGE STATUS (ER ONLY):<br>☐ Red ☐ Yellow ☐ Green ☐ White |
|---|---|---|---|

**PRIMARY PERSON TO CONTACT:**
Name: Eddie Joaquin—husband
Home #: 555-3947
Work #: N/A

### Initial Vital Signs

| TEMP:<br>98.6 | RESP:<br>25 | SAO2: |
|---|---|---|

ORIENTATION TO UNIT: ☒ Call light ☒ Television/telephone
☒ Bathroom ☒ Visiting ☒ Smoking ☒ Meals
☒ Patient rights/responsibilities

| HT:<br>5'0" | WT (lb):<br>140 | B/P:<br>140/80 | PULSE:<br>86 |
|---|---|---|---|

| LAST TETANUS<br>2 years ago | LAST ATE<br>breakfast | LAST DRANK<br>1 hour ago |
|---|---|---|

### CHIEF COMPLAINT/HX OF PRESENT ILLNESS

Having problems urinating

**PERSONAL ARTICLES:** (Check if retained/describe)
☐ Contacts ☐ R ☐ L       ☐ Dentures ☐ Upper ☐ Lower
☐ Jewelry:
☐ Other:

### ALLERGIES: Meds, Food, IVP Dye, Seafood: Type of Reaction

**VALUABLES ENVELOPE:** No
☐ Valuables instructions

### PREVIOUS HOSPITALIZATIONS/SURGERIES

childbirth 5 years ago by midwife

**INFORMATION OBTAINED FROM:**
☒ Patient          ☐ Previous record
☐ Family           ☐ Responsible party

Signature *Enex Joaquin*

### Home Medications (including OTC)      Codes: A=Sent home      B=Sent to pharmacy      C=Not brought in

| Medication | Dose | Frequency | Time of Last Dose | Code | Patient Understanding of Drug |
|---|---|---|---|---|---|
| Glucophage | 850 mg | bid | yesterday | C | no |
|  |  |  |  |  |  |
|  |  |  |  |  |  |
|  |  |  |  |  |  |
|  |  |  |  |  |  |
|  |  |  |  |  |  |
|  |  |  |  |  |  |
|  |  |  |  |  |  |
|  |  |  |  |  |  |
|  |  |  |  |  |  |
|  |  |  |  |  |  |

Do you take all medications as prescribed?   ☐ Yes   ☒ No   If no, why?  I forget.

### PATIENT/FAMILY HISTORY

| | | |
|---|---|---|
| ☐ Cold in past two weeks | ☒ High blood pressure Patient | ☒ Kidney/urinary problems Patient |
| ☐ Hay fever | ☐ Arthritis | ☐ Gastric/abdominal pain/heartburn |
| ☐ Emphysema/lung problems | ☐ Claustrophobia | ☐ Hearing problems |
| ☐ TB disease/positive TB skin test | ☐ Circulation problems | ☐ Glaucoma/eye problems |
| ☐ Cancer | ☐ Easy bleeding/bruising/anemia | ☐ Back pain |
| ☐ Stroke/past paralysis | ☐ Sickle cell disease | ☐ Seizures |
| ☐ Heart attack | ☐ Liver disease/jaundice | ☐ Other |
| ☐ Angina/chest pain | ☐ Thyroid disease | |
| ☐ Heart problems | ☒ Diabetes Patient | |

### RISK SCREENING

Have you had a blood transfusion?   ☐ Yes   ☒ No
Do you smoke?   ☐ Yes   ☒ No
If yes, how many pack(s)   /day for   years
Does anyone in your household smoke?   ☒ Yes   ☐ No
Do you drink alcohol?   ☒ Yes   ☐ No
If yes, how often? weekends  How much? 1-2 beers
When was your last drink?   4 / 7 /
Do you take any recreational drugs?   ☐ Yes   ☒ No
If yes, type:_____   Route
Frequency:_____   Date last used:_____/_____/

**FOR WOMEN Ages 12–52**

Is there any chance you could be pregnant?   ☐ Yes   ☒ No
If yes, expected date (EDC):
Gravida/Para: 1/1

**ALL WOMEN**

Date of last Pap smear: 2/15
Do you perform regular breast self-exams?   ☐ Yes   ☒ No

**ALL MEN**

Do you perform regular testicular exams?   ☐ Yes   ☐ No

Additional comments:

✗ *Marlene Matten, MS, RN*
Signature/Title

**Client name:**  Enez Joaquin
**DOB:**  4/14
**Age:**  24
**Sex:**  Female
**Education:**  High school
**Occupation:**  Secretary
**Hours of work:**  9 am–5 pm
**Household members:**  Husband age 24, type 2 diabetes under control; daughter age 5, in good health
**Ethnic background:**  Pima Indian
**Religious affiliation:**  Catholic
**Referring physician:**  Lourdes Nila, MD (nephrology)

**Chief complaint:**
"My doctor is worried about some of my labs and that I might have kidney problems."

**Patient history:**
Mrs. Joaquin is a 24-yo Native American woman who was diagnosed with type 2 DM when she was 13 years old and has been poorly compliant with prescribed treatment. She is from the Pima Indian tribe of southern Arizona. She is the product of a full-term pregnancy and weighed 11 lb at birth. She lives with her husband and 5-year-old daughter. Her husband also has type 2 DM. He was diagnosed at age 18. Her renal function has been monitored for the past five years. Her glomerular filtration rate has been declining over the past year.
*Onset of disease:* In last year
*Type of Tx:* Medical treatment for type 2 DM
*PMH:* Gravida 1/para 1. Infant weighed 10 lb at birth five years ago.
*Meds:* Glucophage (metformin) 850 mg bid
*Smoker:* No
*Family Hx:* Both mother and father diagnosed with DM

**Physical exam:**
*General appearance:* Overweight Native American female who appears her age
*Vitals:* Temp 98.6°F, BP 140/80 mm Hg, HR 86 bpm, RR 25 bpm
*Heart:* S4, S1, and S2, regular rate and rhythm. I/VI systolic ejection murmur, upper left sternal border
*HEENT:* Normocephalic, equal carotid pulses, neck supple, no bruits
　*Eyes:* PERRLA
　*Ears:* Noncontributory
　*Nose:* Noncontributory
　*Throat:* Noncontributory
*Genitalia:* Normal female
*Neurologic:* Oriented to person, place and time; intact, mild asterixis
*Extremities:* 1+ pitting generalized edema
*Skin:* Dry and yellowish-brown
*Chest/lungs:* Generalized rhonchi with rales that are mild at the bases (Pt breathes with poor effort)
*Peripheral vascular:* Normal pulse (3+) bilaterally
*Abdomen:* Bowel sounds positive, soft; generalized mild tenderness; no rebound

**Nutrition Hx:**
*General:* States that appetite is good and that she doesn't follow any special diet.
*Usual dietary intake:*
| | |
|---|---|
| *Breakfast:* | 1 fried egg |
| | 3 strips fried bacon |
| | Fried potatoes (approx 1 potato) |
| | 1 slice bread with butter |
| *Lunch:* | 2 tamales with chili con carne |
| | Fry bread |
| | 1 can Coke |
| *PM snack:* | Approx 25 potato chips |
| | 1 can Coke |
| *Dinner:* | 3 tacos made with ground beef, chopped tomatoes, chopped onion, chopped lettuce and 3 tortillas (flour) |
| | Coke |
| *HS snacks:* | Crackers and peanut butter |
| *24-hr recall:* | N/A |

*Food allergies/intolerances/aversions:* None
*Previous MNT?* Yes   *If yes, when:* 1200 kcal exchange list diet; 11 years ago when Pt Dx with DM
*Where?* Reservation Health Service
*Food purchase/preparation:* Self
*Vit/min intake:* None
*Current diet order:* 35 kcal/kg, 0.8 g protein/kg, 8–12 mg phosphorus/kg, 2–3 g Na

**Tx plan:**
Captopril 25 mg tid
Glucophage (metformin) 850 mg bid
CBC, chemistry
Closely monitor and document I & O
Daily weights
*Diet Rx:* 0.8 g pro/kg, 30–35 kcal/kg, 3 g Na, 8–12 mg phosphorus/kg
Vitamin/mineral supplement

# U<sub>H</sub> UNIVERSITY HOSPITAL

NAME: E. Joaquin                          DOB: 4/14
AGE: 24                                   SEX: Female
PHYSICIAN: L. Nila


**\*\*\*\*\*\*\*\*\*\*\*\*\*\*\*\*\*\*\*\*\*\*\*\*\*\*\*\*\*\*\*\*\*\*\*\*\*\*\*\*\*\*\*\*CHEMISTRY\*\*\*\*\*\*\*\*\*\*\*\*\*\*\*\*\*\*\*\*\*\*\*\*\*\*\*\*\*\*\*\*\*\*\*\*\*\*\*\*\***

DAY:                                      Admit
DATE:
TIME:
LOCATION:

| | NORMAL | | UNITS |
|---|---|---|---|
| Albumin | 3.6–5 | 3.2 | g/dL |
| Total protein | 6–8 | 6 | g/dL |
| Prealbumin | 19–43 | 19 | mg/dL |
| Transferrin | 200–400 | 250 | mg/dL |
| Sodium | 135–155 | 150 | mmol/L |
| Potassium | 3.5–5.5 | 5.2 | mmol/L |
| Chloride | 98–108 | 100 | mmol/L |
| $PO_4$ | 2.5–4.5 | 4.5 | mmol/L |
| Magnesium | 1.6–2.6 | 2.4 | mmol/L |
| Osmolality | 275–295 | 400 | mmol/kg $H_2O$ |
| Total $CO_2$ | 24–30 | 25 | mmol/L |
| Glucose | 70–120 | 205 | mg/dL |
| BUN | 8–26 | 80.6 | mg/dL |
| Creatinine | 0.6–1.3 | 1.5 | mg/dL |
| Uric acid | 2.6–6 (women) | 6.2 | mg/dL |
| | 3.5–7.2 (men) | | |
| Calcium | 8.7–10.2 | 9.6 | mg/dL |
| Bilirubin | 0.2–1.3 | 1.1 | mg/dL |
| Ammonia ($NH_3$) | 9–33 | 25 | $\mu$mol/L |
| SGPT (ALT) | 10–60 | 45 | U/L |
| SGOT (AST) | 5–40 | 39 | U/L |
| Alk phos | 98–251 | 198 | U/L |
| CPK | 26–140 (women) | 138 | U/L |
| | 38–174 (men) | | |
| LDH | 313–618 | 583 | U/L |
| CHOL | 140–199 | 443 | mg/dL |
| HDL–C | 40–85 (women) | 37 | mg/dL |
| | 37–70 (men) | | |
| VLDL | | | mg/dL |
| LDL | < 130 | 132 | mg/dL |
| LDL/HDL ratio | < 3.22 (women) | | |
| | < 3.55 (men) | | |
| Apo A | 101–199 (women) | | mg/dL |
| | 94–178 (men) | | |
| Apo B | 60–126 (women) | | mg/dL |
| | 63–133 (men) | | |
| TG | 35–160 | 300 | mg/dL |
| $T_4$ | 5.4–11.5 | | $\mu$g/dL |
| $T_3$ | 80–200 | | ng/dL |
| $HbA_{1C}$ | 4.8–7.8 | 8.2 | % |

# U H  *UNIVERSITY HOSPITAL*

NAME: E. Joaquin                    DOB: 4/14
AGE: 24                             SEX: Female
PHYSICIAN: L. Nila

\*\*\*\*\*\*\*\*\*\*\*\*\*\*\*\*\*\*\*\*\*\*\*\*\*\*\*\*\*\*\*\*\*\*\*\*\*\*\*\*\*HEMATOLOGY\*\*\*\*\*\*\*\*\*\*\*\*\*\*\*\*\*\*\*\*\*\*\*\*\*\*\*\*\*\*\*\*\*\*\*\*\*\*

DAY:                                             Admit
DATE:
TIME:
LOCATION:

| | NORMAL | | UNITS |
|---|---|---|---|
| WBC | 4.3–10 | 4.5 | $\times 10^3/mm^3$ |
| RBC | 4–5 (women) | 5 | $\times 10^6/mm^3$ |
| | 4.5–5.5 (men) | | |
| HGB | 12–16 (women) | 9.7 | g/dL |
| | 13.5–17.5 (men) | | |
| HCT | 37–47 (women) | 28.4 | % |
| | 40–54 (men) | | |
| MCV | 84–96 | 90 | fL |
| MCH | 27–34 | 32 | pg |
| MCHC | 31.5–36 | 32.5 | % |
| RDW | 11.6–16.5 | 12.6 | % |
| Plt ct | 140–440 | 250 | $\times 10^3$ |
| Diff TYPE | | | |
| % GRANS | 34.6–79.2 | | % |
| % LYM | 19.6–52.7 | 35.7 | % |
| SEGS | 50–62 | 52 | % |
| BANDS | 3–6 | 4 | % |
| LYMPHS | 25–40 | 35 | % |
| MONOS | 3–7 | 6 | % |
| EOS | 0–3 | 1 | % |
| TIBC | 65–165 (women) | 160 | μg/dL |
| | 75–175 (men) | | |
| Ferritin | 18–160 (women) | 148 | μg/dL |
| | 18–270 (men) | | |
| Vitamin B$_{12}$ | 100–700 | | pg/mL |
| Folate | 2–20 | | ng/mL |
| Total T cells | 812–2318 | | $mm^3$ |
| T-helper cells | 589–1505 | | $mm^3$ |
| T-suppressor cells | 325–997 | | $mm^3$ |
| PT | 11–13 | | sec |

# U<sub>H</sub> UNIVERSITY HOSPITAL

NAME: E. Joaquin                           DOB: 4/14
AGE: 24                                      SEX: Female
PHYSICIAN: L. Nila

\*\*\*\*\*\*\*\*\*\*\*\*\*\*\*\*\*\*\*\*\*\*\*\*\*\*\*\*\*\*\*\*\*\*\*\*\*\*\*\*\*\*\*\*\*\*URINALYSIS\*\*\*\*\*\*\*\*\*\*\*\*\*\*\*\*\*\*\*\*\*\*\*\*\*\*\*\*\*\*\*\*\*\*\*\*\*\*\*\*\*\*\*\*\*\*

| | NORMAL | Admit | 2 | d/c | UNITS |
|---|---|---|---|---|---|
| DAY: | | Admit | 2 | d/c | |
| DATE: | | | | | |
| TIME: | | | | | |
| LOCATION: | | | | | |
| Coll meth | | Random specimen | First morning | First morning | |
| Color | | Straw | Pale yellow | Pale yellow | |
| Appear | | Slightly hazy | Clear | Clear | |
| Sp grv | 1.003-1.030 | 1.1 | | | |
| pH | 5-7 | 7 | | | |
| Prot | NEG | 3+ | | | mg/dL |
| Glu | NEG | POS | | | mg/dL |
| Ket | NEG | POS | | | |
| Occ bld | NEG | NEG | | | |
| Ubil | NEG | NEG | | | |
| Nit | NEG | NEG | | | |
| Urobil | < 1.1 | 0.8 | | | EU/dL |
| Leu bst | NEG | | | | |
| Prot chk | NEG | | | | |
| WBCs | 0-5 | 0 | | | /HPF |
| RBCs | 0-5 | 0 | | | /HPF |
| EPIs | 0 | | | | /LPF |
| Bact | 0 | | | | |
| Mucus | 0 | | | | |
| Crys | 0 | | | | |
| Casts | 0 | | | | /LPF |
| Yeast | 0 | | | | |

## Case Questions

1.  Describe the major functions of the kidneys.

2.  Several biochemical indices are used to diagnose renal disease. One is glomerular filtration rate (GFR). What does GFR measure? What is a normal GFR? What test is usually done to estimate glomerular filtration rate?

3.  Mrs. Joaquin's GFR is 28 mL/min. What does this tell you about her kidney function?

4.  Define the following terms:

    *Acute renal failure (ARF):*

    *Oliguria:*

    *Renal failure:*

    *End-stage renal failure (ESRF):*

    *Uremia:*

    *Azotemia:*

5.  Mrs. Joaquin has a history of type 2 diabetes mellitus. Is it common to develop renal problems with diabetes? Explain.

**6.** Mrs. Joaquin tells you that she is not surprised about the kidney problems—"Many of our family members and neighbors have these same problems." Is this surprising? Why or why not?

**7.** What health problems have been identified in the Pima Indians? In general, describe the epidemiological data regarding these health issues.

**8.** Explain what is meant by the "thrifty gene" theory.

**9.** How does nephropathy affect Pima Indians?

**10.** Which of Mrs. Joaquin's laboratory values are out of normal range? How is that related to her diagnosis?

| Parameter | Normal Value | Pt's Value | Rationale |
|-----------|--------------|------------|-----------|
|           |              |            |           |
|           |              |            |           |
|           |              |            |           |
|           |              |            |           |
|           |              |            |           |
|           |              |            |           |
|           |              |            |           |
|           |              |            |           |
|           |              |            |           |
|           |              |            |           |

**11.** Calculate Mrs. Joaquin's BMI.

**12.** How would you interpret Mrs. Joaquin's BMI?

**13.** What are the energy requirements for renal insufficiency? Would you use Mrs. Joaquin's IBW or adjusted weight? Calculate her energy needs accordingly.

**14.** What are Mrs. Joaquin's protein requirements? What is the rationale?

**15.** What considerations should be made when establishing protein requirements in prerenal ESRD?

**16.** Are there any potential benefits of using different types of protein, such as plant protein rather than animal protein, in the diet for a patient with renal insufficiency? Explain.

**17.** What is likely to happen to Mrs. Joaquin if she does not control her blood glucose levels and HTN?

**18.** Explain the reasons for the following components of Mrs. Joaquin's medical nutrition therapy:

| Medical Nutrition Therapy | Rationale |
| --- | --- |
| 30–40 kcal/kg body wt | |
| 0.8 g protein/kg/day | |
| 8–12 mg $PO_4$/ kg body wt | |
| 3 g Na | |

**19.** What resources would you use to teach Mrs. Joaquin about her diet?

**20.** Using Mrs. Joaquin's typical intake and the prescribed diet, write a sample menu.

| Diet PTA | | Sample Diet |
|---|---|---|
| *Breakfast:* | 1 fried egg | |
| | 3 strips fried bacon | |
| | Fried potato (approx 1) | |
| | 1 slice bread w/ butter | |
| *Lunch:* | 2 tamales w/ chili con carne | |
| | Fry bread | |
| | 1 can Coke | |
| *PM Snack:* | 25 potato chips | |
| | 1 can Coke | |
| *Dinner:* | 3 tacos made w/ground | |
| | beef, chopped tomatoes, | |
| | onions, lettuce, | |
| | 3 flour tortillas | |
| | 1 can Coke | |
| *HS Snack:* | Crackers | |
| | Peanut butter | |

**21.** After evaluating Mrs. Joaquin's typical diet, what other recommendations can you make?

**22.** Mrs. Joaquin has a $PO_4$ restriction. Why?

**23.** What foods have the highest levels of phosphorus?

24.   Mrs. Joaquin tells you that her friend can drink only certain amounts and wants to know if
      that is the case for her. What foods are considered to be fluids? What would you tell her about
      her own case?

25.   If a patient must follow a fluid restriction, what can be done to help reduce his or her thirst?

26.   Write an initial SOAP note for your consultation with Mrs. Joaquin.


      S:


      O:


      A:


      P:

# Bibliography

American Diabetes Association. *Maximizing the Role of Nutrition in Diabetes Management.* Alexandria, VA: American Diabetes Association, 1994.

American Dietetic Association. *Manual of Clinical Dietetics,* 6th ed. Chicago: American Dietetic Association, 2000.

American Dietetic Association and Morrison Health Care. *Medical Nutrition Therapy Across the Continuum of Care.* Chicago: American Dietetic Association, 1998.

Beto J. Which diet for which renal failure: making sense of options. *J Am Diet Assoc.* 1995;95:898–903.

Chronic renal disease. *The Merck Manual.* Available at http://www.merck.com. Accessed December 8, 2000.

Clark CM. Oral therapy in type 2 diabetes: pharmacological properties and clinical use of currently available agents. *Diabetes Spectrum.* 1998;11(4):211–221. Available at http://www.diabetes.org/diabetesspectrum. Accessed May 11, 1999.

Commission on Accreditation for Dietetics Education. Knowledge, skills, and competencies for dietitians. *Accreditation Manual for Dietetics Education Programs,* rev. 4th ed. Chicago: American Dietetic Association, 2000.

Coyne DW. Renal diseases. In Lee HH, Carey CF, Woeltje KF (eds.), *The Washington Manual of Medical Therapeutics,* 29th ed. Philadelphia: Lippincott, Williams & Wilkins, 1998.

Escott-Stump S. *Nutrition and Diagnosis-Related Care,* 4th ed. Baltimore: Williams & Wilkins, 1998.

Fischbach F. *Manual of Laboratory and Diagnostic Tests,* 6th ed. Philadelphia: Lippincott, 2000.

Gedney F. Renal disease education. *Virtual Hospital.* Available at http://www.vh.org. Accessed October 3, 2000.

Graber MA, Martinez-Bianchi V. Genitourinary and renal disease: proteinuria, nephritic syndrome, and nephritic urine. *Virtual Hospital.* Available at http://www.vh.org. Accessed October 3, 2000.

Graber MA, Martinez-Bianchi V. Genitourinary and renal disease: renal failure. *Virtual Hospital.* Available at http://www.vh.org. Accessed October 3, 2000.

Hemodialysis. *The Merck Manual.* Available at http://www.merck.com. Accessed December 8, 2000.

Huether SE. Alterations of renal and urinary tract function. In McCance KL, Huether SE (eds.), *Pathophysiology: The Biologic Basis for Disease in Adults and Children,* 3rd ed. St. Louis: Mosby, 1998.

Humes HD. Limiting acute renal failure. *Hospital Practice.* Available at http://www.hospract.com. Accessed December 4, 2000.

Kittler PG, Sucher KP. *Cultural Foods: Traditions and Trends.* Belmont, CA: Wadsworth Thompson Learning, 2000.

MedlinePlus Health information. Available at http://www.nlm.nih.gov/medlineplus/druginformation.html. Accessed September 8, 2000.

Morrison G, Stover J. Renal disease. In Morrison G, Hark L (eds.), *Medical Nutrition and Disease,* 2nd ed. Malden, MA: Blackwell Science, 1999.

Nelson R, et al. Development and progression of renal disease in Pima Indians with non-insulin dependent diabetes mellitus. *N Engl J Med.* 1996;335:1636.

National Institute of Diabetes and Digestive and Kidney Diseases (NIDDK). Diabetes in American Indians and Alaska Natives. Available at http://www.niddk.nih.gov. Accessed December 6, 2000.

National Institute of Diabetes and Digestive and Kidney Diseases (NIDDK). High rates of kidney disease in American Indians. Available at http://www.niddk.nih.gov. Accessed December 6, 2000.

National Institute of Diabetes and Digestive and Kidney Diseases (NIDDK). The Pima Indians. Obesity and diabetes. Available at http://www.niddk.nih.gov. Accessed March 3, 2001.

O'Connell BS. Early renal disease in diabetes: a brief review. *Diabetes Care and Education.* 2001;22(1):7–11.

The Pima Indians: Pathfinders for Health. Available at http://www.niddk.nih.gov/health/diabetes/pima/pathfind/pathfind.htm. Accessed December 6, 2000.

Romano MM. Renal conditions. In Lysen LK (ed.), *Quick Reference to Clinical Dietetics.* Gaithersburg, MD: ASPEN, 1997.

Sevilla G. Life changed after river stopped. *Arizona Republic,* October 31, 1999. Available at http://www.azcentral.com/news/specials/pima/1031history.shtml. Accessed March 3, 2001.

Sharp AR. Nutritional implications of new medications to treat diabetes. *On the Cutting Edge.* 1998;19(2):8–10.

Stanford EK. Management of chronic renal failure. Available at http://www.medical-library.org. Accessed September 22, 2000.

Swearingen PL, Ross DG. Endocrine disorders. *Manual of Medical-Surgical Nursing Care,* 4th ed. St. Louis: Mosby, 1999.

Wilkens KG. Medical nutrition therapy for renal disease. In Mahan LK, Escott-Stump S (eds.), *Krause's Food, Nutrition, and Diet Therapy,* 10th ed. Philadelphia: Saunders, 2000.

## Case 32

# End-Stage Renal Disease Treated with Hemodialysis

*Introductory Level*

## Objectives

After completing this case, the student will be able to:

1. Describe the pathophysiology of end-stage renal disease.
2. Integrate physiology of hemodialysis into MNT recommendations.
3. Interpret laboratory parameters for nutritional implications and significance.
4. Individualize fluid and electrolyte requirements for the patient with end-stage renal disease.
5. Manage monitoring of patient's food and nutrient intake.
6. Use nutrition assessment data and interpretation of pathophysiology to determine appropriate nutrient requirements.

7. Identify appropriate MNT goals.
8. Make appropriate documentation in the medical record.
9. Integrate knowledge of sociocultural and ethnic food consumption issues into nutrition care plan.

Mrs. Joaquin, who was introduced in Case 31, is admitted two years later in end-stage renal failure. During this admission, she will have an arteriovenous fistula placed and will begin treatment with hemodialysis.

 **UNIVERSITY HOSPITAL**

## ADMISSION DATABASE

Name: Enez Joaquin
DOB: 4/14  age 26
Physician: L. Nila

| BED#<br>2 | DATE:<br>3/5 | TIME:<br>1830 | TRIAGE STATUS (ER ONLY):<br>☐ Red  ☐ Yellow  ☐ Green  ☐ White |
|---|---|---|---|

**PRIMARY PERSON TO CONTACT:**
Name: Eddie Joaquin—husband
Home #: 555-3947
Work #: 554-2100

### Initial Vital Signs

| TEMP:<br>98.6 | RESP:<br>25 | | SAO2: |
|---|---|---|---|
| HT:<br>5'0" | WT (lb):<br>170 | B/P:<br>220/80 | PULSE:<br>84 |
| LAST TETANUS<br>4 years ago | | LAST ATE<br>2 days ago | LAST DRANK<br>4 hours—water |

**ORIENTATION TO UNIT:** ☒ Call light ☒ Television/telephone
☒ Bathroom ☒ Visiting ☒ Smoking ☒ Meals
☒ Patient rights/responsibilities

### CHIEF COMPLAINT/HX OF PRESENT ILLNESS

N/V

**PERSONAL ARTICLES:** (Check if retained/describe)
☐ Contacts ☐ R ☐ L        ☐ Dentures ☐ Upper ☐ Lower
☐ Jewelry:
☐ Other:

### ALLERGIES: Meds, Food, IVP Dye, Seafood: Type of Reaction

None

**VALUABLES ENVELOPE:**
☐ Valuables instructions

### PREVIOUS HOSPITALIZATIONS/SURGERIES

childbirth 7 years ago

**INFORMATION OBTAINED FROM:**
☒ Patient           ☐ Previous record
☐ Family            ☐ Responsible party

Signature  *Enez Joaquin*

| Home Medications (including OTC) | | Codes: A=Sent home | | B=Sent to pharmacy | | C=Not brought in |
|---|---|---|---|---|---|---|
| Medication | Dose | Frequency | Time of Last Dose | Code | Patient Understanding of Drug | |
| Glucophage | 850 mg | bid | ? | C | no | |
| | | | | | | |
| | | | | | | |
| | | | | | | |
| | | | | | | |
| | | | | | | |
| | | | | | | |
| | | | | | | |
| | | | | | | |

Do you take all medications as prescribed?   ☐ Yes   ☒ No   If no, why? They make me feel tired.

### PATIENT/FAMILY HISTORY

| | | |
|---|---|---|
| ☐ Cold in past two weeks | ☒ High blood pressure Patient | ☒ Kidney/urinary problems Patient |
| ☐ Hay fever | ☐ Arthritis | ☒ Gastric/abdominal pain/heartburn Patient |
| ☐ Emphysema/lung problems | ☐ Claustrophobia | ☐ Hearing problems |
| ☐ TB disease/positive TB skin test | ☐ Circulation problems | ☐ Glaucoma/eye problems |
| ☐ Cancer | ☐ Easy bleeding/bruising/anemia | ☐ Back pain |
| ☐ Stroke/past paralysis | ☐ Sickle cell disease | ☐ Seizures |
| ☐ Heart attack | ☐ Liver disease/jaundice | ☐ Other |
| ☐ Angina/chest pain | ☐ Thyroid disease | |
| ☐ Heart problems | ☒ Diabetes Patient | |

### RISK SCREENING

Have you had a blood transfusion?   ☐ Yes   ☒ No
Do you smoke?   ☐ Yes   ☒ No
If yes, how many pack(s)      /day for      years
Does anyone in your household smoke?   ☒ Yes   ☐ No
Do you drink alcohol?   ☒ Yes   ☐ No
If yes, how often? daily  How much? 12 oz beer
When was your last drink?    3 / 4 /
Do you take any recreational drugs?   ☐ Yes   ☒ No
If yes, type:_____  Route
Frequency:_____  Date last used:_____/_____/

**FOR WOMEN Ages 12–52**

Is there any chance you could be pregnant?   ☐ Yes   ☒ No
If yes, expected date (EDC):
Gravida/Para: 1/1

**ALL WOMEN**

Date of last Pap smear: 1/25
Do you perform regular breast self-exams?   ☐ Yes   ☒ No

**ALL MEN**

Do you perform regular testicular exams?   ☐ Yes   ☐ No

Additional comments:

**✗** *Liz Romero, RN*
Signature/Title

**Client name:** Enez Joaquin
**DOB:** 4/14
**Age:** 26
**Sex:** Female
**Education:** High school
**Occupation:** Secretary
**Hours of work:** 9 AM–5 PM
**Household members:** Husband age 26, type 2 diabetes under control; daughter age 7, in good health
**Ethnic background:** Pima Indian
**Religious affiliation:** Catholic
**Referring physician:** Lourdes Nila, MD (nephrology)

## Chief complaint:

Pt complains of anorexia; N/V; four kg weight gain in the past two weeks, edema in extremities, face and eyes; malaise; progressive SOB with three-pillow orthopnea; pruritus; muscle cramps; and inability to urinate

## Patient history:

Mrs. Joaquin is a 26-yo Native American woman with renal insufficiency secondary to diabetes mellitus. She is from the Pima Indian tribe. She is the product of a full-term pregnancy and weighed 11 lb at birth. She lives with her husband and 7-year-old daughter. Mrs. Joaquin was diagnosed with type 2 DM when she was 13 years old and has been poorly compliant with prescribed treatment. Her husband also has type 2 DM. He was diagnosed at the age of 18. Her renal function has been monitored for the past seven years. Progressive decompensation of renal function has been documented by declining GRF, increasing creatinine and urea concentrations, elevated serum phosphate, and normochromic, normocytic anemia. She is being admitted for placement of AV fistula to begin hemodialysis treatment.

*Onset of disease:* Diagnosed with impaired renal function two years ago. Her acute symptoms have developed over the last two weeks.

*Type of Tx:* Control BP, create access for hemodialysis, nutrition consult.

*PMH:* Gravida 1/para 1. Infant weighed 10 lb at birth seven years ago. Pt admits that she has recently stopped taking a prescribed hypoglycemic agent, and she has never filled her prescription for antihypertensive medication.

*Meds:* Glucophage (metformin) 850 mg bid

*Smoker:* No

*Family Hx:* Both mother and father diagnosed with DM

## Physical exam:

*General appearance:* Overweight Native American female who appears her age. Lethargic, complaining of N/V.

*Vitals:* Temp 98.6°F, BP 220/80 mm Hg, HR 84 bpm, RR 25 bpm

*Heart:* S4, S1, and S2, regular rate and rhythm. I/VI systolic ejection murmur, upper left sternal border.

*HEENT:* Normocephalic, equal carotid pulses, neck supple, no bruits
  *Eyes:* PERRLA
  *Ears:* Noncontributory

*Nose:* Noncontributory

*Throat:* Noncontributory

*Genitalia:* Normal female

*Neurologic:* Oriented to person, place, and time; intact, mild asterixis

*Extremities:* Muscle weakness; 3+ pitting edema to the knees, no cyanosis

*Skin:* Dry and yellowish-brown

*Chest/lungs:* Generalized rhonchi with rales that are mild at the bases (Pt breathes with poor effort)

*Peripheral vascular:* Normal pulse (3+) bilaterally

*Abdomen:* Bowel sounds positive, soft; generalized mild tenderness; no rebound

## Nutrition Hx:

*General:* Intake has been poor due to anorexia, N&V. Patient states that she tried to follow the diet that she was taught two years ago. "It went pretty well for awhile but it was hard to keep up with."

*Usual dietary intake:*

| | |
|---|---|
| *Breakfast:* | Cold cereal |
| | Bread or fried potatoes |
| | Fried egg (occasionally) |
| *Lunch:* | Bologna sandwich |
| | Potato chips |
| | Coke |
| *Dinner:* | Chopped meat |
| | Fried potatoes |
| *Snacks:* | Crackers and peanut butter |

*Food allergies/intolerances/aversions:* None

*Previous MNT?* Yes    *If yes, when:* 2 years ago when Pt Dx with impaired renal function    *Where?* Reservation Health Service

*Food purchase/preparation:* Self

*Vit/min intake:* None

*Current diet order:* 35 kcal/kg, 0.8 g protein/kg, 8–12 mg phosphorus/kg, 2–3 g Na

## Tx plan:

Surgical placement of AV graft

Captopril

Erythropoietin (r-HuEPO) 30 units/kg

Vitamin/mineral supplement

Calcitriol 0.25 ϕg/d po

35 kcal/kg, 1.2 g protein/kg, 2 g K, 1 g phosphorus, 2 g Na

1000 mL fluid + urine output per day

Glucophage (metformin) 850 mg bid

CBC, chemistry

Phos Lo

Stool softener

Sodium bicarbonate, 2 g q d

Occult fecal blood

$\underline{\mathbf{U_H}\ UNIVERSITY\ HOSPITAL}$

NAME: E. Joaquin                         DOB: 4/14
AGE: 26                                  SEX: Female
PHYSICIAN: L. Nila

\*\*\*\*\*\*\*\*\*\*\*\*\*\*\*\*\*\*\*\*\*\*\*\*\*\*\*\*\*\*\*\*\*\*\*\*\*\*\*\*\*\*\*\*\*\*CHEMISTRY\*\*\*\*\*\*\*\*\*\*\*\*\*\*\*\*\*\*\*\*\*\*\*\*\*\*\*\*\*\*\*\*\*\*\*\*\*\*\*\*\*\*\*\*\*

| DAY: | | Admit | d/c | |
|---|---|---|---|---|
| DATE: | | | | |
| TIME: | | | | |
| LOCATION: | | | | |
| | NORMAL | | | UNITS |
| Albumin | 3.6–5 | 3.7 | 3.4 | g/dL |
| Total protein | 6–8 | | | g/dL |
| Prealbumin | 19–43 | | | mg/dL |
| Transferrin | 200–400 | | | mg/dL |
| Sodium | 135–155 | 130 | 134 | mEq/L |
| Potassium | 3.5–5.5 | 5.8 | 5.6 | mEq/L |
| Chloride | 98–108 | 91 | 100 | mEq/L |
| $PO_4$ | 2.5–4.5 | 9.5 | 7.2 | mEq/L |
| Magnesium | 1.6–2.6 | 2.9 | 2.7 | mEq/L |
| Osmolality | 275–295 | | | mmol/kg $H_2O$ |
| Total $CO_2$ | 24–30 | 20 | 23 | mmol/L |
| Glucose | 70–120 | 282 | 200 | mg/dL |
| BUN | 8–26 | 69 | 55 | mg/dL |
| Creatinine | 0.6–1.3 | 12.0 | 8.5 | mg/dL |
| Uric acid | 2.6–6 (women) | | | mg/dL |
| | 3.5–7.2 (men) | | | |
| Calcium | 8.7–10.2 | 8.2 | 8.6 | mg/dL |
| Bilirubin | 0.2–1.3 | | | mg/dL |
| Ammonia ($NH_3$) | 9–33 | | | μmol/L |
| SGPT (ALT) | 10–60 | 26 | | U/L |
| SGOT (AST) | 5–40 | 28 | | U/L |
| Alk phos | 98–251 | 131 | | U/L |
| CPK | 26–140 (women) | | | U/L |
| | 38–174 (men) | | | |
| LDH | 313–618 | 315 | | U/L |
| CHOL | 140–199 | 220 | | mg/dL |
| HDL-C | 40–85 (women) | | | mg/dL |
| | 37–70 (men) | | | |
| VLDL | | | | mg/dL |
| LDL | < 130 | | | mg/dL |
| LDL/HDL ratio | < 3.22 (women) | | | |
| | < 3.55 (men) | | | |
| Apo A | 101–199 (women) | | | mg/dL |
| | 94–178 (men) | | | |
| Apo B | 60–126 (women) | | | mg/dL |
| | 63–133 (men) | | | |
| TG | 35–160 | 200 | | mg/dL |
| $T_4$ | 5.4–11.5 | | | μg/dL |
| $T_3$ | 80–200 | | | ng/dL |
| $HbA_{1C}$ | 4.8–7.8 | 8.9 | | % |

# U<sub>H</sub> UNIVERSITY HOSPITAL

```
NAME: E. Joaquin                    DOB: 4/14
AGE: 26                             SEX: Female
PHYSICIAN: L. Nila
```

\*\*\*\*\*\*\*\*\*\*\*\*\*\*\*\*\*\*\*\*\*\*\*\*\*\*\*\*\*\*\*\*\*\*\*\*\*\*\*\*\*\*URINALYSIS\*\*\*\*\*\*\*\*\*\*\*\*\*\*\*\*\*\*\*\*\*\*\*\*\*\*\*\*\*\*\*\*\*\*\*\*\*\*\*\*\*\*

| DAY: | | Admit | Postop | d/c | |
|---|---|---|---|---|---|
| DATE: | | | | | |
| TIME: | | | | | |
| LOCATION: | | | | | |
| | NORMAL | | | | UNITS |
| Coll meth | | Random specimen | First morning | First morning | |
| Color | | Straw | Straw | Pale yellow | |
| Appear | | Hazy | Slightly hazy | Slightly hazy | |
| Sp grv | 1.003–1.030 | 1.010 | | | |
| pH | 5–7 | 7.9 | | | |
| Prot | NEG | 2+ | | | mg/dL |
| Glu | NEG | | | | mg/dL |
| Ket | NEG | | | | |
| Occ bld | NEG | | | | |
| Ubil | NEG | | | | |
| Nit | NEG | | | | |
| Urobil | < 1.1 | | | | EU/dL |
| Leu bst | NEG | | | | |
| Prot chk | NEG | | | | |
| WBCs | 0–5 | 20 | | | /HPF |
| RBCs | 0–5 | | | | /HPF |
| EPIs | 0 | | | | /LPF |
| Bact | 0 | | | | |
| Mucus | 0 | | | | |
| Crys | 0 | | | | |
| Casts | 0 | | | | /LPF |
| Yeast | 0 | | | | |

# UH UNIVERSITY HOSPITAL

Name: E. Joaquin
Physician: L. Nila

## PATIENT CARE SUMMARY SHEET

Date: 3/5   Room: 324   Wt Yesterday:   lb   Today: 170 lb   Postdialysis:   lb

| Temp °F | NIGHTS | | | | | | | | DAYS | | | | | | | | EVENINGS | | | | | | | |
|---|---|---|---|---|---|---|---|---|---|---|---|---|---|---|---|---|---|---|---|---|---|---|---|---|
| | 00 | 01 | 02 | 03 | 04 | 05 | 06 | 07 | 08 | 09 | 10 | 11 | 12 | 13 | 14 | 15 | 16 | 17 | 18 | 19 | 20 | 21 | 22 | 23 |
| 105 | | | | | | | | | | | | | | | | | | | | | | | | |
| 104 | | | | | | | | | | | | | | | | | | | | | | | | |
| 103 | | | | | | | | | | | | | | | | | | | | | | | | |
| 102 | | | | | | | | | | | | | | | | | | | | | | | | |
| 101 | | | | | | | | | | | | | | | | | | | | | | | | |
| 100 | | | | | | | | | | | | | | | | | | | | | | | | |
| 99 | | | | | | | | | | | | | | | | | | | | | | | | |
| 98 | | | | | | | | | | | | | | | | | | | | | | | | |
| 97 | | | | | | | | | | | | | | | | | | | | | | | | |
| 96 | | | | | | | | | | | | | | | | | | | | | | | | |
| Pulse | | | | | | | | | | | | | | | | | | | 84 | | | | | 82 |
| Respiration | | | | | | | | | | | | | | | | | | | 25 | | | | | 24 |
| BP | | | | | | | | | | | | | | | | | | | 220/80 | | | | | 210/78 |
| Blood Glucose | | | | | | | | | | | | | | | | | | | 210 | | | | | |
| Appetite/Assist | | | | | | | | | | | | | | | | | | | 0 | | | | | |
| INTAKE | | | | | | | | | | | | | | | | | | | | | | | | |
| Oral | | | | | | | | | | | | | | | | | | | 0 | 50 | | | | |
| IV | | | | | | | | | | | | | | | | | | | 0 | | | | | |
| TF Formula/Flush | | | | | | | | | | | | | | | | | | | 0 | | | | | |
| Shift Total | | | | | | | | | | | | | | | | | | | | | | | | |
| OUTPUT | | | | | | | | | | | | | | | | | | | | | | | | |
| Void | | | | | | | | | | | | | | | | | | | N/A | | | | | 100 |
| Cath. | | | | | | | | | | | | | | | | | | | | | | | | |
| Emesis | | | | | | | | | | | | | | | | | | | | | 50 | | | |
| BM | | | | | | | | | | | | | | | | | | | | | | | | |
| Drains | | | | | | | | | | | | | | | | | | | | | | | | |
| Shift Total | | | | | | | | | | | | | | | | | | | | | | | | |
| Gain | | | | | | | | | | | | | | | | | | | | | | | | 50 cc |
| Loss | | | | | | | | | | | | | | | | | | | | | | | | 150 cc |
| Signatures | | | | | | | | | | | | | | | | | | Sandy Dunn, RN | | | | | | |

 UNIVERSITY HOSPITAL

Name: E. Joaquin

Physician: L. Nila

# PATIENT CARE SUMMARY SHEET

Date: 3/6    Room: 324    Wt Yesterday: 170 lb    Today: 165 lb    Postdialysis: 165 lb

| Temp °F | NIGHTS | | | | | | | | DAYS | | | | | | | | EVENINGS | | | | | | | |
|---|---|---|---|---|---|---|---|---|---|---|---|---|---|---|---|---|---|---|---|---|---|---|---|---|
| | 00 | 01 | 02 | 03 | 04 | 05 | 06 | 07 | 08 | 09 | 10 | 11 | 12 | 13 | 14 | 15 | 16 | 17 | 18 | 19 | 20 | 21 | 22 | 23 |
| 105 | | | | | | | | | | | | | | | | | | | | | | | | |
| 104 | | | | | | | | | | | | | | | | | | | | | | | | |
| 103 | | | | | | | | | | | | | | | | | | | | | | | | |
| 102 | | | | | | | | | | | | | | | | | | | | | | | | |
| 101 | | | | | | | | | | | | | | | | | | | | | | | | |
| 100 | | | | | | | | | | | | | | | | | | | | | | | | |
| 99 | | | | | | | | 99 | | | | | | | | | | | | | | | | |
| 98 | | | | | | | | | | | | | | | | | | | | | | | | |
| 97 | | | | | | | | | | | | | | | | | | | | | | | | |
| 96 | | | | | | | | | | | | | | | | | | | | | | | | |
| Pulse | | | | | | | | 80 | | | 84 | | | | | | | | | | | | | |
| Respiration | | | | | | | | 23 | | | 25 | | | | | | | | | | | | | |
| BP | | | | | | | | 200/75 | | | 220/80 | | | | | | | | | | | | | |
| Blood Glucose | | | | | | | | 170 | | | 200 | | | | | | | | | | | | | |
| Appetite/Assist | | | | | | | | NPO | | | NPO | | | | | | | | | | | | | |
| INTAKE | | | | | | | | | | | | | | | | | | | | | | | | |
| Oral | | | | | | | | 0 | | | 0 | | | | | | | | | | | | | |
| IV | | | | | | | | | | | | | | | | | | | | | | | | |
| TF Formula/Flush | | | | | | | | | | | | | | | | | | | | | | | | |
| Shift Total | | | | | | | | | | | | | | | | | | | | | | | | |
| OUTPUT | | | | | | | | | | | | | | | | | | | | | | | | |
| Void | | | | | | | | 200 | | | | | | | | | | | | | 300 | | | |
| Cath. | | | | | | | | | | | | | | | | | | | | | | | | |
| Emesis | | | 100 | | | | | | | | | | | | 50 | | | | | | | | | |
| BM | | | | | | | | X 1 | | | | | | | | | | | | | | | | |
| Drains | | | | | | | | | | | | | | | | | | | | | | | | |
| Shift Total | | | | | | | | | | | | | | | | | | | | | | | | |
| Gain | NPO | | | | | | | | | | | | | | | | | | | | | | | |
| Loss | 300 cc | | | | | | | | | | | | | | | | | | | | | | | |
| Signatures | *Bill Salazar, RN* | | | | | | | | *Sandy Dunn, RN* | | | | | | | | *Marie Seymour, RN* | | | | | | | |

## Case Questions

1. Describe the physiological function of the kidneys.

2. What diseases/conditions can lead to kidney failure?

3. Signs, symptoms, and laboratory abnormalities distinguish disease pathophysiology. What is the difference between a sign and a symptom? What are the signs and symptoms of chronic kidney failure?

4. From your reading of Mrs. Joaquin's history and physical, what signs and symptoms did Mrs. Joaquin have?

5. Evaluate Mrs. Joaquin's chemistry report. What labs support the diagnosis of ESRD?

6. What are the treatment options for kidney failure?

7. Mrs. Joaquin is admitted to the hospital to have an AV fistula placed for use in hemodialysis. Explain the concept of hemodialysis.

8. What is an AV fistula? Are there other types of access for hemodialysis?

9. The treatment plan for Mrs. Joaquin consists of three major components: medical nutrition therapy, pharmacological treatment, and the actual dialysis treatment. Identify goals for each component of her care.

   *MNT:*

   *Meds:*

   *Dialysis prescription:*

10. Mrs. Joaquin was diagnosed with type 2 DM when she was 13 years old. Does she fit the "profile" for someone with type 2 DM? Why or why not?

11. Evaluate Mrs. Joaquin's chemistry values. Why would you expect to see each of the laboratory value discrepancies, and what could be done nutritionally to affect each value?

| Parameter | Normal Value | Pt's Value | Nutritional Implications |
|---|---|---|---|
|  |  |  |  |
|  |  |  |  |
|  |  |  |  |
|  |  |  |  |
|  |  |  |  |
|  |  |  |  |
|  |  |  |  |
|  |  |  |  |
|  |  |  |  |
|  |  |  |  |

12. Explain why the following medications were prescribed.

| Medication | Indications/Mechanism | Nutritional Implications |
|---|---|---|
| Captopril |  |  |
| Erythropoietin |  |  |
| Vitamin/mineral supplement |  |  |
| Calcitriol |  |  |
| Glucophage |  |  |
| Sodium bicarbonate |  |  |
| Phos Lo |  |  |

**13.**   Assess the patient's anthropometric values.

**14.**   Is the prescribed kilocalorie level based on ABW, UBW, or IBW? (Explain rationale.)

**15.**   What is "dry weight"?

**16.**   Determine the rest of Mrs. Joaquin's medical nutrition therapy prescription. Be sure to include protein, Na, $K^+$, $PO_4$, calcium, and fluid. Check the nursing flow sheet for calculation of fluid restriction. Give the rationale for each component of the prescription.

**17.**   Energy and protein recommendations increased after Mrs. Joaquin began hemodialysis. Explain why.

**18.**   Why is it recommended for patients to have at least 50% of their protein from sources that have high biological value?

**19.**   The MD ordered daily use of multivitamin/mineral supplement containing B-complex but not fat-soluble vitamins. Why are these restrictions specified?

**20.**   Using the MNT plan, plan a day's menu using renal diet choices.

**21.**   Examine the patient care summary sheet for hospital day 2. What was Mrs. Joaquin's weight postdialysis? Why did it change? Which of Mrs. Joaquin's other symptoms would you expect to begin to improve? What other information would be important to examine from this record?

**22.**   How often are a dialysis patient's nutritional goals adjusted?

**23.**   How are they adjusted?

**24.**    Check Mrs. Joaquin's chemistry labs as she is being discharged. What has changed? Why? Do you need to make adjustments in her MNT prescription?

| Chemistry Lab Value | Change | Rationale |
| --- | --- | --- |
|  |  |  |
|  |  |  |
|  |  |  |
|  |  |  |
|  |  |  |
|  |  |  |
|  |  |  |

**25.**    How does the dietetic professional follow a dialysis patient?

**26.**    After examining Mrs. Joaquin's labs after her first dialysis, would you make any adjustments in her medical nutrition therapy prescription?

**27.**    Review Mrs. Joaquin's admission record. Is anything noted in her database, but not mentioned in her nutrition history, that the RD should note?

**28.**    Write an initial SOAP note for your consultation with Mrs. Joaquin.

S:

O:

A:

P:

# Bibliography

American Diabetes Association. Maturity-onset diabetes of the young (MODY). Available at http://www.diabetes.org. Accessed May 26, 2001.

American Diabetes Association. *Maximizing the Role of Nutrition in Diabetes Management.* Alexandria, VA: American Diabetes Association, 1994.

American Dietetic Association. *Manual of Clinical Dietetics,* 6th ed. Chicago: American Dietetic Association, 2000.

American Dietetic Association and Morrison Health Care. *Medical Nutrition Therapy Across the Continuum of Care.* Chicago: American Dietetic Association, 1998.

Chronic renal disease. *The Merck Manual.* Available at http://www.merck.com. Accessed December 8, 2000.

Clark CM. Oral therapy in type 2 diabetes: pharmacological properties and clinical use of currently available agents. *Diabetes Spectrum.* 1998;11(4):211–221. Available at http://www.diabetes.org/diabetesspectrum. Accessed May 11, 1999.

Commission on Accreditation for Dietetics Education. Knowledge, skills, and competencies for dietitians. *Accreditation Manual for Dietetics Education Programs,* rev. 4th ed. Chicago: American Dietetic Association, 2000.

Coyne DW. Renal diseases. In Lee HH, Carey CF, Woeltje KF (eds.), *The Washington Manual of Medical Therapeutics,* 29th ed. Philadelphia: Lippincott, Williams & Wilkins, 1998.

Diabetes research highlight: Scientists pinpoint genes in diabetes of the young. *National Center for Research Resources Reporter,* March–April 1997. Available at http://www.ncrr.nih.gov. Accessed May 26, 2001.

Fischbach F. *Manual of Laboratory and Diagnostic Tests,* 6th ed. Philadelphia: Lippincott, 2000.

Gedney F. Renal disease education. *Virtual Hospital.* Available at http://www.vh.org. Accessed October 3, 2000.

Graber MA, Martinez-Bianchi V. Genitourinary and renal disease: proteinuria, nephritic syndrome, and nephritic urine. *Virtual Hospital.* Available at http://www.vh.org. Accessed October 3, 2000.

Graber MA, Martinez-Bianchi V. Genitourinary and renal disease: renal failure. *Virtual Hospital.* Available at http://www.vh.org. Accessed October 3, 2000.

Hemodialysis. *The Merck Manual.* Available at http://www.merck.com. Accessed December 8, 2000.

Humes HD. Limiting acute renal failure. *Hospital Practice.* Available at http://www.hospract.com. Accessed December 4, 2000.

Kittler PG, Sucher KP. *Cultural Foods: Traditions and Trends.* Belmont, CA: Wadsworth Thomson Learning, 2000.

MedlinePlus Health information. Available at http://www.nlm.nih.gov/medlineplus/druginformation.html. Accessed September 8, 2000.

Morrison G, Stover J. Renal disease. In Morrison G and Hark L (eds.), *Medical Nutrition and Disease,* 2nd ed. Malden, MA: Blackwell Science, 1999.

National Institute of Diabetes, Digestive and Kidney Diseases. Diabetes mellitus: Challenges and opportunities final report and recommendations. Available at http://www.niddk.nih.gov. Accessed May 26, 2001.

National Institute of Diabetes, Digestive and Kidney Diseases. High rates of kidney disease in American Indians. Available at http://www.niddk.nih.gov. Accessed December 6, 2000.

*The Pima Indians: Pathfinders for Health.* Available at http://www.niddk.nih.gov/health/diabetes/pima/pathfind/pathfind.htm. Accessed December 6, 2000.

Romano MM. Renal conditions. In Lysen LK (ed.), *Quick Reference to Clinical Dietetics.* Gaithersburg, MD: ASPEN, 1997.

Sharp AR. Nutritional implications of new medications to treat diabetes. *On the Cutting Edge.* 1998;19(2):8–10.

Stanford EK. Management of chronic renal failure. Available at http://www.medical-library.org. Accessed September 22, 2000.

Swearingen PL, Ross DG. Endocrine disorders. *Manual of Medical-Surgical Nursing Care,* 4th ed. St. Louis: Mosby, 1999.

Wilkens KG. Medical nutrition therapy for renal disease. In Mahan LK, Escott-Stump S (eds.), *Krause's Food, Nutrition, and Diet Therapy,* 10th ed. Philadelphia: Saunders, 2000.

# Unit Nine

# MEDICAL NUTRITION THERAPY FOR HYPERMETABOLISM, INFECTION, AND TRAUMA

## Introduction

The physiological response to stress, trauma, and infection has been an important area of nutrition research for the past several decades. This metabolic response is characterized by catabolism of stored nutrients to meet the increased energy requirements. Unlike other situations when the body faces increased energy requirements, the stress response demands a preferential use of glucose for fuel. Because glycogen stores are quickly depleted, the body turns to lean body mass for glucose produced via gluconeogenesis. Under the influence of counterregulatory hormones such as glucagon, epinephrine, norepinephrine, and cortisol, as well as cytokines such as interleukin and tumor necrosis factor, the body shifts its normal tendency of anabolism to catabolism. All sources of fuel metabolism are affected by the stress response and the subsequent control of counter-regulatory hormones. Despite increased lipolysis, the use of fatty acids and glycerol as fuel appear not to significantly increase oxidation of those fuel sources. The body's inability to keep up with the rate of protein catabolism results in significant loss of skeletal muscle and high urinary losses of nitrogen. The liver's rate of gluconeogenesis is increased, and hyperglycemia is common. In addition, many tissues—especially skeletal tissue—develop insulin resistance, which contributes to the hyperglycemic state.

Nutrition support during these conditions is challenging, to say the least. Research indicates that both overfeeding and underfeeding can harm the patient. Advances in enteral and par-

enteral feeding have allowed refinement of this nutrition support practice, and today medical nutrition therapy can certainly support the trauma patient appropriately and adequately.

The cases in this section allow you to assess a patient with a closed head injury from a motor vehicle accident. Closed head injuries are an excellent example of the posttraumatic, hypermetabolic state. Determining nutritional needs, prescribing appropriate nutrition support, and monitoring of daily progress are all involved in this case. Other examples of metabolic stress include sepsis and burns. These situations also demand close attention to nutrition support to minimize complications of protein–calorie malnutrition and to optimize recovery through medical nutrition therapy. You can easily transfer the same concepts for nutrition assessment and support to other individual cases you may encounter.

The remaining cases in this section involve HIV and AIDS. These conditions may seem very different from closed head injury, but in many ways the metabolic response is similar. Viral load, opportunistic infections, and the presence of wasting syndrome all can increase energy expenditure and shift substrate metabolism. Other issues for HIV and AIDS are included in these cases. Drug–nutrient interactions, biochemical indices of viral load, and appropriate nutrition education are all crucial aspects of medical nutrition therapy for the patient with HIV and AIDS.

# Closed Head Injury: Metabolic Stress with Nutrition Support

*Introductory Level*

## Objectives

After completing this case, the student will be able to:

1. Describe the pathophysiology of head injury.
2. Delineate the metabolic response to stress and trauma.
3. Identify nutrition goals during both acute care and for discharge.
4. Interpret medical terminology associated with a traumatic brain injury.
5. Explain the roles of members of the health care team involved in both the acute care and rehabilitation of the patient.
6. Determine nutrient, fluid, and electrolyte requirements for children.
7. Demonstrate the ability to calculate enteral formulation appropriate for the diagnosis and individual patient requirements.

Chelsea Montgomery is a 9-year-old girl admitted through the emergency room after being injured as a restrained front-seat passenger in a motor vehicle accident. She is transferred to the neurointensive care unit with a closed head injury.

## ADMISSION DATABASE

Name: Chelsea Montgomery
DOB: 1/12  age 9
Physician: E. Mantio, MD

| BED#<br>1 | DATE:<br>5/24 | TIME:<br>1400 | TRIAGE STATUS (ER ONLY):<br>☒ Red ☐ Yellow ☐ Green ☐ White |
|---|---|---|---|

**PRIMARY PERSON TO CONTACT:**
Name: Jacob and Melanie Montgomery
Home #: 334-421-5689
Work #: 334-351-3200

### Initial Vital Signs

| TEMP:<br>97 | RESP:<br>27 | SAO2: |
|---|---|---|

| HT:<br>4'4" | WT (lb):<br>61 | B/P:<br>138/90 | PULSE:<br>100 |
|---|---|---|---|

**ORIENTATION TO UNIT:** ☐ Call light ☐ Television/telephone
☐ Bathroom ☐ Visiting ☐ Smoking ☐ Meals
☐ Patient rights/responsibilities

| LAST TETANUS<br>6 months ago | LAST ATE<br>lunch today | LAST DRANK<br>? |
|---|---|---|

### CHIEF COMPLAINT/HX OF PRESENT ILLNESS

Admitted through ER–victim of high-speed MVA with head-on collision with truck. Restrained front seat passenger.

**PERSONAL ARTICLES:** (Check if retained/describe)
☐ Contacts ☐ R ☐ L   ☐ Dentures ☐ Upper ☐ Lower
☐ Jewelry:
☐ Other:

### ALLERGIES: Meds, Food, IVP Dye, Seafood: Type of Reaction

NKA

**VALUABLES ENVELOPE:**
☐ Valuables instructions

### PREVIOUS HOSPITALIZATIONS/SURGERIES

**INFORMATION OBTAINED FROM:**
☐ Patient          ☐ Previous record
☒ Family           ☐ Responsible party

Signature: *Melanie Montgomery*

| Home Medications (including OTC) | Codes: A=Sent home | | B=Sent to pharmacy | | C=Not brought in |
|---|---|---|---|---|---|
| Medication | Dose | Frequency | Time of Last Dose | Code | Patient Understanding of Drug |
| multivitamin | 1 | qd | | | |
| | | | | | |
| | | | | | |
| | | | | | |
| | | | | | |
| | | | | | |
| | | | | | |
| | | | | | |
| | | | | | |
| | | | | | |
| | | | | | |

Do you take all medications as prescribed?  ☐ Yes  ☐ No  If no, why?

### PATIENT/FAMILY HISTORY

| | | |
|---|---|---|
| ☐ Cold in past two weeks | ☒ High blood pressure Maternal grandmother | ☐ Kidney/urinary problems |
| ☐ Hay fever | ☒ Arthritis Maternal grandmother | ☐ Gastric/abdominal pain/heartburn |
| ☐ Emphysema/lung problems | ☐ Claustrophobia | ☐ Hearing problems |
| ☐ TB disease/positive TB skin test | ☐ Circulation problems | ☐ Glaucoma/eye problems |
| ☐ Cancer | ☐ Easy bleeding/bruising/anemia | ☐ Back pain |
| ☐ Stroke/past paralysis | ☐ Sickle cell disease | ☐ Seizures |
| ☒ Heart attack Paternal grandfather | ☐ Liver disease/jaundice | ☐ Other |
| ☐ Angina/chest pain | ☐ Thyroid disease | |
| ☒ Heart problems Paternal grandfather | ☒ Diabetes Sibling | |

### RISK SCREENING

Have you had a blood transfusion?  ☐ Yes  ☒ No
Do you smoke?  ☐ Yes  ☐ No
If yes, how many pack(s) _____ /day for _____ years
Does anyone in your household smoke?  ☐ Yes  ☐ No
Do you drink alcohol?  ☐ Yes  ☐ No
If yes, how often?_____  How much?
When was your last drink? _____/_____/_____
Do you take any recreational drugs?  ☐ Yes  ☐ No
If yes, type:_____  Route
Frequency:_____  Date last used:_____/_____/_____

**FOR WOMEN Ages 12–52**

Is there any chance you could be pregnant?  ☐ Yes  ☐ No
If yes, expected date (EDC):_____/_____/_____
Gravida/Para:

**ALL WOMEN**

Date of last Pap smear:_____/_____/_____
Do you perform regular breast self-exams?  ☐ Yes  ☐ No

**ALL MEN**

Do you perform regular testicular exams?  ☐ Yes  ☐ No

Additional comments:

✗ *Ginger Syler, RN*
Signature/Title

**Client name:** Chelsea Montgomery
**DOB:** 1/12
**Age:** 9
**Sex:** Female
**Education:** Less than high school    *What grade/level?* 3rd grade
**Occupation:** Student
**Hours of work:** N/A
**Household members:** Mother age 36, well; father age 37, well; brother age 11, type 1 DM
**Ethnic background:** Caucasian
**Religious affiliation:** Catholic
**Referring physician:** Elizabeth Mantio, MD (intensive care)

**Chief complaint:**
Admitted through ER after high-speed MVA head-on with truck. She was a restrained front seat passenger.

**Patient history:**
*Onset of disease:* N/A
*Type of Tx:* N/A
*PMH:* Full-term infant weighing 9 lb 1 oz, delivered via cesarean. Healthy except for severe myopia. Good student; competitive gymnast, softball player, and participant in Girl Scouts and after-school program
*Meds:* None
*Smoker:* No
*Family Hx: What?* CAD

**Physical exam:**
*General appearance:* 9-year-old female child alternating crying and unconsciousness
*Vitals:* Temp: 97°F,  BP 138/90, HR 100 bpm, RR 27 bpm
*Heart:* RRR, nl S1–S2, tachycardia, no murmur
*HEENT:*
    *Eyes:* Pupils 4 mm reactive; no battle/raccoon signs
    *Ears:* WNL
    *Nose:* WNL
    *Throat:* WNL
*Genitalia:* + rectal tone − heme negative
*Neurologic:* GCS = 10 E4 V2 M4. Obtundation and L-sided hemiparesis. No verbal responses. Withdrawal and moaning when touched
*Extremities:* DTR symmetric, WNL. 3+ lower extremities; 2+ R biceps; 1+ L biceps. 2-cm laceration on R knee
*Skin:* WNL
*Chest/lungs:* Breath sounds bilaterally
*Peripheral vascular:* No ankle edema
*Abdomen:* Soft bowel sounds ↓, linear mark in LUQ, + guarding throughout

**Nutrition Hx:**
*General:* Parents indicate that patient had normal growth and appetite PTA

*Usual dietary intake:*
| | |
|---|---|
| *Breakfast:* | Cereal, juice, milk, toast |
| *Lunch:* | At school cafeteria |
| *Snacks:* | Prior to gymnastics or softball practice, cookies, fruit, juice, or milk |
| *Dinner:* | Meat, pasta or potatoes, rolls or bread. Likes only green beans, corn, and salad as vegetables. Will eat any fruit |
| *24-hr recall:* | NPO |

*Food allergies/intolerances/aversions:* NKA
*Previous MNT?* No
*Food purchase/preparation:* Parents
*Vit/min intake:* General multivitamin with iron

**Nutrition consult** (excerpt from nutrition assessment note):
Recommendations for enteral feeding. Nutren Jr with fiber @ 25 cc/hr. ↑10 cc 4–6 hrs to goal rate 85 cc/hr via continuous drip × 16 hours then gradually switch to bolus as patient tolerates. Start bolus q 4 hours @ 60 cc; then ↑ 120 cc; then ↑ 340 cc. Suggest to ↓ IVF as TF↑
(Signed) P. Marietta, MS, RD

**Tx plan:**
Admit to Neurointensive Care Unit
$D_5$0.9 NS with 10 mEq KCl
Zantac 25 mg q6 hr; Tylenol 450 mg q 6 hr; ibuprofen 200 mg q 6 hr; Zofran 2 mg IV q 6 hr
NPO
NG to LIGS
$O_2$ to keep sat >95%
I/O
Foley to gravity

**Hospital course:**
By day 4, aroused easily—automatic speech of "No-No-No." One-level commands followed. Oriented to parents but not place or time. CT and MRI completed. Rehabilitation consult. Nutrition consult on day 3 for nutrition support recommendations. Patient began PO on hospital day 14. Weaned from enteral feeding completely on hospital day 17. During hospitalization, patient had extensive physical, speech, and occupational therapy. Patient discharged on hospital day 21 with orders for patient direct supervision 24 hours/day, 7 days a week, with gradual removal of restrictions as clinically indicated. Patient to receive PT weekly; OT 3–5 ×/week and speech therapy 3–5 ×/week.

# U̲H̲ ̲U̲N̲I̲V̲E̲R̲S̲I̲T̲Y̲ ̲H̲O̲S̲P̲I̲T̲A̲L̲

NAME: Chelsea Montgomery        DOB: 1/12
AGE: 9        SEX: F
PHYSICIAN: Dr. E. Mantio

\*\*\*\*\*\*\*\*\*\*\*\*\*\*\*\*\*\*\*\*\*\*\*\*\*\*\*\*\*\*\*\*\*\*\*\*\*\*\*\*\*\*\*\*CHEMISTRY\*\*\*\*\*\*\*\*\*\*\*\*\*\*\*\*\*\*\*\*\*\*\*\*\*\*\*\*\*\*\*\*\*\*\*\*\*\*\*\*\*\*

| | NORMAL | DAY: 1<br>DATE: 5/24<br>TIME:<br>LOCATION: | 10<br>6/3 | UNITS |
|---|---|---|---|---|
| Albumin | 3.6-5 | 3.7 | 3.3 | g/dL |
| Total protein | 6-8 | 6.4 | | g/dL |
| Prealbumin | 19-43 | | 19 | mg/dL |
| Transferrin | 200-400 | | | mg/dL |
| Sodium | 135-155 | 142 | 139 | mmol/L |
| Potassium | 3.5-5.5 | 3.9 | 3.6 | mmol/L |
| Chloride | 98-108 | 110 | 113 | mmol/L |
| $PO_4$ | 2.5-4.5 | | | mmol/L |
| Magnesium | 1.6-2.6 | | | mmol/L |
| Osmolality | 275-295 | 286 | 279 | mmol/kg $H_2O$ |
| Total $CO_2$ | 24-30 | | 22 | mmol/L |
| Glucose | 70-120 | 189 | 115 | mg/dL |
| BUN | 8-26 | 6 | 4 | mg/dL |
| Creatinine | 0.6-1.3 | 0.4 | 0.4 | mg/dL |
| Uric acid | 2.6-6 (women)<br>3.5-7.2 (men) | | | mg/dL |
| Calcium | 8.7-10.2 | 8.5 | 9.2 | mg/dL |
| Bilirubin | 0.2-1.3 | 0.3 | | mg/dL |
| Ammonia ($NH_3$) | 9-33 | | | $\mu$mol/L |
| SGPT (ALT) | 10-60 | 141 | | U/L |
| SGOT (AST) | 5-40 | 153 | | U/L |
| Alk phos | 98-251 | 222 | | U/L |
| CPK | 26-140 (women)<br>38-174 (men) | | | U/L |
| LDH | 313-618 | | | U/L |
| CHOL | 140-199 | | | mg/dL |
| HDL-C | 40-85 (women)<br>37-70 (men) | | | mg/dL |
| VLDL | | | | mg/dL |
| LDL | < 130 | | | mg/dL |
| LDL/HDL ratio | < 3.22 (women)<br>< 3.55 (men) | | | |
| Apo A | 101-199 (women)<br>94-178 (men) | | | mg/dL |
| Apo B | 60-126 (women)<br>63-133 (men) | | | mg/dL |
| TG | 35-160 | | | mg/dL |
| $T_4$ | 5.4-11.5 | | | $\mu$g/dL |
| $T_3$ | 80-200 | | | ng/dL |
| $HbA_{1C}$ | 4.8-7.8 | | | % |

Name: Chelsea Montgomery

Physician: Dr. E. Mantio

# PATIENT CARE SUMMARY SHEET

| Date: 6/5 | Room: NICU Bed 3 | Wt. Yesterday: 25.5 kg | Today: 25.2 kg | Postdialysis: | 1b |

| Temp °F | NIGHTS | | | | | | | | DAYS | | | | | | | | EVENINGS | | | | | | | |
|---|---|---|---|---|---|---|---|---|---|---|---|---|---|---|---|---|---|---|---|---|---|---|---|---|
| | 00 | 01 | 02 | 03 | 04 | 05 | 06 | 07 | 08 | 09 | 10 | 11 | 12 | 13 | 14 | 15 | 16 | 17 | 18 | 19 | 20 | 21 | 22 | 23 |
| 105 | | | | | | | | | | | | | | | | | | | | | | | | |
| 104 | | | | | | | | | | | | | | | | | | | | | | | | |
| 103 | | | | | | | | | | | | | | | | | | | | | | | | |
| 102 | | | | | | | | | | | | | | | | | | | | | | | | |
| 101 | | | | | | | | | | | | | | | | | | | | | | | | |
| 100 | | | | | | | | | | | | | | | | | | | | | | | | |
| 99 | | | | | | | | | | | | | | | | | | | | | | | | |
| 98 | | | | | | | | | | | | | | | | | | | | | | | | |
| 97 | | | | | | | | | | | | | | | | | | | | | | | | |
| 96 | | | | | | | | | | | | | | | | | | | | | | | | |
| Pulse | 108 | | | | | | | | 94 | | | | | | | | 100 | | | | | | | |
| Respiration | 20 | | | | | | | | 20 | | | | | | | | 24 | | | | | | | |
| BP | 100/51 | | | | | | | | 121/62 | | | | | | | | 124/72 | | | | | | | |
| Blood Glucose | | | | | | | | | | | | | | | | | | | | | | | | |
| Appetite/Assist | NG | | | | | | | | NG | | | | | | | | NG | | | | | | | |
| INTAKE | | | | | | | | | | | | | | | | | | | | | | | | |
| Oral | | | | | | | | | | | | | | | | | | | | | | | | |
| IV | | | | | | | | | | | | | | | | | | | | | | | | |
| TF Formula/Flush | 85 | 85 | 85 | 85 | 85 | 85 | 85 | 85/30 | 85 | 85 | 85 | 85 | 85 | 85 | 85 | 85 | 85 | 85 | * | | | | | 50 |
| Shift Total | 680 TF + 30 flush | | | | | | | | 680 TF | | | | | | | | 220 | | | | | | | |
| OUTPUT | | | | | | | | | INC | | | | | | | | | | | | | | | |
| Void | | 50 | | | | | | | 50 | | | | | 100 | | | 70 | | 100 | | | | | |
| Cath | | | | | | | | | | | | | | | | | | | | | | | | |
| Emesis | | | | | | | | | | | | | | | | | | | | | | | | |
| BM | | | | | | | | 1-soft | | | | | | | | | | | | | | | | |
| Drains | | | | | | | | | | | | | | | | | | | | | | | | |
| Shift Total | INC | | | | | | | | INC | | | | | | | | INC | | | | | | | |
| Gain | | | | | | | | | | | | | | | | | | | | | | | | |
| Loss | | | | | | | | | | | | | | | | | | | | | | | | |
| Signatures | C. Taylor, RN | | | | | | | | B. Phillips, RN | | | | | | | | M. Frazier, RN | | | | | | | |
| | | | | | | | | | | | | | | | | | | | | | | | | |
| | | | | | | | | | | | | | | | | | | | | | | | | |

* Held for residual.

**DEPARTMENT OF RADIOLOGY**

CT Report

Date: 5/24

Patient: Chelsea Montgomery

DOB: 1/12   age 9

Physician: Elizabeth Mantio, MD

Two areas of increased density in L frontal lobe near vertex and possibly left central modality.

*Victoria Roundtree, MD*

Department of Radiology

---

**DEPARTMENT OF RADIOLOGY**

MRI Report

Date: 5/29

Patient: Chelsea Montgomery

DOB: 1/12   age 9

Physician: Elizabeth Mantio, MD

MRI showed areas of hemorrhagic edema in deep white matter of L frontal lobe anteriorly. Additionally heme and edema found in the splenium of corpus callosum. 3.4 cm x 4.2 cm x 1.0 cm representing areas of shearing injury.

*James Morgan, MD*

Department of Radiology

5/29

---

**DEPARTMENT OF SPEECH PATHOLOGY**

RE: Interpretation of video fluoroscopy and speech/swallowing evaluation

Date: 6/3 Hospital day 10

Patient: Chelsea Montgomery

DOB: 1/12   age 9

Physician: Elizabeth Mantio, MD

Patient accepted macaroni and cheese with appropriate tongue lateralization and chewing skills but choked after 5-7 ice chips. Oral skills appropriate. Showed significant signs of fatigue and decreased cooperation after a few swallows, which therefore inhibited PO feeding. Video swallow studies showed no evidence of penetration or aspiration.

*Carol Davie, MS, SLP*

## Case Questions

1. What is the GCS (Glasgow Coma Scale)?

2. What was Chelsea's initial GCS score? Is anything in the initial physical assessment consistent with this score? Explain.

3. Define the following terms found in the admitting history and physical:

   a. *Intensivist:*

   b. *L-sided hemiparesis:*

4. Read the CT scan and MRI report. The CT scan report was very general, noting density in the frontal lobe. The MRI indicated more localized areas of edema and blood in the frontal lobe. It also discusses a shearing injury.

   a. What causes edema and bleeding in a traumatic brain injury?

   b. What general functions occur in the frontal lobe? How might Chelsea's injury affect her in the long term?

5. What factors place the patient with traumatic brain injury at nutritional risk?

6. Chelsea's height is 132 cm, and her weight on admission is 27.7 kg. At 9 years of age, what is the most appropriate method to evaluate her height and weight? Assess her height and weight.

7. What method should you use to determine Chelsea's energy and protein requirements? After specifying your method, determine her energy and protein needs.

8. Chelsea was to receive a goal rate of Nutren Jr with fiber @ 85 cc/hour. How much energy and protein would this provide? Show your calculations. Does it meet her needs?

9.   Using the patient care summary sheet, answer the following:

   a.  What was the total volume of feeding she received on June 5?

   b.  What was the nutritional value of her feeding for that day? Calculate the total energy and protein.

   c.  What percentage of her needs was met?

   d.  There is a note on the evening shift that the feeding was held for high residual. What does that mean?

   e.  What is the usual procedure for handling a high gastric residual? How do you think Chelsea's situation was handled?

   f.  What other information would you assess on the daily flow sheet to determine her tolerance to the enteral feeding?

   g.  Look at the additional information on the patient care summary sheet. Are there any factors of concern? Explain.

10.  Evaluate Chelsea's laboratory data. Note any changes from admission day labs to June 3. Are any changes of nutritional concern?

11.  On June 6, a 24-hour urine sample was collected for nitrogen balance. On this day, she received 1650 cc of Nutren Jr. Her total nitrogen output was 14 grams.

   a.  Calculate her nitrogen balance from this information. Show all your calculations.

   b.  How would you assess this information? Explain your response in the context of her hypermetabolism.

    **c.** Are there any factors that may affect the accuracy of this test?

    **d.** The intern taking care of Chelsea pages you when he reads your note regarding her negative nitrogen balance. He asks whether he should change the enteral formula to one higher in nitrogen. Explain the results in the context of the metabolic stress response.

**12.** Chelsea has worked with occupational therapy, speech therapy, and physical therapy. Summarize the training that each of these professionals receives and what their role might be for Chelsea's rehabilitation.

**13.** The speech pathologist saw Chelsea for a swallowing evaluation on hospital day 10. (See p. 383.)

    **a.** What is a video fluoroscopy?

    **b.** What factors were noted that support the need for enteral feeding at this time?

**14.** As Chelsea's recovery proceeds, she begins a PO mechanical soft diet. Her calorie counts are as follows:

(10/14)
Oatmeal ¼ c
Brown sugar 2 T
Whole milk 1 c
240 cc Carnation Instant Breakfast (CIB) prepared with 2% milk
Mashed potatoes 1 c
Gravy 2 T
(10/15)
Cheerios 1 c
Whole milk 1 c
240 cc CIB prepared with 2% milk
Grilled cheese sandwich (2 slices bread, 1 oz American cheese, 1 t margarine)
Jell-o 1 c
240 cc CIB prepared with 2% milk

    **a.** Calculate her intake and average for these two days of calorie counts.

    **b.** What recommendations would you make regarding her enteral feeding?

# Bibliography

Bloch A, Mueller C. Enteral and Parenteral Nutrition Support. In Mahan LK, Escott-Stump S (eds.), *Krause's Food, Nutrition, and Diet Therapy,* 10th ed. Philadelphia: Saunders, 2000:463–482.

Brain Trauma Foundation. *Guidelines for the prehospital management of traumatic brain injury.* New York: Brain Trauma Foundation, 2000. Available at www.braintrauma.org. Accessed December 5, 2000.

Brain Trauma Task Force. Management and prognosis of severe traumatic brain injury. *J Neurotrauma.* 2000; 17:451–553.

Commission on Accreditation for Dietetics Education. Knowledge, skills, and competencies for dietitians. *Accreditation Manual for Dietetics Education Programs,* rev. 4th ed. Chicago: American Dietetic Association, 2000.

Ghajar J. Traumatic brain injury. *Lancet.* 2000;356: 923–929.

Gould BE. *Pathophysiology for the Health-Related Professions.* Philadephia: Saunders, 1997:320–376.

Hall CA. Patient management in head injury care: a nursing perspective. *Intensive and Crit Care Nurs.* 1997;13:329–337.

Marshall LF, Gautille T, Klauber MR, et al. The outcome of severe closed head injury. *Neurosurg.* 1991; 75:S28–S36.

Murray CJL, Lopez AD. Global mortality, disability and the contribution of risk factors: global burden of disease study. *Lancet.* 1997;349:1436–1442,1498–1504.

Rimel RW, Giordani B, Barth JT, et al. Disability caused by mild head injury. *Neurosurg.* 1981;9:3221–3228.

Sosin DM, Sniezek JE, Thurman DJ. Incidence of mild and moderate brain injury in the United States, 1991. *Brain Injury.* 1996;10:47–54.

Winkler MF, Manchester S. Medical nutrition therapy for metabolic stress: sepsis, trauma, burns, and surgery. In Mahan LK, Escott-Stump S (eds.), *Krause's Food, Nutrition, and Diet Therapy,* 10th ed. Philadelphia: Saunders, 2000:722–741.

# Case 34

# Human Immunodeficiency Virus (HIV)

*Introductory Level*

## Objectives

After completing this case, the student will be able to:

1. Describe the pathophysiology of the human immunodeficiency virus.
2. Identify the relationship of transmission during pregnancy and breastfeeding.
3. Determine nutritional risk factors for the patient with HIV.
4. Complete nutrition screening and nutrition assessment for the patient with HIV.
5. Interpret laboratory and anthropometric measurements for medical and nutritional significance.

6. Explain HAART pharmacological treatment for HIV and potential drug–nutrient interactions.
7. Develop nutrition care plan for a patient with HIV.

Ms. Kimberly Nicholson, a 23-year-old woman, is admitted to University Hospital with nausea, vomiting, and diarrhea. Ms. Nicholson is HIV positive and has recently been started on a new triple-therapy drug regimen.

# ADMISSION DATABASE

Name: Kimberly Nicholson
DOB: 6/23   age 23
Physician: S. Zhargam, MD

| BED# 1 | DATE: 5/22 | TIME: 0900 | TRIAGE STATUS (ER ONLY): ☐ Red ☐ Yellow ☐ Green ☐ White |
|---|---|---|---|

### Initial Vital Signs

| TEMP: 98.2 | RESP: 23 | SAO2: |
|---|---|---|

| HT: 5'3" | WT (lb): 205 | B/P: 135/70 | PULSE: 82 |
|---|---|---|---|

| LAST TETANUS 8 years ago | LAST ATE last pm | LAST DRANK last pm |
|---|---|---|

**PRIMARY PERSON TO CONTACT:**
Name: Janet Johnson (mother)
Home #: 222-7865
Work #: n/a

**ORIENTATION TO UNIT:** ☒ Call light  ☒ Television/telephone
☒ Bathroom  ☒ Visiting  ☒ Smoking  ☒ Meals
☒ Patient rights/responsibilities

### CHIEF COMPLAINT/HX OF PRESENT ILLNESS

"I've been throwing up a lot and I have diarrhea since I started on these new medications."

**ALLERGIES: Meds, Food, IVP Dye, Seafood: Type of Reaction**

NKA

### PREVIOUS HOSPITALIZATIONS/SURGERIES

Cesarean section 16 months ago

**PERSONAL ARTICLES:** (Check if retained/describe)
☐ Contacts ☐ R ☐ L      ☐ Dentures ☐ Upper ☐ Lower
☐ Jewelry:
☐ Other:

**VALUABLES ENVELOPE:**
☐ Valuables instructions

**INFORMATION OBTAINED FROM:**
☒ Patient     ☐ Previous record
☒ Family      ☐ Responsible party

Signature: *Kimberly Nicholson*

| Home Medications (including OTC) | Codes: A=Sent home | | B=Sent to pharmacy | | C=Not brought in |
|---|---|---|---|---|---|
| Medication | Dose | Frequency | Time of Last Dose | Code | Patient Understanding of Drug |
| Epivir | | | | c | no |
| Zerit | | | | c | no |
| Crixivan | | | | c | no |
| | | | | | |
| | | | | | |
| | | | | | |
| | | | | | |
| | | | | | |
| | | | | | |
| | | | | | |

Do you take all medications as prescribed?  ☐ Yes  ☒ No   If no, why? I try, but this change in medications has been hard.

### PATIENT/FAMILY HISTORY

| | | |
|---|---|---|
| ☐ Cold in past two weeks | ☒ High blood pressure Mother | ☐ Kidney/urinary problems |
| ☐ Hay fever | ☒ Arthritis Maternal grandmother | ☐ Gastric/abdominal pain/heartburn |
| ☐ Emphysema/lung problems | ☐ Claustrophobia | ☐ Hearing problems |
| ☐ TB disease/positive TB skin test | ☐ Circulation problems | ☐ Glaucoma/eye problems |
| ☒ Cancer Maternal grandmother | ☐ Easy bleeding/bruising/anemia | ☐ Back pain |
| ☐ Stroke/past paralysis | ☐ Sickle cell disease | ☐ Seizures |
| ☐ Heart attack | ☐ Liver disease/jaundice | ☐ Other |
| ☐ Angina/chest pain | ☐ Thyroid disease | |
| ☐ Heart problems | ☒ Diabetes Mother | |

### RISK SCREENING

Have you had a blood transfusion?  ☐ Yes  ☒ No
Do you smoke?  ☒ Yes  ☐ No
If yes, how many pack(s) _1_ /day for _8_ years
Does anyone in your household smoke?  ☒ Yes  ☐ No
Do you drink alcohol?  ☒ Yes  ☐ No
If yes, how often? 3-4 x week     How much? 3-4 beers
When was your last drink? _5_/_19_/
Do you take any recreational drugs?  ☒ Yes  ☐ No
If yes, type: marijuana     Route smoking
Frequency: 1x week     Date last used: 5/15

**FOR WOMEN Ages 12–52**

Is there any chance you could be pregnant?  ☐ Yes  ☒ No
If yes, expected date (EDC):_____/_____/
Gravida/Para: 1/1

**ALL WOMEN**

Date of last Pap smear: _1_/_12_/last year
Do you perform regular breast self-exams?  ☐ Yes  ☒ No

**ALL MEN**

Do you perform regular testicular exams?  ☐ Yes  ☐ No

Additional comments:

**✗** *Mary Webb, RN, BSN*
Signature/Title

**Client name:** Kimberly Nicholson
**DOB:** 6/23
**Age:** 23
**Sex:** Female
**Education:** Less than high school    *What grade/level?* 11th grade
**Occupation:** Unemployed
**Hours of work:** N/A
**Household members:** Mother age 38, sister age 19, niece age 8 months, daughter age 16 months
**Ethnic background:** Caucasian
**Religious affiliation:** Baptist
**Referring physician:** Steve Zargham, MD (infectious disease)

## Chief complaint:
"I've been throwing up a lot and I have diarrhea since I started on these new medications."

## Patient history:
*Onset of disease:* Patient was diagnosed as HIV seropositive three years ago. Her previous boyfriend was also HIV positive. She began treatment with AZT at 14th week of recent pregnancy.
*Type of Tx:* Recently has begun combination therapy with Epivir, Zerit, and Crixivan
*PMH:* Gravida 1/para 1, 38-week-gestation, female infant weighing 6 lb 4 oz; no other contributing conditions
*Meds:* Epivir, Zerit, Crixivan
*Smoker:* Yes
*Family Hx: What?* CAD    *Who?* Maternal grandparents

## Physical exam:
*General appearance:* Obese young woman in no acute distress
*Vitals:* Temp 98.2°F, BP 135/70, HR 82 bpm/normal, RR 23 bpm
*Heart:* Regular rate and rhythm without murmurs or gallops
*HEENT:*
    *Head:* Exam normal
    *Eyes:* PERRLA
    *Ears:* Clear
    *Nose:* Clear
    *Throat:* Thyroid nonpalpable
*Genitalia:* WNL
*Neurologic:* Alert and oriented; strength 5/5 throughout
*Extremities:* Without edema
*Skin:* Warm, dry; no petechia or ecchymoses
*Chest/lungs:* Lungs clear to auscultation and percussion
*Abdomen:* Obese with bowel sounds in all four quadrants; no abdominal bruits, tenderness, masses, or organomegaly; well-healed cesarean incision

## Nutrition Hx:
*General:* Appetite fair. Client relates that she wakes up about 7 AM to take first meds and then goes back to bed. She has first food in the middle of the day. Her schedule thereafter is sporadic, and she generally eats when she feels like it. States that she hates to exercise and does not do any physical

activity except care for her daughter. Had cesarean section for delivery of her daughter, weighing 6 lb 4 oz at 38 weeks gestation. Patient did not breastfeed her daughter because she thought she might pass the virus to her that way.

*Usual dietary intake:* Drinks Kool-Aid and soda throughout day. Intake is generally sandwiches or fast food. Cooks about twice a week and eats out all other meals from McDonalds, Kentucky Fried Chicken, and other fast-food restaurants.

*24-hr recall:*

| | |
|---|---|
| *9 am:* | 8 oz Kool-Aid with sugar, 1 oz ham, 1 oz bologna on 1 slice white bread |
| *4:30 pm:* | McDonald's double cheeseburger, large fries, 22 oz orange soda |
| *1 am:* | 2 sausage patties, 1 slice bread, 2 12-oz cans Pepsi |

*Food allergies/intolerances/aversions (specify):* Does not like vegetables
*Previous MNT?* No
*Food purchase/preparation:* Self, mother, and sister
*Vit/min intake:* None
*Anthropometric data:* Ht 5′3″, Wt 205#, prepregnancy wt 193# with total weight gain during pregnancy of 27#. TSF 34 mm; MAC—380 mm

## Tx plan:

*Diagnosis:* HIV—R/O gastroenteritis, dehydration
*Condition:* Stable
*Vitals:* Routine/Q shift; allergies: NKDA; diet: clear liquids advance as tolerated
*Activity:* Ad lib
*Lab:* CBC, CD4, CD8, and HIV viral load, SMA, U/A, stool sample for O&P
*Meds:* Epivir 225 mg bid; Zerit 75 mg tid; Crixivan 600 mg tid
*PRN Meds:* MOM, Mylanta, Tylenol, benzodiazepine as needed
Nutrition, social services, and pharmacy consults

## Hospital course:

Patient was discharged 48 hours after admission with diagnosis of viral gastroenteritis. The patient did not experience any further vomiting or diarrhea since admission.

# UH *UNIVERSITY HOSPITAL*

NAME: Kimberly Nicholson                    DOB: 6/23
AGE: 23                                     SEX: F
PHYSICIAN: S. Zhargam, MD


\*\*\*\*\*\*\*\*\*\*\*\*\*\*\*\*\*\*\*\*\*\*\*\*\*\*\*\*\*\*\*\*\*\*\*\*\*\*\*\*\*\*\*\*\*CHEMISTRY\*\*\*\*\*\*\*\*\*\*\*\*\*\*\*\*\*\*\*\*\*\*\*\*\*\*\*\*\*\*\*\*\*\*\*\*\*\*\*\*\*\*\*\*

DAY:                                            1
DATE:
TIME:
LOCATION:

| | NORMAL | | UNITS |
|---|---|---|---|
| Albumin | 3.6–5 | 4.2 | g/dL |
| Total protein | 6–8 | 6 | g/dL |
| Prealbumin | 19–43 | 40 | mg/dL |
| Transferrin | 200–400 | 250 | mg/dL |
| Sodium | 135–155 | 138 | mmol/L |
| Potassium | 3.5–5.5 | 4.3 | mmol/L |
| Chloride | 98–108 | 101 | mmol/L |
| $PO_4$ | 2.5–4.5 | 3.2 | mmol/L |
| Magnesium | 1.6–2.6 | 2.1 | mmol/L |
| Osmolality | 275–295 | 286 | mmol/kg $H_2O$ |
| Total $CO_2$ | 24–30 | 27 | mmol/L |
| Glucose | 70–120 | 119 | mg/dL |
| BUN | 8–26 | 12 | mg/dL |
| Creatinine | 0.6–1.3 | 0.8 | mg/dL |
| Uric acid | 2.6–6 (women) | 3.2 | mg/dL |
| | 3.5–7.2 (men) | | |
| Calcium | 8.7–10.2 | 9.1 | mg/dL |
| Bilirubin | 0.2–1.3 | 0.8 | mg/dL |
| Ammonia ($NH_3$) | 9–33 | | $\mu$mol/L |
| SGPT (ALT) | 10–60 | 15 | U/L |
| SGOT (AST) | 5–40 | 21 | U/L |
| Alk phos | 98–251 | 101 | U/L |
| CPK | 26–140 (women) | | U/L |
| | 38–174 (men) | | |
| LDH | 313–618 | | U/L |
| CHOL | 140–199 | 185 | mg/dL |
| HDL-C | 40–85 (women) | 45 | mg/dL |
| | 37–70 (men) | | |
| VLDL | | | mg/dL |
| LDL | < 130 | 129 | mg/dL |
| LDL/HDL ratio | < 3.22 (women) | 2.8 | |
| | < 3.55 (men) | | |
| Apo A | 101–199 (women) | | mg/dL |
| | 94–178 (men) | | |
| Apo B | 60–126 (women) | | mg/dL |
| | 63–133 (men) | | |
| TG | 35–160 | 155 | mg/dL |
| $T_4$ | 5.4–11.5 | | $\mu$g/dL |
| $T_3$ | 80–200 | | ng/dL |
| $HbA_{1C}$ | 4.8–7.8 | | % |

# UH UNIVERSITY HOSPITAL

NAME: Kimberly Nicholson             DOB: 6/23
AGE: 23                              SEX: F
PHYSICIAN: S. Zhargham, MD

\*\*\*\*\*\*\*\*\*\*\*\*\*\*\*\*\*\*\*\*\*\*\*\*\*\*\*\*\*\*\*\*\*\*\*\*\*\*\*HEMATOLOGY\*\*\*\*\*\*\*\*\*\*\*\*\*\*\*\*\*\*\*\*\*\*\*\*\*\*\*\*\*\*\*\*\*\*\*\*\*\*\*

DAY:                                             1
DATE:
TIME:
LOCATION:

| | NORMAL | | UNITS |
|---|---|---|---|
| WBC | 4.3–10 | 5.6 | $\times 10^3/mm^3$ |
| RBC | 4–5 (women) | 4.2 | $\times 10^6/mm^3$ |
| | 4.5–5.5 (men) | | |
| HGB | 12–16 (women) | 12.5 | g/dL |
| | 13.5–17.5 (men) | | |
| HCT | 37–47 (women) | 37 | % |
| | 40–54 (men) | | |
| MCV | 84–96 | 88 | fL |
| MCH | 27–34 | 30 | pg |
| MCHC | 31.5–36 | 33 | % |
| RDW | 11.6–16.5 | 12.5 | % |
| Plt ct | 140–440 | 320 | $\times 10^3$ |
| Diff TYPE | | | |
| % GRANS | 34.6–79.2 | | % |
| % LYM | 19.6–52.7 | | % |
| SEGS | 50–62 | | % |
| BANDS | 3–6 | | % |
| LYMPHS | 25–40 | | % |
| MONOS | 3–7 | | % |
| EOS | 0–3 | | % |
| TIBC | 65–165 (women) | | $\mu g/dL$ |
| | 75–175 (men) | | |
| Ferritin | 18–160 (women) | | $\mu g/dL$ |
| | 18–270 (men) | | |
| Vitamin $B_{12}$ | 100–700 | | pg/mL |
| Folate | 2–20 | | ng/mL |
| HIV viral load | 0 | 10,000 | $mm^3$ |
| CD4 | 589–1505 | 425 | $mm^3$ |
| CD8 | 325–997 | 300 | $mm^3$ |
| PT | 11–13 | | sec |

## Case Questions

1.  Describe the human immunodeficiency virus.

2.  How is HIV transmitted?

3.  Identify the stages of infection for HIV.

4.  The patient's history indicates that Ms. Nicholson was HIV seropositive prior to her pregnancy. What tests are used to measure seropositivity? What is the risk that her daughter will also be HIV positive? What is the probable relationship between this risk and the initiation of drug treatment during Ms. Nicholson's pregnancy?

5.  Was the patient correct in choosing not to breastfeed? What are the current recommendations regarding breastfeeding for women who are HIV positive?

6.  Identify the mechanism of action for each medication that Ms. Nicholson is currently taking. In addition, determine any nutrient interactions.

| Medication | Action | Nutrient Interaction/ Nutritional Side Effects | Nutrition Recommendations |
| --- | --- | --- | --- |
|  |  |  |  |
|  |  |  |  |
|  |  |  |  |
|  |  |  |  |

7.  Calculate Ms. Nicholson's healthy weight, % healthy weight, %UBW, and BMI. What do these figures tell you about her weight?

**8.** Evaluate and interpret her other anthropometric information that is available to you.

**9.** Calculate Ms. Nicholson's energy and protein needs.

**10.** Evaluate Ms. Nicholson's biochemical indices.

    **a.** Which labs would you use to monitor Ms. Nicholson's HIV status? What do they indicate about her medical status?

    **b.** Interpret Ms. Nicholson's labs for nutritional significance.

**11.** Can she be classified as having AIDS? Explain.

**12.** How would you assess Ms. Nicholson's dietary information?

**13.** Is Ms. Nicholson's diet adequate? Explain.

**14.** Identify at least three nutritional risk factors for Ms. Nicholson.

**15.** For each risk factor, determine the appropriate interventions and the expected outcome.

| Risk Factor | Intervention | Outcome |
| --- | --- | --- |
|  |  |  |
|  |  |  |
|  |  |  |
|  |  |  |
|  |  |  |
|  |  |  |

**16.** You note that Ms. Nicholson uses marijuana. Is this information significant? How would you use this information in planning your nutritional care?

**17.** What are the important factors to include when planning nutrition education for Ms. Nicholson?

**18.** Complete an initial SOAP note for Ms. Nicholson:

S:


O:


A:


P:

## Bibliography

Castaneda D. HIV/AIDS-related services for women and the rural community context. *AIDS Care.* 2000; 12(5):549–565.

Commission on Accreditation for Dietetics Education. Knowledge, skills, and competencies for dietitians. *Accreditation Manual for Dietetics Education Programs,* rev. 4th ed. Chicago: American Dietetic Association, 2000.

Fenton M, Silverman E. Medical nutrition therapy for human immunodeficiency virus infection and acquired immunodeficiency syndrome. In Mahan LK, Escott-Stump S (eds.), *Krause's Food, Nutrition, and Diet Therapy,* 10th ed. Philadelphia: Saunders, 2000: 887–911.

Fields-Gardner C, Thomson CA, Rhodes SS. *A Clinician's Guide to Nutrition in HIV and AIDS.* Chicago: American Dietetic Association, 1997.

Gerbert B, Bronstone A, Clanon K, Abercrombie P, Bangsberg D. Combination antiretroviral therapy: health care providers confront emerging dilemmas. *AIDS Care.* 2000;12(4):409–421.

*HIV Clinical Management.* Available at http://www.medscape.com/Medscape/HIV/ClinicalMgmt/CM.drug/CM.drug05.html. Accessed March 10, 2001.

Lert F. Advances in HIV treatment and prevention: should treatment optimism lead to prevention pessimism? *AIDS Care.* 2000;12(6):745–755.

Meystre-Agustoni G, Dubois-Arber F, Cochand P, Telenti A. Antiretroviral therapies from the patient's perspective. *AIDS Care.* 2000;12(6):717–721.

Miller S, Exner TM, Williams SP, Ehrhardt AA. A gender-specific intervention for at-risk women in the USA. *AIDS Care.* 2000;12(5):603–612.

Simoni JM, Ng MT. Trauma, coping, and depression among women with HIV/AIDS in New York City. *AIDS Care.* 2000;12(5):567–580.

**Client name:** Terry Long
**DOB:** 5/12
**Age:** 32
**Sex:** Male
**Education:** Bachelor's degree
**Occupation:** Currently on disability but previously worked as nurse in dialysis clinic
**Hours of work:** N/A
**Household members:** Father age 69, mother age 66, both well
**Ethnic background:** African American
**Religious affiliation:** AME (African Methodist Episcopal)
**Referring physician:** Agnes Fremont, MD (family medicine/internal medicine)

## Chief complaint:

"I was diagnosed with HIV four years ago when I was living in St. Louis. I just recently moved back home because I am not able to work right now. I have not been treated before but I am pretty sure I will need to be. I feel exhausted all the time—I have a really sore mouth and throat. I have lost a lot of weight. I think I've just been denying that I may have AIDS. But a lot of people I know are doing OK on drugs, so I came to this new physician. The case manager at the Health Department set it up for me. The doc thinks that I may have pneumonia as well, so she admitted me for a full workup."

## Patient history:

*Onset of disease:* Seropositive for HIV-1 confirmed by ELISA and Western Blot four years previously. Etiology of contraction not known but was employed in high-risk environment. Admits to intercourse with multiple partners but denies same-sex intercourse.
*Type of Tx:* None
*PMH:* Tonsillectomy age 6; appendectomy age 18
*Meds:* Multivitamin, vitamin E, vitamin C, ginseng, milk thistle, echinacea
*Smoker:* No—quit five years ago
*Family Hx: What?* CAD, HTN *Who?* Father

## Physical exam:

*General appearance:* Thin African American male in no acute distress
*Vitals:* Temp 98.6°F, BP 120/84, HR 92 bpm/normal, RR 18 bpm
*Heart:* Regular rate and rhythm—normal heart sounds
*HEENT:*
   *Eyes:* PERRLA
   *Ears:* Unremarkable
   *Nose:* Mucosa pink without drainage
   *Throat:* Erythematous with white, patchy exudate
*Genitalia:* Rectal exam normal. Stool: heme negative
*Neurologic:* Oriented × 3, no focal motor or sensory deficits, cranial nerves intact, DTR +2 in all groups
*Extremities:* Good pulses, no edema
*Skin:* Warm, dry, with flaky patches
*Chest/lungs:* Rhonchi in lower left lung
*Abdomen:* Nondistended, nontender, hyperactive bowel sounds

**Nutrition Hx:**

*General:* Patient describes appetite as OK but not normal. "I have always been a picky eater. There are a lot of foods that I don't like. But in the last few days it is the sores in my mouth and throat that have made the biggest difference. It hurts pretty badly, and I can hardly even drink. I have been reading about nutrition and HIV on the Internet—I've been trying to do some research. That's when I started taking more supplements. I thought if I wasn't eating like I should that I could at least take supplements. They are expensive, though, so I don't have them every day like I probably should. My highest weight ever was about 175#, which was during college almost 10 years ago. But I have never been this thin as an adult."

MAC 10"; TSF 7 mm; % body fat 12.5%, Ht 6'1", Wt 151#, UBW 160–165#

*Usual dietary intake* (before mouth sores):

| | |
|---|---|
| *Breakfast/lunch:* | ("I usually don't get up before noon because I stay up really late.") cold cereal 1–2 c, ½ c whole milk |
| *Supper:* | Meat, potatoes or rice, tea or soda |
| *Snacks:* | Pizza, candy bar, or cookies with tea or soda. Drinks 1–2 beers or glass of wine several times a week |
| *24-hr recall:* | Sips of apple juice, yogurt 1 c, rice and gravy 1 c, iced tea with sugar—sips throughout the day |

*Food allergies/intolerances/aversions (specify):* Can only tolerate small amount of milk at a time; does not like beef, coffee, or vegetables (except salad)

*Previous MNT?* No

*Food purchase/preparation:* Parent(s), self

*Vit/min intake:* Multivitamin 1 qd, vitamin E 1500 IU qd, vitamin C 500 mg qid, ginseng 500 mg bid, milk thistle 200 mg bid, echinacea 3 capsules daily (88.5 mg per capsule)

**Tx plan:**

Admit: R/O progression to AIDS, oral candidasis, R/O pneumonia; CXR, WBC with diff, CD4, and viral load; begin D5 ½ NS @ 100 cc/hr; fluconazole IV

**Hospital course:**

*Dx:* AIDS—clinical category C2 with oral thrush; no clinical evidence of pneumonia; antiretroviral regimen initiated of AZT (zidovudine) 200 mg q 8 hr, Crixivan (indinavir) 800 mg q 8 hr, and 3TC Epivir (lamivudine) 150 mg q 12 hr; d/c on antiretroviral regimen and oral Diflucan

# $U_H$ _UNIVERSITY HOSPITAL_

NAME: Terry Long                    DOB: 5/12
AGE: 32                             SEX: M
PHYSICIAN: A. Fremont, MD

\*\*\*\*\*\*\*\*\*\*\*\*\*\*\*\*\*\*\*\*\*\*\*\*\*\*\*\*\*\*\*\*\*\*\*\*\*\*\*\*\*\*\*\*\*CHEMISTRY\*\*\*\*\*\*\*\*\*\*\*\*\*\*\*\*\*\*\*\*\*\*\*\*\*\*\*\*\*\*\*\*\*\*\*\*\*\*\*\*\*\*\*\*\*

DAY:                                                1
DATE:                                               10/17
TIME:
LOCATION:

| | NORMAL | | UNITS |
|---|---|---|---|
| Albumin | 3.6–5 | 3.6 | g/dL |
| Total protein | 6–8 | 6.0 | g/dL |
| Prealbumin | 19–43 | 17 | mg/dL |
| Transferrin | 200–400 | 201 | mg/dL |
| Sodium | 135–155 | 142 | mmol/L |
| Potassium | 3.5–5.5 | 3.6 | mmol/L |
| Chloride | 98–108 | 101 | mmol/L |
| $PO_4$ | 2.5–4.5 | 3.2 | mmol/L |
| Magnesium | 1.6–2.6 | 1.8 | mmol/L |
| Osmolality | 275–295 | 292 | mmol/kg $H_2O$ |
| Total $CO_2$ | 24–30 | 27 | mmol/L |
| Glucose | 70–120 | 75 | mg/dL |
| BUN | 8–26 | 11 | mg/dL |
| Creatinine | 0.6–1.3 | 0.8 | mg/dL |
| Uric acid | 2.6–6 (women) | 5.2 | mg/dL |
| | 3.5–7.2 (men) | | |
| Calcium | 8.7–10.2 | 9.1 | mg/dL |
| Bilirubin | 0.2–1.3 | 0.9 | mg/dL |
| Ammonia ($NH_3$) | 9–33 | | $\mu$mol/L |
| SGPT (ALT) | 10–60 | 12 | U/L |
| SGOT (AST) | 5–40 | 17 | U/L |
| Alk phos | 98–251 | 102 | U/L |
| CPK | 26–140 (women) | 110 | U/L |
| | 38–174 (men) | | |
| LDH | 313–618 | 642 | U/L |
| CHOL | 140–199 | 150 | mg/dL |
| HDL–C | 40–85 (women) | 42 | mg/dL |
| | 37–70 (men) | | |
| VLDL | | | mg/dL |
| LDL | < 130 | 114 | mg/dL |
| LDL/HDL ratio | < 3.22 (women) | 2.7 | |
| | < 3.55 (men) | | |
| Apo A | 101–199 (women) | | mg/dL |
| | 94–178 (men) | | |
| Apo B | 60–126 (women) | | mg/dL |
| | 63–133 (men) | | |
| TG | 35–160 | 78 | mg/dL |
| $T_4$ | 5.4–11.5 | | $\mu$g/dL |
| $T_3$ | 80–200 | | ng/dL |
| $HbA_{1C}$ | 4.8–7.8 | | % |

# U H *UNIVERSITY HOSPITAL*

NAME: Terry Long                    DOB: 5/12
AGE: 32                             SEX: M
PHYSICIAN: A. Fremont, MD

\*\*\*\*\*\*\*\*\*\*\*\*\*\*\*\*\*\*\*\*\*\*\*\*\*\*\*\*\*\*\*\*\*\*\*\*\*\*\*\*\*\*\*HEMATOLOGY\*\*\*\*\*\*\*\*\*\*\*\*\*\*\*\*\*\*\*\*\*\*\*\*\*\*\*\*\*\*\*\*\*\*\*\*\*\*\*\*\*\*\*

DAY:                                        1
DATE:                                     10/17
TIME:                                     0400
LOCATION:                                  UMC

| | NORMAL | | UNITS |
|---|---|---|---|
| WBC | 4.3–10 | 8.5 | $\times 10^3/mm^3$ |
| RBC | 4–5 (women) | 5.2 | $\times 10^6/mm^3$ |
| | 4.5–5.5 (men) | | |
| HGB | 12–16 (women) | 14.2 | g/dL |
| | 13.5–17.5 (men) | | |
| HCT | 37–47 (women) | 40 | % |
| | 40–54 (men) | | |
| MCV | 84–96 | 96 | fL |
| MCH | 27–34 | 34.2 | pg |
| MCHC | 31.5–36 | 35.5 | % |
| RDW | 11.6–16.5 | 16.3 | % |
| Plt ct | 140–440 | 220 | $\times 10^3$ |
| Diff TYPE | | | |
| % GRANS | 34.6–79.2 | 82 | % |
| % LYM | 19.6–52.7 | 3 | % |
| SEGS | 50–62 | 51 | % |
| BANDS | 3–6 | 4 | % |
| LYMPHS | 25–40 | 3 | % |
| MONOS | 3–7 | 8 | % |
| EOS | 0–3 | 3 | % |
| TIBC | 65–165 (women) | | µg/dL |
| | 75–175 (men) | | |
| Ferritin | 18–160 (women) | | µg/dL |
| | 18–270 (men) | | |
| Vitamin $B_{12}$ | 100–700 | | pg/mL |
| Folate | 2–20 | | ng/mL |
| Viral load | 0 | 29000 | $mm^3$ |
| CD4 | 1500–4000 | 12 | $mm^3$ |
| CD8 | 325–997 | 157 | $mm^3$ |
| PT | 11–13 | 11.9 | sec |

## Case Questions

1.  How is HIV transmitted? After reading Mr. Long's history and physical, what risk factors would you say he has had for contracting HIV?

2.  Mr. Long says he found out he was HIV positive four years ago. Why is he only symptomatic now?

3.  The history and physical indicate that he is seropositive. What does that mean? The Western Blot and ELISA confirmed that he was seropositive. Describe these tests.

4.  What is thrush, and why might Mr. Long have this condition?

5.  What are common nutritional complications of HIV and AIDS? After reading Mr. Long's history and physical, can you identify any of these complications in him?

6.  After this admission Mr. Long was diagnosed with AIDS, Category C2. What information can you see from his medical record that confirms this diagnosis?

7.  Evaluate the patient's anthropometric information.

    a.  Calculate %UBW and BMI.

    b.  Compare the TSF to population standards. What does this comparison mean? Is this a viable comparison? Explain.

    c.  Using MAC and TSF, calculate upper arm muscle area. What can you infer from this calculation?

    d.  Mr. Long's body fat percentage is 12.5%. What does this mean? Compare to standards.

    e.  Summarize Mr. Long's nutritional risk (if any) from your assessment of anthropometric information.

8.  Evaluate Mr. Long's dietary information. What tools could you use to evaluate his dietary intake? Does he seem to be consuming adequate amounts of food? Can you identify anything from his history that indicates he is having difficulty eating? Explain.

9.  Mr. Long states that he consumes alcohol several times a week. Are there any contraindications for alcohol consumption for him?

10. Using this patient's laboratory values, identify those labs used to monitor his HIV. What do these specifically measure, and how would you interpret them for him?

11. What laboratory values can be used to evaluate nutritional status? Are there any for Mr. Long that identify nutritional risk?

12. Calculate Mr. Long's optimal energy and protein intake. How does this compare to his diet history?

13. What other information would you want to obtain from Mr. Long in order to fully evaluate his nutritional status?

14. Mr. Long was started on three medications that he will be discharged on.

    a. Identify these medications and the purpose of each.

    b. Are there any specific drug–nutrient interactions to be concerned about? Explain.

    c. Is there specific information you would want Mr. Long to know about taking these medications?

15. Mr. Long is taking several vitamin and herbal supplements. Find out why someone with AIDS might take each of the supplements. What would you tell Mr. Long about these supplements? Do they pose any risk? Use the following table to organize your answers.

| Supplement | Proposed Use in HIV/AIDS | Potential Risk |
|---|---|---|
| Vitamin C | | |
| Vitamin E | | |
| Ginseng | | |
| Milk thistle | | |
| Echinacea | | |
| Multivitamin | | |

16. After evaluating Mr. Long's medical record, identify three specific nutritional problems that he is experiencing. Then identify appropriate goal(s) for each problem. Outline a minimum of one intervention you would recommend to assist this patient in meeting each goal.

| Problem | Goal | Intervention |
|---|---|---|
| | | |
| | | |
| | | |
| | | |

17. Patients with AIDS are at increased risk for infection. What nutritional practices would you teach Mr. Long to help him prevent illness related to food or water intake?

18. Why is exercise important as a component of the nutritional care plan? What general recommendations could you give to Mr. Long regarding physical activity?

## Bibliography

Center for Disease Control and Prevention. *Cryptosporidosis: A Guide for Persons with HIV/AIDS.* Atlanta, GA: CDC, 1997. Available at http://www.cdc.gov/mmwr. Accessed November 15, 2000.

Commission on Accreditation for Dietetics Education. Knowledge, skills, and competencies for dietitians. *Accreditation Manual for Dietetics Education Programs,* rev. 4th ed. Chicago: American Dietetic Association, 2000.

Escott-Stump S. *Nutrition and Diagnosis Related Care,* 4th ed. Baltimore: Williams & Wilkins, 1998.

Fenton M, Silverman E. Medical nutrition therapy for human immunodeficiency virus infection and acquired immunodeficiency syndrome. In Mahan LK, Escott-Stump S (eds.), *Krause's Food, Nutrition, and Diet Therapy,* 10th ed. Philadelphia: Saunders, 2000: 887–911.

Fields-Gardner C, Thomson CA, Rhodes SS. *A Clinician's Guide to Nutrition in HIV and AIDS.* Chicago: American Dietetic Association, 1997.

*HIV Clinical Management.* Available at: http://www.medscape.com/Medscape/HIV/ClinicalMgmt/CM.drug/CM.drug05.html. Accessed November 15, 2000.

Mathai K. Integrative medicine and herbal therapy. In Mahan LK, Escott-Stump S (eds.), *Krause's Food, Nutrition, and Diet Therapy,* 10th ed. Philadelphia: Saunders, 2000:415–430.

# MEDICAL NUTRITION THERAPY FOR HEMATOLOGY–ONCOLOGY

## Introduction

The layperson often uses *cancer* as a name for one disease. The term *cancer* or *neoplasm* actually describes any condition where cells proliferate at a rapid rate and in an unrestrained manner. Each type of cancer is a different disease with different origins and responses to therapy. It is difficult to discuss the role of nutrition and cancer, because each diagnosis is truly an individual case. However, it is very obvious to any clinician participating in the care of cancer patients that nutrition problems are common. More than 80% of patients with cancer experience some degree of malnutrition.

Nutrition problems may be some of the first symptoms the patient recognizes. Unexplained weight loss, changes in ability to taste, or decrease in appetite are often present at diagnosis. The malignancy itself may affect not only energy requirements but also the metabolism of nutrients.

As the patient begins therapy for a malignancy—surgery, radiation therapy, chemotherapy, immunotherapy, or bone marrow transplant—nutritional side effects occur that can affect nutritional status. Can nutrition make a difference? Adequate nutrition helps prevent surgical complications, helps meet increased energy and protein requirements, and helps repair and rebuild tissues, which cancer therapies often damage. Furthermore, good nutrition allows increased tolerance of therapy and helps maintain the patient's quality of life. And finally, as with many medical conditions, cancer patients also face significant psychosocial issues.

All these factors must be considered when planning nutritional and medical care. The cases in this section allow you to plan nutritional care for some of the most common problems during cancer diagnosis and therapy. In addition, the cases let you practice nutrition support and tackle psychosocial issues in alternative and complementary therapy.

# Lymphoma: Part One

*Introductory Level*

## Objectives

After completing this case, the student will be able to:

1. Use nutrition assessment knowledge to assess baseline nutritional status.
2. Evaluate laboratory indices for nutritional and medical significance.
3. Identify the diagnostic procedures for solid tumors.
4. Assess dietary information for nutritional adequacy.
5. Evaluate the medical and nutritional side effects of chemotherapy.
6. Determine appropriate nutrition interventions for medical and nutritional side effects of cancer and chemotherapy.

Ms. Denise Mitchell, a 21-year-old college student, is admitted for evaluation of viral illness in which she has experienced night sweats, fevers, and weight loss. A chest X-ray indicates a possible mass. After chest CT, MRI, bone marrow biopsy, and biopsy of suspect lymph nodes, she is diagnosed with stage II diffuse large B-cell lymphoma with mediastinal disease and positive lymph nodes.

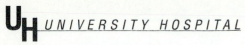

# UNIVERSITY HOSPITAL

## ADMISSION DATABASE

Name: D. Mitchell
DOB: 2/18  age 21
Physician: S. Miller, MD

| BED#<br>1 | DATE:<br>3/8 | TIME:<br>0300 | TRIAGE STATUS (ER ONLY):<br>☐ Red  ☐ Yellow  ☐ Green  ☐ White |
|---|---|---|---|

### Initial Vital Signs

| TEMP:<br>100.5 | RESP:<br>18 | | SAO2: |
|---|---|---|---|
| HT:<br>5'6" | WT (lb): 120<br>(UBW:130) | B/P:<br>95/70 | PULSE:<br>85 |
| LAST TETANUS | | LAST ATE<br>this am | LAST DRANK<br>this am |

### CHIEF COMPLAINT/HX OF PRESENT ILLNESS

"I have continued to feel sick since I had the flu. I still have a fever sometimes, and the cough won't go away."

### ALLERGIES: Meds, Food, IVP Dye, Seafood: Type of Reaction

NKA

### PREVIOUS HOSPITALIZATIONS/SURGERIES

Tonsillectomy–age 5

---

**PRIMARY PERSON TO CONTACT:**
Name: Mrs. Mitchell (mother)
Home #: 212-555-1322
Work #: same

ORIENTATION TO UNIT: ☒ Call light  ☒ Television/telephone
☒ Bathroom  ☒ Visiting  ☒ Smoking  ☒ Meals
☒ Patient rights/responsibilities

PERSONAL ARTICLES: (Check if retained/describe)
☒ Contacts ☒ R ☒ L      ☐ Dentures ☐ Upper ☐ Lower
☐ Jewelry:
☐ Other:

VALUABLES ENVELOPE:
☒ Valuables instructions

INFORMATION OBTAINED FROM:
☒ Patient      ☐ Previous record
☒ Family       ☐ Responsible party

Signature  *Denise Mitchell*

---

| Home Medications (including OTC) | | Codes: A=Sent home | B=Sent to pharmacy | | C=Not brought in |
|---|---|---|---|---|---|
| Medication | Dose | Frequency | Time of Last Dose | Code | Patient Understanding of Drug |
| Dimetapp | | occ | 9 pm yesterday | A | yes |
| | | | | | |
| | | | | | |
| | | | | | |
| | | | | | |
| | | | | | |
| | | | | | |
| | | | | | |
| | | | | | |
| | | | | | |

Do you take all medications as prescribed?  ☐ Yes  ☐ No   If no, why?

### PATIENT/FAMILY HISTORY

| | | |
|---|---|---|
| ☒ Cold in past two weeks Patient | ☒ High blood pressure Paternal grandfather | ☐ Kidney/urinary problems |
| ☐ Hay fever | ☐ Arthritis | ☐ Gastric/abdominal pain/heartburn |
| ☐ Emphysema/lung problems | ☐ Claustrophobia | ☐ Hearing problems |
| ☐ TB disease/positive TB skin test | ☐ Circulation problems | ☐ Glaucoma/eye problems |
| ☒ Cancer Maternal grandmother | ☐ Easy bleeding/bruising/anemia | ☐ Back pain |
| ☐ Stroke/past paralysis | ☐ Sickle cell disease | ☐ Seizures |
| ☒ Heart attack Paternal grandfather | ☐ Liver disease/jaundice | ☐ Other |
| ☒ Angina/chest pain Paternal grandfather | ☐ Thyroid disease | |
| ☒ Heart problems Paternal grandfather | ☐ Diabetes | |

### RISK SCREENING

Have you had a blood transfusion?  ☐ Yes  ☒ No
Do you smoke?  ☐ Yes  ☒ No
If yes, how many pack(s)        /day for        years
Does anyone in your household smoke?  ☐ Yes  ☒ No
Do you drink alcohol?  ☐ Yes  ☒ No
If yes, how often?        How much?
When was your last drink?        /        /
Do you take any recreational drugs?  ☐ Yes  ☒ No
If yes, type:_____   Route
Frequency:_____   Date last used:_____/____/____

**FOR WOMEN Ages 12–52**

Is there any chance you could be pregnant?  ☐ Yes  ☒ No
If yes, expected date (EDC):
Gravida/Para:

**ALL WOMEN**

Date of last Pap smear: 08/20/
Do you perform regular breast self-exams?  ☒ Yes  ☐ No

**ALL MEN**

Do you perform regular testicular exams?  ☐ Yes  ☐ No

---

Additional comments:

✗ *S. Smith, RN, BSN*
Signature/Title

**Client name:**  Denise Mitchell
**DOB:**  2/18
**Age:**  21
**Sex:**  Female
**Education:**  College student
**Occupation:**  Student
**Hours of work:**  N/A
**Household members:**  Mother age 45; father age 50; brothers ages 9, 12, and 16
**Ethnic background:**  Caucasian
**Religious affiliation:**  Methodist
**Referring physician:**  Simon Miller, MD (hematology/oncology)

## Chief complaint:

"I have continued to feel run down since I had the flu. I still have a fever sometimes and the cough won't go away."

## Patient history:

*Onset of disease:* Ms. Mitchell is a 21-year-old female who currently is a sophomore at an Ivy League university. She has had an uneventful medical history with no significant illness until the past 2–3 months. Patient describes having the "flu" and feeling run down ever since. She has continued to have fevers, especially at night; she describes having to change her nightgown and bedclothes due to excessive sweating. She now presents for admission on referral from her family physician.
*Type of Tx:* None at present
*PMH:* Tonsillectomy age 5
*Meds:* OTC cough medicine
*Smoker:* No
*Family Hx:* Noncontributory

## Physical exam:

*General appearance:* Patient is a thin, pale young woman who appears tired.
*Vitals:* Temp 100.5°F, BP 95/70 mm Hg, HR 85 bpm, RR 18 bpm
*Heart:* Regular rate and rhythm, no gallops or rubs, point of maximal impulse at the fifth intercostal space in the midclavicular line.
*HEENT:*
  *Head:* Normocephalic
  *Eyes:* Extraocular movements intact; wears glasses for myopia; fundi grossly normal bilaterally
  *Ears:* Tympanic membranes normal
  *Nose:* Dry mucous membranes without lesions
  *Throat:* Slightly dry mucous membranes without exudates or lesions; abnormal lymph nodes
*Genitalia:* Normal without lesions
*Neurologic:* Alert and oriented, cranial nerves II–XII grossly intact, strength 5/5 throughout, sensation to light touch intact, normal gait, and normal reflexes
*Extremities:* Normal muscular tone with normal ROM, nontender
*Skin:* Warm and dry without lesions
*Chest/lungs:* Respirations are shallow, dullness present to percussion

*Peripheral vascular:* Pulse +2 bilaterally, warm and nontender

*Abdomen:* Normal active bowel sounds, soft and nontender without masses or organomegaly

## Nutrition Hx:

*General:* Appetite decreased. No nausea, vomiting, constipation, or diarrhea

*Usual dietary intake:*

| | |
|---|---|
| AM: | Cold cereal, toast or doughnut, skim milk, juice |
| Lunch: | (In college cafeteria) Sandwich or salad, frozen yogurt, chips or pretzels, soda |
| PM: | Meat (eats only chicken and fish), 1–2 vegetables including a salad, iced tea, or skim milk |
| Snack: | Popcorn, occasionally pizza, soda, juice, iced tea |

*24-hr recall:*

| | |
|---|---|
| AM: | 1 slice dry toast, plain hot tea |
| Lunch: | ½ cup ice cream, ¼ cup fruit cocktail, few bites of other foods on tray |
| Dinner: | Few bites of chicken, 2 T mashed potatoes, ½ c Jell-o, plain hot tea |

*Food allergies/intolerances/aversions:* NKA

*Previous MNT?* No

*Food purchase/preparation:* Self, parents, college cafeteria

*Vit/min intake:* None

## Hospital course:

Chest X-ray indicated possible mass. After chest CT, MRI, bone marrow biopsy, and biopsy of suspect lymph nodes

*Dx:* Stage II diffuse large B-cell lymphoma with mediastinal disease and positive lymph nodes. Bone marrow and other organs show no indication of disease.

## Tx plan:

A chemotherapy regimen of cyclophosphamide, doxorubicin, vincristine, and prednisone (CHOP) is prescribed. Prednisone will be administered orally on the first five days of each cycle, and the other chemotherapeutic medications will be given intravenously on the first day of each 21-day cycle. Radiotherapy is planned to start three weeks after the third cycle of CHOP.

## Case Questions

1. What type of cancer is lymphoma?

2. What symptoms found in her history and physical are consistent with the classic signs of lymphoma?

3. Ms. Mitchell's diagnosis stated that she had stage II lymphoma. What does this mean, and how does her physical examination support this?

4. Generally, patients with cancer are treated with surgery, radiation therapy, chemotherapy, immunotherapy, or a combination of therapies. Ms. Mitchell's medical plan indicates that she will have both chemotherapy and radiation therapy. How does chemotherapy act to treat malignant cells? How does radiation therapy act to treat the malignancy?

5. Radiation and chemotherapy may also affect healthy tissues.

   a. What other cells in the body may be affected by either or both of these treatments?

   b. What symptoms may the patient experience from the destruction of these cells?

6. Identify each of the drugs that the patient is prescribed, and note the possible nutritional side effects of each. In general, what might you tell this patient to expect from receiving her chemotherapy?

| Drug | Possible Nutritional Side Effect(s) |
|---|---|
| Cyclophosphamide | |
| Doxorubicin | |
| Vincristine | |
| Prednisone | |

7.  Calculate this patient's body mass index and the percent usual body weight. How do they differ? Which is the most appropriate to determine nutritional risk for this patient?

8.  Using the Harris-Benedict equation, calculate the patient's energy requirements. Which weight (IBW, UBW, or current body weight) should you use to accurately calculate the patient's energy needs?

9.  Which labs can be used to assess protein status?

    a.  Which labs will reflect acute changes in protein status versus chronic changes? Why?

    b.  Which are available for this patient? Considering her diagnosis, which labs would *not* be appropriate to use to evaluate protein status?

    c.  Determine the nutritional risk associated with this patient's laboratory value. Would you request any additional nutrition assessment labs?

10. Calculate the patient's protein requirements.

11. How would you assess the dietary information gathered for usual nutritional intake?

12. What additional information would you ask the patient to provide regarding her usual intake?

13. Using one of the methods you have identified, determine whether this patient's usual intake is adequate to meet her needs. Explain.

14. What method would you use to assess her 24-hour recall? Is it adequate to meet her needs? Explain.

**15.** What common side effects of her illness may affect her dietary intake and subsequently her nutritional status?

**16.** What physical symptom(s) is this patient experiencing that might affect her dietary intake?

**17.** For each symptom, identify at least two interventions.

| Symptom | Intervention |
|---------|--------------|
|         |              |
|         |              |

**18.** How would you follow up or evaluate the interventions you have determined?

**19.** What types of nutrition education would be important to provide for this patient? When would it be appropriate to provide this education? What factors might interfere with the patient's reception of nutrition education?

# Bibliography

*Ann Arbor Staging System.* Available at: http:// www.1.nai.med.Kyushu-u.acjp/~skoba/hematools/ annarbor.html. Accessed February 26, 2001.

Blackburn GL, et al. Nutritional and metabolic assessment of the hospitalized patient. *J Parenter Enteral Nutr.* 1997;1(1):11.

Commission on Accreditation for Dietetics Education. Knowledge, skills, and competencies for dietitians. *Accreditation Manual for Dietetics Education Programs,* rev. 4th ed. Chicago: American Dietetic Association, 2000.

Darbinian J, Coulston A. Impact of chemotherapy on the nutrition status of the cancer patient. In Bloch A (ed.), *Nutrition Management of the Cancer Patient.* Rockville, MD: Aspen, 1990:161–171.

Hammond K. Dietary and clinical assessment. In Mahan LK, Escott-Stump S (eds.), *Krause's Food, Nutrition, and Diet Therapy,* 10th ed. Philadelphia: Saunders, 2000:353–379.

Nahikian-Nelms M. General feeding problems. In Bloch A (ed.), *Nutrition Management of the Cancer Patient.* Rockville, MD: Aspen, 1990:41–52.

Ross BT. Cancer's impact on nutrition status. In Bloch A (ed.), *Nutrition Management of the Cancer Patient.* Rockville, MD: Aspen, 1990:10–13.

Ross BT. Impact of radiation therapy on the nutrition status of the cancer patient: an overview. In Bloch A (ed.), *Nutrition Management of the Cancer Patient.* Rockville, MD: Aspen, 1990:173–180.

Shils M. Nutrition needs of cancer patients. In Bloch A (ed.), *Nutrition Management of the Cancer Patient.* Rockville, MD: Aspen, 1990:3–10.

Tyler V. *Herbs of Choice: The Therapeutic Use of Phytomedicinals.* New York: Haworth Press, 1994.

Tyler V. *The Honest Herbal,* 3rd ed. Binghamton, NY: Pharmaceutical Products Press, 1993.

# Lymphoma: Part Two

*Advanced Practice*

## Objectives

After completing this case, the student will be able to:

1. Use nutrition assessment knowledge to assess baseline nutritional status.
2. Evaluate laboratory indices for nutritional and medical significance.
3. Assess dietary information for nutritional adequacy.
4. Evaluate the medical and nutritional side effects of chemotherapy.
5. Determine appropriate nutrition interventions for medical and nutritional side effects of cancer and chemotherapy.
6. Evaluate literature supporting efficacy of alternative therapies for cancer.
7. Determine appropriate strategies for counseling cancer patients using nontraditional medical care.

Denise Mitchell is readmitted with a probable fungal infection 20 days after receiving her first round of chemotherapy. She is immunosuppressed and dehydrated.

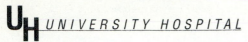 **UNIVERSITY HOSPITAL**

## ADMISSION DATABASE

Name: D. Mitchell
DOB: 2/18 age 21
Physician: S. Miller, MD

| BED# 2 | DATE: 3/19 | TIME: 0300 | TRIAGE STATUS (ER ONLY): ☐ Red ☐ Yellow ☐ Green ☐ White |
|---|---|---|---|

**Initial Vital Signs**

| TEMP: 102.5 | RESP: 22 | SAO2: |
|---|---|---|

| HT: 5'6" | WT (lb): 108 | B/P: 95/70 | PULSE: 90 |
|---|---|---|---|

| LAST TETANUS | LAST ATE last pm | LAST DRANK last pm |
|---|---|---|

**PRIMARY PERSON TO CONTACT:**
Name: Mrs. Mitchell (mother)
Home #: 212-555-1322
Work #: same

**ORIENTATION TO UNIT:** ☒ Call light ☒ Television/telephone ☒ Bathroom ☒ Visiting ☒ Smoking ☒ Meals ☒ Patient rights/responsibilities

### CHIEF COMPLAINT/HX OF PRESENT ILLNESS

Pt's mother states that Ms. Mitchell's mouth hurts so badly that she can hardly talk. She has had only puréed foods, vegetable juices, and a little water over the last 3 days.

### ALLERGIES: Meds, Food, IVP Dye, Seafood: Type of Reaction

NKA

**PERSONAL ARTICLES:** (Check if retained/describe)
☐ Contacts ☐ R ☐ L       ☐ Dentures ☐ Upper ☐ Lower
☐ Jewelry:
☒ Other: glasses

**VALUABLES ENVELOPE:** none
☐ Valuables instructions

### PREVIOUS HOSPITALIZATIONS/SURGERIES

Biopsy last admission for diagnosis of stage II diffuse large B-cell lymphoma

**INFORMATION OBTAINED FROM:**
☒ Patient       ☐ Previous record
☒ Family        ☐ Responsible party

Signature  *Denise Mitchell*

**Home Medications (including OTC)**       Codes: A=Sent home       B=Sent to pharmacy       C=Not brought in

| Medication | Dose | Frequency | Time of Last Dose | Code | Patient Understanding of Drug |
|---|---|---|---|---|---|
| Tylox | 250 mg | q 6 hours | this am | C | yes |
| | | | | | |
| | | | | | |
| | | | | | |
| | | | | | |
| | | | | | |
| | | | | | |
| | | | | | |
| | | | | | |
| | | | | | |

Do you take all medications as prescribed?   ☒ Yes   ☐ No   If no, why?

### PATIENT/FAMILY HISTORY

| | | |
|---|---|---|
| ☐ Cold in past two weeks | ☒ High blood pressure Paternal grandfather | ☐ Kidney/urinary problems |
| ☐ Hay fever | ☐ Arthritis | ☐ Gastric/abdominal pain/heartburn |
| ☐ Emphysema/lung problems | ☐ Claustrophobia | ☐ Hearing problems |
| ☐ TB disease/positive TB skin test | ☐ Circulation problems | ☐ Glaucoma/eye problems |
| ☒ Cancer Patient | ☐ Easy bleeding/bruising/anemia | ☐ Back pain |
| ☐ Stroke/past paralysis | ☐ Sickle cell disease | ☐ Seizures |
| ☒ Heart attack Paternal grandfather | ☐ Liver disease/jaundice | ☐ Other |
| ☒ Angina/chest pain Paternal grandfather | ☐ Thyroid disease | |
| ☒ Heart problems Paternal grandfather | ☐ Diabetes | |

### RISK SCREENING

Have you had a blood transfusion?   ☐ Yes   ☒ No
Do you smoke?   ☐ Yes   ☒ No
If yes, how many pack(s)    /day for    years
Does anyone in your household smoke?   ☐ Yes   ☒ No
Do you drink alcohol?   ☐ Yes   ☒ No
If yes, how often?   How much?
When was your last drink?    /    /
Do you take any recreational drugs?   ☐ Yes   ☒ No
If yes, type:_____   Route
Frequency:_____   Date last used:_____/_____/_____

**FOR WOMEN Ages 12–52**

Is there any chance you could be pregnant?   ☐ Yes   ☒ No
If yes, expected date (EDC):
Gravida/Para:

**ALL WOMEN**

Date of last Pap smear: 8/20
Do you perform regular breast self-exams?   ☐ Yes   ☒ No

**ALL MEN**

Do you perform regular testicular exams?   ☐ Yes   ☐ No

Additional comments: Patient is day 20 post initiation of chemotherapy regimen of cyclophosphamide, doxorubicin, vincristine, and prednisone. She is scheduled to begin radiation therapy after third course of chemo.

✗ *Sherryl Jones, RN*
Signature/Title

**Client name:** Denise Mitchell
**DOB:** 2/18
**Age:** 21
**Sex:** Female
**Education:** Some college
**Occupation:** Student
**Hours of work:** N/A
**Household members:** Mother age 45; father age 50; brothers ages 9, 12, and 16
**Ethnic background:** Caucasian
**Religious affiliation:** Methodist
**Referring physician:** Simon Miller, MD, Hematology/Oncology

## Chief complaint:

Ms. Mitchell's mother states that patient's mouth hurts so badly that she can hardly talk. She has had only puréed fruits, vegetable juices, and a little water over the last 3 days. Patient's mother states that the patient's aunt and uncle introduced them to a special "anticancer" diet that "will help her cure her disease." Ms. Mitchell was told to only eat natural foods without any meat or dairy products. She was also started on 5000 mg of vitamin C, 20,000 μg RE of vitamin A, and 1000 mg of vitamin E. The patient is also taking a mixture of echinacea, ginger, and ginseng. The patient tells you privately that she doesn't know what to believe about "this diet stuff" but doesn't want to make anyone mad. She also states that the vegetable juices and fruits make her mouth burn worse. "I'd really like some sherbet, but they said I couldn't have any."

## Patient history:

*Onset of disease:* Ms. Mitchell is a 21-year-old female recently diagnosed with stage II diffuse large B-cell lymphoma with mediastinal disease and positive lymph nodes. She is day 20 s/p her first chemotherapy cycle of cyclophosphamide, doxorubicin, vincristine, and prednisone (CHOP).
*PMH:* Tonsillectomy at age 5
*Meds:* Prescribed Tylox for pain, which she started taking yesterday
*Smoker:* No

## Physical exam:

*General appearance:* Patient is a thin, pale young woman in obvious distress.
*Vitals:* Temp 102.5°F, BP 95/70 mmHg, HR 90 bpm, RR 22 bpm
*Heart:* Regular rate and rhythm, no gallops or rubs, point of maximal impulse at the fifth intercostal space in the midclavicular line.
*HEENT:*
  *Head:* Normocephalic
  *Eyes:* Extraocular movements intact; wears glasses for myopia; fundi grossly normal bilaterally
  *Ears:* Tympanic membranes normal
  *Nose:* Dry mucous membranes without lesions
  *Throat:* Dry, bright red mucous membranes with white exudate; abnormal lymph nodes noted
*Genitalia:* Normal without lesions
*Neurologic:* Alert and oriented, cranial nerves II–XII grossly intact, strength 5/5 throughout, sensation to light touch intact, normal gait, and normal reflexes
*Extremities:* Normal muscular tone with normal ROM, nontender

*Skin:* Warm and dry without lesions
*Chest/lungs:* Respirations are shallow; dullness present to percussion
*Peripheral vascular:* Pulse +2 bilaterally, warm and nontender
*Abdomen:* Normal active bowel sounds, soft and nontender without masses or organomegaly

**Nutrition Hx:**
*General:* Appetite poor. Nausea, vomiting during the actual chemotherapy and several times since, no constipation, slight diarrhea since chemotherapy

*24-hr recall* (prior to admission):
AM:        2 T applesauce, 1 oz orange juice
Lunch:     Nothing
Dinner:    Carrot, tomatoes, and broccoli blended into juice—few sips

*Food allergies/intolerances/aversions (specify):* NKA
*Previous MNT?* Yes    *If yes, when:* Last admission    *Where?* At this hospital. RD discussed high-calorie, high-protein diet.
*Food purchase/preparation:* Self, parents
*Vit/min intake:* Vitamins C, A, E, and herbal supplements

**Tx plan:**
*Dx:* Immunosuppression, candidiasis fungal infection, and dehydration, R/O pneumonia
IVF D5 1/2 NS with 20 mEq KCL @ 50cc/hr
Ancef 500 mg q 6 hr
ABGs q 6 hr CXR—EPA/LAT
Sputum cultures and Gram stain
Begin nystatin swish and swallow
Nutrition consult

# U<sub>H</sub> _UNIVERSITY HOSPITAL_

NAME: Denise Mitchell                    DOB: 2/18
AGE: 21                                  SEX: F
PHYSICIAN: S. Miller, MD

\*\*\*\*\*\*\*\*\*\*\*\*\*\*\*\*\*\*\*\*\*\*\*\*\*\*\*\*\*\*\*\*\*\*\*\*\*\*\*\*\*\*\*\*CHEMISTRY\*\*\*\*\*\*\*\*\*\*\*\*\*\*\*\*\*\*\*\*\*\*\*\*\*\*\*\*\*\*\*\*\*\*\*\*\*\*\*\*\*\*\*

| DAY: | | Admit 1 | Admit 2 | Admit 2 | |
|---|---|---|---|---|---|
| DATE: | | | Day 1 | Day 2 | |
| TIME: | | | | | |
| LOCATION: | | | | | |
| | NORMAL | | | | UNITS |
| Albumin | 3.6–5 | 3.3 | 3.5 | 3.0 | g/dL |
| Total protein | 6–8 | 5.5 | 6.0 | 5.4 | g/dL |
| Prealbumin | 19–43 | | | | mg/dL |
| Transferrin | 200–400 | | | | mg/dL |
| Sodium | 135–155 | 141 | 149 | 138 | mmol/L |
| Potassium | 3.5–5.5 | 3.8 | 4.8 | 3.6 | mmol/L |
| Chloride | 98–108 | 100 | 108 | 102 | mmol/L |
| $PO_4$ | 2.5–4.5 | 3.9 | 4.4 | 4.1 | mmol/L |
| Magnesium | 1.6–2.6 | | | | mmol/L |
| Osmolality | 275–295 | 292 | 301 | 285 | mmol/kg $H_2O$ |
| Total $CO_2$ | 24–30 | | | | mmol/L |
| Glucose | 70–120 | 105 | 110 | 98 | mg/dL |
| BUN | 8–26 | 14 | 20 | 12 | mg/dL |
| Creatinine | 0.6–1.3 | 0.6 | 1.1 | 0.4 | mg/dL |
| Uric acid | 2.6–6 (women) | | | | mg/dL |
| | 3.5–7.2 (men) | | | | |
| Calcium | 8.7–10.2 | 9.2 | 9.1 | 8.8 | mg/dL |
| Bilirubin | 0.2–1.3 | 0.8 | 1.2 | 1.0 | mg/dL |
| Ammonia ($NH_3$) | 9–33 | | | | μmol/L |
| SGPT (ALT) | 10–60 | 15 | 13 | 11 | U/L |
| SGOT (AST) | 5–40 | 10 | 20 | 13 | U/L |
| Alk phos | 98–251 | 110 | 180 | 150 | U/L |
| CPK | 26–140 (women) | | | | U/L |
| | 38–174 (men) | | | | |
| LDH | 313–618 | | | | U/L |
| CHOL | 140–199 | 141 | | 120 | mg/dL |
| HDL-C | 40–85 (women) | | | | mg/dL |
| | 37–70 (men) | | | | |
| VLDL | | | | | mg/dL |
| LDL | < 130 | | | | mg/dL |
| LDL/HDL ratio | < 3.22 (women) | | | | |
| | < 3.55 (men) | | | | |
| Apo A | 101–199 (women) | | | | mg/dL |
| | 94–178 (men) | | | | |
| Apo B | 60–126 (women) | | | | mg/dL |
| | 63–133 (men) | | | | |
| TG | 35–160 | | | | mg/dL |
| $T_4$ | 5.4–11.5 | | | | μg/dL |
| $T_3$ | 80–200 | | | | ng/dL |
| $HbA_{1c}$ | 4.8–7.8 | | | | % |

**U<sub>H</sub>** _UNIVERSITY HOSPITAL_

NAME: Denise Mitchell                    DOB: 2/18
AGE: 21                                  SEX: F
PHYSICIAN: S. Miller, MD

\*\*\*\*\*\*\*\*\*\*\*\*\*\*\*\*\*\*\*\*\*\*\*\*\*\*\*\*\*\*\*\*\*\*\*\*\*\*\*\*\*\*\*\*HEMATOLOGY\*\*\*\*\*\*\*\*\*\*\*\*\*\*\*\*\*\*\*\*\*\*\*\*\*\*\*\*\*\*\*\*\*\*\*\*\*\*\*\*\*\*\*\*

| | NORMAL | Admit 1 | Admit 2 Day 2 | UNITS |
|---|---|---|---|---|
| DAY: | | Admit 1 | Admit 2 | |
| DATE: | | | Day 2 | |
| TIME: | | | | |
| LOCATION: | | | | |
| WBC | 4.3–10 | 12.0 | .01100 | $\times\ 10^3/mm^3$ |
| RBC | 4–5 (women) | 4.2 | 3.1 | $\times\ 10^6/mm^3$ |
| | 4.5–5.5 (men) | | | |
| HGB | 12–16 (women) | 11 | 9 | g/dL |
| | 13.5–17.5 (men) | | | |
| HCT | 37–47 (women) | 31 | 29 | % |
| | 40–54 (men) | | | |
| MCV | 84–96 | 70 | 65 | fL |
| MCH | 27–34 | 28 | 22 | pg |
| MCHC | 31.5–36 | 27 | 21 | % |
| RDW | 11.6–16.5 | | | % |
| Plt ct | 140–440 | | | $\times\ 10^3$ |
| Diff TYPE | | | | |
| % GRANS | 34.6–79.2 | | | % |
| % LYM | 19.6–52.7 | 23 | n/a | % |
| SEGS | 50–62 | | | % |
| BANDS | 3–6 | | | % |
| LYMPHS | 25–40 | | | % |
| MONOS | 3–7 | | | % |
| EOS | 0–3 | | | % |
| TIBC | 65–165 (women) | | | $\mu$g/dL |
| | 75–175 (men) | | | |
| Ferritin | 18–160 (women) | | | $\mu$g/dL |
| | 18–270 (men) | | | |
| Vitamin $B_{12}$ | 100–700 | | | pg/mL |
| Folate | 2–20 | | | ng/mL |
| Total T cells | 812–2318 | | | $mm^3$ |
| T-helper cells | 589–1505 | | | $mm^3$ |
| T-suppressor cells | 325–997 | | | $mm^3$ |
| PT | 11–13 | | | sec |

## Case Questions

1. Evaluate the patient's physical examination. What is abnormal? Are they consistent with her admitting diagnoses?

2. In the physical examination, what do you think is the first clue that the patient may have a fungal infection?

3. Evaluate the patient's weight. How has this changed from her first admission? How does this weight differ from her usual body weight? Are the changes (if any) significant? Why or why not?

4. Review Ms. Mitchell's chemistry labs. The labs have changed from day 1 to day 2 of this admission. What is the most likely cause?

5. How has this patient's albumin changed? Should you be concerned?

6. What conclusions can you draw about her nutritional status?

7. This patient has a decreased Hgb and Hct. What is the most likely cause for the variation in these values? What tests could be run to determine if the patient has an iron deficiency anemia?

8. Evaluate her white blood cell count. Why is this patient's white blood cell count decreased? What is this condition called? How is this related to her diagnosis for this admission?

9. What precautions need to be followed for this patient while her white blood cell count is low?

10.   What questions would you ask the patient regarding her supplement intake? Is it important to know the form of the supplement she is consuming? For example, will it matter if the patient is taking beta-carotene or retinol as her vitamin A supplement? Are there any risks for this patient when taking the current dosages she has indicated to you? Explain.

11.   What are the supplements that Ms. Mitchell is taking? For what conditions are they commonly prescribed? Is there any indication that her current herbal supplements may be beneficial? Are there any potential risks?

12.   How would you advise this patient regarding adherence to this "anticancer" diet? Why may cancer patients be especially vulnerable to nutrition quackery?

13.   What diet would you recommend for Ms. Mitchell?

14.   Are there other suggestions you could make to increase her oral intake?

15.   How would you monitor the effectiveness of the nutrition therapy you have recommended? What nutritional parameters would you follow?

16.   Complete a SOAP note for this patient's readmission.

      S:

      O:

      A:

      P:

# Bibliography

*Ann Arbor Staging System.* Available at: http://www.1.nai.med.Kyushu-u.acjp/~skoba/hematools/annarbor.html. Accessed February 26, 2001.

Blackburn GL, et al. Nutritional and metabolic assessment of the hospitalized patient. *J Parenter Enteral Nutr.* 1997;1(1):11.

Commission on Accreditation for Dietetics Education. Knowledge, skills, and competencies for dietitians. *Accreditation Manual for Dietetics Education Programs,* rev. 4th ed. Chicago: American Dietetic Association, 2000.

Darbinian J, Coulston A. Impact of chemotherapy on the nutrition status of the cancer patient. In Bloch A (ed.), *Nutrition Management of the Cancer Patient.* Rockville, MD: Aspen, 1990:161–171.

Hammond K. Dietary and clinical assessment. In Mahan LK, Escott-Stump S (eds.), *Krause's Food, Nutrition, and Diet Therapy,* 10th ed. Philadelphia: Saunders, 2000:353–379.

Nahikian-Nelms M. General feeding problems. In Bloch A (ed.), *Nutrition Management of the Cancer Patient.* Rockville, MD: Aspen, 1990:41–52.

Ross BT. Cancer's impact on nutrition status. In Bloch A (ed.), *Nutrition Management of the Cancer Patient.* Rockville, MD: Aspen, 1990:10–13.

Ross BT. Impact of radiation therapy on the nutrition status of the cancer patient: an overview. In Bloch A (ed.), *Nutrition Management of the Cancer Patient.* Rockville, MD: Aspen, 1990:173–180.

Shils M. Nutrition needs of cancer patients. In Bloch A (ed.), *Nutrition Management of the Cancer Patient.* Rockville, MD: Aspen, 1990:3–10.

Tyler V. *Herbs of Choice: The Therapeutic Use of Phytomedicinals.* New York: Haworth Press, 1994.

Tyler V. *The Honest Herbal,* 3rd ed. Binghamton, NY: Pharmaceutical Products Press, 1993.

# Acute Leukemia Treated with Total Body Irradiation, Chemotherapy, and Bone Marrow Transplant

*Advanced Practice*

*Deborah Cohen, MMSc, RD, CNSD*

## Objectives

After completing this case, the student will be able to:

1. Use nutrition knowedge to assess baseline nutrition status.
2. Evaluate laboratory indices for nutritional and medical significance.
3. Explain the rationale for using high-dose chemotherapy, total body irradiation, and bone marrow rescue in the treatment of hematologic malignancies.
4. Describe the side effects of high-dose chemotherapy and total body irradiation and their nutritional implications.
5. Describe the nutritional side effects of the pharmacologic agents used in conjunction with BMT.
6. Discuss the role of parenteral nutrition support after BMT and explain the rationale for macro- and micronutrient changes to the parenteral formula/solution.
7. Explain the mechanism of graft versus host disease (GVHD).
8. State the nutritional implications of graft versus host disease and the drugs used for its treatment.
9. Discuss the rationale for limiting oral intake during and after a bout with gastrointestinal GVHD.

Mrs. Rachel Dean, a 25-year-old chemical engineer, is admitted to University Hospital for an allogenic bone marrow transplant. She recently received chemotherapy for treatment of acute myelogenous leukemia, which is now in complete remission.

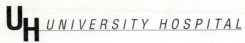

## ADMISSION DATABASE

Name: Rachel Dean
DOB: 5/24  age 25
Physician: M. Hansen

| BED#<br>A | DATE:<br>4/15 | TIME:<br>0730 | TRIAGE STATUS (ER ONLY):<br>☐ Red  ☐ Yellow  ☐ Green  ☐ White |
|---|---|---|---|

**PRIMARY PERSON TO CONTACT:**
Name: Jeffrey Dean
Home #: 402-555-7789
Work #: 402-555-2207

### Initial Vital Signs

| TEMP:<br>98.6 | RESP:<br>15 | SAO2: |
|---|---|---|

**ORIENTATION TO UNIT:** ☒ Call light  ☒ Television/telephone
☒ Bathroom  ☒ Visiting  ☒ Smoking  ☒ Meals
☒ Patient rights/responsibilities

| HT:<br>5'3" | WT (lb):<br>115 | B/P:<br>110/70 | PULSE:<br>65 |
|---|---|---|---|

| LAST TETANUS | | LAST ATE<br>1230 | LAST DRANK<br>1230 |
|---|---|---|---|

### CHIEF COMPLAINT/HX OF PRESENT ILLNESS

"I had a sinus infection for about 3 months that didn't clear up with antibiotics so my doctor drew some blood and found leukemia; now I am here for a bone marrow transplant."

**PERSONAL ARTICLES:** (Check if retained/describe)
☐ Contacts ☐ R ☐ L          ☐ Dentures ☐ Upper ☐ Lower
☐ Jewelry:
☐ Other:

### ALLERGIES: Meds, Food, IVP Dye, Seafood: Type of Reaction

**VALUABLES ENVELOPE:**
☒ Valuables instructions

### PREVIOUS HOSPITALIZATIONS/SURGERIES

bone marrow biopsy

induction chemotherapy Ara-C, hospitalized for 10 days

double lumen catheter insertion

**INFORMATION OBTAINED FROM:**
☒ Patient     ☐ Previous record
☐ Family      ☐ Responsible party

Signature  *Rachel Dean*

### Home Medications (including OTC)     Codes: A=Sent home     B=Sent to pharmacy     C=Not brought in

| Medication | Dose | Frequency | Time of Last Dose | Code | Patient Understanding of Drug |
|---|---|---|---|---|---|
| Tri-Norinyl | ? | one pill per day | last pm | | yes |
| Ambien | 5 mg | 2-3 times per week | last pm | | yes |
| | | | | | |
| | | | | | |
| | | | | | |
| | | | | | |
| | | | | | |
| | | | | | |
| | | | | | |
| | | | | | |
| | | | | | |

Do you take all medications as prescribed?  ☒ Yes  ☐ No  If no, why?

### PATIENT/FAMILY HISTORY

| | | |
|---|---|---|
| ☐ Cold in past two weeks | ☐ High blood pressure | ☐ Kidney/urinary problems |
| ☐ Hay fever | ☐ Arthritis | ☒ Gastric/abdominal pain/heartburn Mother |
| ☐ Emphysema/lung problems | ☐ Claustrophobia | ☐ Hearing problems |
| ☐ TB disease/positive TB skin test | ☐ Circulation problems | ☐ Glaucoma/eye problems |
| ☐ Cancer | ☐ Easy bleeding/bruising/anemia | ☐ Back pain |
| ☐ Stroke/past paralysis | ☐ Sickle cell disease | ☐ Seizures |
| ☒ Heart attack Maternal grandfather | ☐ Liver disease/jaundice | ☐ Other |
| ☐ Angina/chest pain | ☐ Thyroid disease | |
| ☒ Heart problems Maternal grandfather | ☒ Diabetes Paternal grandmother | |

### RISK SCREENING

Have you had a blood transfusion?  ☒ Yes  ☐ No
Do you smoke?  ☐ Yes  ☒ No
If yes, how many pack(s) ____ /day for ____ years
Does anyone in your household smoke?  ☐ Yes  ☐ No
Do you drink alcohol?  ☒ Yes  ☐ No
If yes, how often? 1×/week   How much? 2-3 glasses of wine
When was your last drink?  ____ / ____ /
Do you take any recreational drugs?  ☐ Yes  ☒ No
If yes, type:_____   Route
Frequency:_____   Date last used:_____/____/

**FOR WOMEN Ages 12–52**

Is there any chance you could be pregnant?  ☐ Yes  ☒ No
If yes, expected date (EDC):
Gravida/Para:

**ALL WOMEN**

Date of last Pap smear: 12/18/
Do you perform regular breast self-exams?  ☒ Yes  ☐ No

**ALL MEN**

Do you perform regular testicular exams?  ☐ Yes  ☐ No

Additional comments:

x *Joanne Yardley, RN*
Signature/Title

**Client name:** Rachel Dean
**DOB:** 5/24
**Age:** 25
**Sex:** Female
**Education:** Some graduate school   *What grade/level?* BS, chemical engineering
**Occupation:** Chemical engineer
**Hours of work:** 45–50 per week
**Household members:** Husband age 27
**Ethnic background:** Caucasian
**Religious affiliation:** Methodist
**Referring physician:** Michela Hansen, MD (hematology)

## Chief complaint:

Patient c/o chronic sinus congestion for several weeks before finally seeing her physician, who prescribed antibiotics for a presumed sinus infection. Her symptoms failed to respond after 3 months, and therefore her physician ordered a chemistry panel and CBC and discovered elevated WBC and blasts.

## Patient history:

*Onset of disease:* 4 months PTA
*Type of Tx:* Induction chemotherapy consisting of cytarabine and idarubicin; 1st CR achieved
*PMH:* Appendectomy age 7, otherwise unremarkable
*Meds:* Tri-Norinyl 1 tab q d, Ambien 5 mg prn
*Smoker:* No
*Family Hx: What?* CAD   *Who?* Maternal grandfather

## Physical exam:

*General appearance:* Slender, anxious-appearing young woman in no apparent distress
*Vitals:* Temp 98.6°F, BP 110/70, HR 65 bpm, RR 15 bpm
*Heart:* RRR, no gallops, rubs, or murmurs heard
*HEENT:*
  *Head:* Normocephalic, alopecia
  *Eyes:* Good visual acuity all fields, pupils round and reactive to light
  *Ears:* Ear canals clear, tympanic membranes intact, hearing good, no tinnitus or vertigo
  *Nose:* Nares patent bilaterally, no discharge or lesions noted, mild sinus congestion
*Genitalia:* Exam deferred
*Neurologic:* Denies seizures or sensory loss; alert and oriented × 3, cranial nerves grossly intact; normal gait; superficial and deep tendon reflexes normal
*Chest/lungs:* Lungs clear to auscultation, normal breath sounds, no cough noted
*Abdomen:* Soft and flat, no masses felt, no tenderness on palpation; +BS in all four quadrants, normal spleen percussed, no hepatomegaly noted

## Nutrition Hx:

*General:* Appetite has been fair; somewhat decreased the last 2–3 days PTA due to anxiety concerning BMT.

*Usual dietary intake:*

| | |
|---|---|
| *Breakfast:* | Raisin bran, skim milk, orange juice, coffee |
| *Lunch:* | Turkey or tuna fish sandwich, chips, diet soda, 2–3 Fig Newtons |
| *Dinner:* | Chicken or fish, vegetables, bread/rolls, lemonade or iced tea, frozen yogurt or sherbet |
| *Snack:* | Popcorn or cookies |
| *24-hr recall:* | Not done |

*Food allergies/intolerances/aversions:* None
*Previous MNT?* No
*Food purchase/preparation:* Self
*Vitamin/mineral intake:* Takes a vitamin/mineral supplement daily; vitamin C 1000 mg/day

## Dx:

Acute leukemia in CR. Admitted for bone marrow transplant.

## Tx plan:

Preparative regimen consisting of cyclophosphamide (day −6, day −5) and total body irradiation (days −3, −2, −1) followed by infusion of HLA-matched sibling bone marrow (day 0).
Methotrexate (IV) for graft versus host disease (GVHD); prophylaxis will be given on days +1, +3, +6, and +11.
Intravenous cyclosporine for GVHD prophylaxis will be instituted on day −2 and given until the patient is tolerating p.o. and bowel movements are normal, at which time the dose will be converted to p.o. until day +180.
Prophylactic antifungal (fluconazole) and antiviral (acyclovir) infusions to begin on day +1.
Hematopoietic growth factors (GM-CSF) will be administered intravenously on days 0, +1, +2, and +3.
Antibiotics will be administered at the onset of fever.
Pain will be managed by IV morphine.
*Antiemetic regimen:* Lorazepam and ondansetron

# U**H** *UNIVERSITY HOSPITAL*

NAME: Rachel Dean                          DOB: 5/24
AGE: 25                                     SEX: F
PHYSICIAN: M. Hansen, MD

\*\*\*\*\*\*\*\*\*\*\*\*\*\*\*\*\*\*\*\*\*\*\*\*\*\*\*\*\*\*\*\*\*\*\*\*\*\*\*\*\*\*\*CHEMISTRY\*\*\*\*\*\*\*\*\*\*\*\*\*\*\*\*\*\*\*\*\*\*\*\*\*\*\*\*\*\*\*\*\*\*\*\*\*\*\*\*\*\*\*

| | DAY: | −7 | 0 | +7 | +16 | +20 | |
|---|---|---|---|---|---|---|---|
| | DATE: | | | | | | |
| | TIME: | | | | | | |
| | LOCATION: | | | | | | |
| | NORMAL | | | | | | UNITS |
| Albumin | 3.6–5 | 3.9 | 3.0 | 2.8 | 2.5 | 2.6 | g/dL |
| Total protein | 6–8 | 6.3 | 6.0 | 5.8 | 5.2 | 5.3 | g/dL |
| Prealbumin | 19–43 | 40 | | | | | mg/dL |
| Transferrin | 200–400 | | | | | | mg/dL |
| Sodium | 135–155 | 145 | 140 | 136 | 132 | 135 | mmol/L |
| Potassium | 3.5–5.5 | 4.2 | 3.8 | 3.2 | 2.8 | 3.3 | mmol/L |
| Chloride | 98–108 | 102 | 100 | 97 | 95 | 96 | mmol/L |
| $PO_4$ | 2.5–4.5 | 4.0 | 4.8 | 2.5 | 2.1 | 2.4 | mmol/L |
| Magnesium | 1.6–2.6 | 2.3 | 2.0 | 1.4 | 1.3 | 1.4 | mmol/L |
| Osmolality | 275–295 | | | | | | mmol/kg $H_2O$ |
| Total $CO_2$ | 24–30 | 27 | 26 | 26 | 26 | | mmol/L |
| Glucose | 70–120 | 110 | 97 | 84 | 132 | 375 | mg/dL |
| BUN | 8–26 | 12 | 10 | 14 | 17 | 16 | mg/dL |
| Creatinine | 0.6–1.3 | 0.7 | 1.0 | 1.3 | 1.6 | 1.6 | mg/dL |
| Uric acid | 2.6–6 (women) | 3.1 | 4.3 | 2.2 | | | mg/dL |
| | 3.5–7.2 (men) | | | | | | |
| Calcium | 8.7–10.2 | 9.5 | 9.0 | 8.7 | 8.8 | 9.0 | mg/dL |
| Bilirubin | 0.2–1.3 | 0.3 | 0.5 | 1.4 | 2.1 | 1.8 | mg/dL |
| Ammonia ($NH_3$) | 9–33 | | | | | | μmol/L |
| SGPT (ALT) | 10–60 | 55 | 53 | 50 | | | U/L |
| SGOT (AST) | 5–40 | 35 | 32 | 43 | | | U/L |
| Alk phos | 98–251 | 200 | 210 | 205 | | | U/L |
| CPK | 26–140 (women) | | | | | | U/L |
| | 38–174 (men) | | | | | | |
| LDH | 313–618 | | | | | | U/L |
| CHOL | 140–199 | 175 | | | | | mg/dL |
| HDL-C | 40–85 (women) | | | | | | mg/dL |
| | 37–70 (men) | | | | | | |
| VLDL | | | | | | | mg/dL |
| LDL | < 130 | | | | | | mg/dL |
| LDL/HDL ratio | < 3.22 (women) | | | | | | |
| | < 3.55 (men) | | | | | | |
| Apo A | 101–199 (women) | | | | | | mg/dL |
| | 94–178 (men) | | | | | | |
| Apo B | 60–126 (women) | | | | | | mg/dL |
| | 63–133 (men) | | | | | | |
| TG | 35–160 | | | | | | mg/dL |
| $T_4$ | 5.4–11.5 | | | | | | μg/dL |
| $T_3$ | 80–200 | | | | | | ng/dL |
| $HbA_{1C}$ | 4.8–7.8 | | | | | | % |

# U_H UNIVERSITY HOSPITAL

NAME: Rachel Dean                    DOB: 5/24
AGE: 25                              SEX: F
PHYSICIAN: M. Hansen, MD


\*\*\*\*\*\*\*\*\*\*\*\*\*\*\*\*\*\*\*\*\*\*\*\*\*\*\*\*\*\*\*\*\*\*\*\*\*\*\*\*\*\*\*\*\*HEMATOLOGY\*\*\*\*\*\*\*\*\*\*\*\*\*\*\*\*\*\*\*\*\*\*\*\*\*\*\*\*\*\*\*\*\*\*\*\*\*\*\*\*\*\*\*\*\*

| DAY: | | −7 | 0 | +7 | |
|------|--------|------|------|------|------|
| DATE: | | | | | |
| TIME: | | | | | |
| LOCATION: | | | | | |
| | NORMAL | | | | UNITS |
| WBC | 4.3–10 | 4.4 | 0.1 | 0.4 | $\times 10^3/mm^3$ |
| RBC | 4–5 (women) | 4.0 | 2.8 | 2.6 | $\times 10^6/mm^3$ |
| | 4.5–5.5 (men) | | | | |
| HGB | 12–16 (women) | 10 | 8 | 9 | g/dL |
| | 13.5–17.5 (men) | | | | |
| HCT | 37–47 (women) | 34 | 27 | 28 | % |
| | 40–54 (men) | | | | |
| MCV | 84–96 | 85 | 70 | 62 | fL |
| MCH | 27–34 | 30 | 20 | 19 | pg |
| MCHC | 31.5–36 | 34 | 30 | .7 | g/dL |
| RDW | 11.6–16.5 | | | | % |
| Plt ct | 140–440 | 127 | 64 | 33 | $\times 10^3$ |
| Diff TYPE | | | | | |
| % GRANS | 34.6–79.2 | 35.2 | 15 | 7 | % |
| % LYM | 19.6–52.7 | 20 | 0 | 1 | % |
| SEGS | 50–62 | 51 | 2 | 2 | % |
| BANDS | 3–6 | 4 | 2 | 0 | % |
| LYMPHS | 25–40 | 23 | 2 | 0 | % |
| MONOS | 3–7 | 2 | 1 | 0 | % |
| EOS | 0–3 | 2 | 0 | 0 | % |
| TIBC | 65–165 (women) | | | | $\mu$g/dL |
| | 75–175 (men) | | | | |
| Ferritin | 18–160 (women) | | | | $\mu$g/dL |
| | 18–270 (men) | | | | |
| Vitamin $B_{12}$ | 100–700 | | | | pg/mL |
| Folate | 2–20 | | | | ng/mL |
| Total T cells | 812–2318 | | | | $mm^3$ |
| T-helper cells | 589–1505 | | | | $mm^3$ |
| T-suppressor cells | 325–997 | | | | $mm^3$ |
| PT | 11–13 | | | | sec |

## Case Questions

1.  What is acute myelogenous leukemia (AML), and how are the symptoms at diagnosis related to its pathology?

2.  What are the primary goals of high-dose chemotherapy and total body irradiation?

3.  What is the rationale for using allogeneic (as opposed to autologous) bone marrow transplant for the treatment of AML?

4.  Why is bacterial infection common in the immediate posttransplant period (day +1 to day +30) or until engraftment?

5.  Describe the expected medical side effects of the cyclophosphamide/total body irradiation (TBI) regimen, cyclosporine, methotrexate, fluconazole, acyclovir, antibiotics, and GM-CSF.

| Treatment | Medical Side Effects |
|---|---|
| Cyclophosphamide | |
| TBI | |
| Cyclosporine | |
| Methotrexate | |
| Fluconazole | |
| Acyclovir | |
| Antibiotics | |
| GM-CSF | |

**6.** Describe the expected nutritional side effects of the cyclophosphamide/TBI regimen, cyclosporine, methotrexate, fluconazole, acyclovir, antibiotics, and GM-CSF.

| Treatment | Nutritional Side Effects |
|---|---|
| Cyclophosphamide | |
| TBI | |
| Cyclosporine | |
| Methotrexate | |
| Fluconazole | |
| Acyclovir | |
| Antibiotics | |
| GM-CSF | |

**7.** By day +7, Mrs. Dean's mucositis has become severe and she is unable to tolerate ice chips or Jell-o. In addition, she has diarrhea amounting to approximately 650 cc/day. Her weight has dropped to 107 lb. What type of nutritional support would you recommend, and why?

**8.** Calculate Mrs. Dean's energy requirements based on her current body weight.

**9.** Calculate Mrs. Dean's protein requirements.

**10.** Devise a parenteral nutrition support regimen—include macronutrients as well as any additional micronutrients, electrolytes, fluids the patient may need.

**11.**   Assess Mrs. Dean's nutrition-related labs, and state how they relate to the treatment and to her nutritional status.

**12.**   How would you monitor her tolerance to the TPN and her nutritional status?

**13.**   Mrs. Dean's absolute neutrophil count reached 1100 by day +11. Her mucositis began to improve, stool output decreased to 250–300 cc/day, and her N/V seemed under control; however, she continued to be anorexic and unable to tolerate p.o. By day +16 Mrs. Dean developed a maculopapular rash on her palms and trunk, her bilirubin climbed to 2.1 mg/dL, and her stool output increased to 1200 cc / 24 hours. Mrs. Dean was diagnosed with stage I/II acute graft versus host disease (GVHD) and was immediately started on methylprednisolone 2mg/kg/day and diphenoxylate HCl. Explain the mechanism of GVHD, and describe its clinical manifestations. How is nutritional status affected?

**14.**   What adjustments need to be made to the parenteral nutrition formula to accommodate for her metabolic and physiologic changes (increased stool output, electrolyte imbalances)? Does her protein intake need to be restricted? Explain.

**15.**   By day +20, Mrs. Dean's blood glucose was averaging 350 mg/dL. What could be contributing to the sudden onset of hyperglycemia? What changes could you make in her parenteral nutrition regimen to help alleviate this problem? What changes need to be made to the electrolytes in the TPN? Explain.

**16.**   State the long-term nutritional implications associated with the chronic use of corticosteroids for the treatment of GVHD.

**17.**   Mrs. Dean responded well to glucocorticoid therapy, and her stool output eventually decreased to <150 cc/day, bilirubin stabilized at 1.7 mg/dL, and her rash diminished. After tolerating a clear liquid diet for three days, her diet was advanced to a GVHD diet (lactose free, residue/fiber free, low fat). She tolerated this well with minimal increase in stool output, and therefore her diet was liberalized to the second-phase GVHD diet (low lactose, low fiber). TPN was weaned and discontinued by day +32. IVF was also discontinued. Mrs. Dean was instructed to follow the GVHD diet (2nd phase) until day +100. Which nutrients may be inadequate on this diet and therefore may need to be supplemented?

## Bibliography

Commission on Accreditation for Dietetics Education. Knowledge, skills, and competencies for dietitians. *Accreditation Manual for Dietetics Education Programs,* rev. 4th ed. Chicago: American Dietetic Association, 2000.

Lenssen P. Bone marrow and stem cell transplantation. In Matarese L, Gottschlich MM (eds.), *Contemporary Nutrition Support Practice: A Clinical Guide.* Philadelphia: Saunders, 1998.

McMahno MM. What to do about hyperglycemia in hospitalized patients. *Nutr Clin Pract.* 1997;12:35–38.

Ringwald-Smith K, Williams R, Horwitz E, Schmidt M. Determination of energy expenditure in the bone marrow transplant patient. *Nutr Clin Pract.* 1998;13:215–218.

Rodino MA, Shane E. Osteoporosis after organ transplantation. *Am J Med.* 1998;104:459–469.

Shapiro TW, Davison DB, Rust DM. *A Clinical Guide to Stem Cell and Bone Marrow Transplantation.* Boston: Jones and Bartlett, 1997.

Tallman MS. Therapy of acute myeloid leukemia. *Cancer Control: Journal of the Moffitt Cancer Center.* 2001;8(1):62–78.

Ziegler TR, et al. Clinical and metabolic efficacy of glutamine-supplemented parenteral nutrition after bone marrow transplantation: a randomized, double-blind, controlled study. *Ann Intern Med.* 1992;116:821–828.

# COMMON MEDICAL ABBREVIATIONS

| | | | | |
|---|---|---|---|---|
| ab lib | at pleasure; as desired (*ab libitum*) | CA+ | calcium |
| ACTH | adrenocorticotropic hormone | CABG | coronary artery bypass graft |
| ac | before meals | CAD | coronary artery disease |
| AD | Alzheimer's Disease | CAPD | continuous ambulatory peritoneal dialysis |
| ADA | American Dietetic Association, American Diabetes Association | cath | catheter, catheterize |
| ADH | antidiuretic hormone | CAVH | continuous arteriovenous hemofiltration |
| ad lib | as desired (*ad libitum*) | | |
| ADL | activities of daily living | CBC | complete blood count |
| AGA | antigliadin antibody | cc | cubic centimeter |
| AIDS | acquired immunodeficiency syndrome | CCK | cholecystokinin |
| | | CCU | coronary care unit |
| ALP (Alk phos) | alkaline phosphatase | CDAI | Crohn's disease activity index |
| ALS | amyotrophic lateral sclerosis | CDC | Centers for Disease Control |
| ALT | alanine aminotransferase | CHD | coronary heart disease |
| amp | ampule | CHF | congestive heart failure |
| ANC | absolute neutrophil count | CHI | closed head injury |
| ANCA | antisacchromyces antibodies | CHO | carbohydrate |
| AP | anterior posterior | CHOL | cholesterol |
| ARDS | adult respiratory distress syndrome | cm | centimeter |
| ARF | acute renal failure, acute respiratory failure | CNS | central nervous system |
| | | c/o | complains of |
| ASA | acetylsalicylic acid, aspirin | COPD | chronic obstructive pulmonary disease |
| ASCA | antineutrophil cytoplasmic antibodies | | |
| ASHD | arteriosclerotic heart disease | CPK | creatinine phosphokinase |
| AV | arteriovenous | Cr | creatinine |
| BANDS | neutrophils | CR | complete remission |
| BCAA | branched chain amino acids | CSF | cerebrospinal fluid |
| BE | barium enema | CT | computed tomography |
| BEE | basal energy expenditure | CVA | cerebrovascular accident |
| BG | blood glucose | CVP | central venous pressure |
| b.i.d. | twice a day | CXR | chest X-ray |
| bili | bilirubin | DBW | desirable body weight |
| BM | bowel movement | d/c | discharge |
| BMI | body mass index | D/C | discontinue |
| BMR | basal metabolic rate | DCCT | Diabetes Control and Complications Trial |
| BMT | bone marrow transplant | | |
| BP (B/P) | blood pressure | DKA | diabetic ketoacidosis |
| BPD | bronchopulmonary dysplasia | dL | deciliter |
| BPH | benign prostate hypertrophy | DM | Diabetes Mellitus |
| bpm | beats per minute, breaths per minute | $D_5NS$ | Dextrose, 5% in normal saline |
| BS | bowel sounds, breath sounds, or blood sugar | $D_5W$ | Dextrose, 5% in water |
| | | DRI | dietary reference intake |
| BSA | body surface area | DTR | deep tendon reflex |
| BUN | blood urea nitrogen | DTs | delirium tremens |
| c | with | DVT | deep vein thrombosis |
| c. | cup | Dx | diagnosis |
| C | centigrade | ECF | extracellular fluid |
| CA | cancer; carcinoma | ECG/EKG | electrocardiogram |

*Note:* Abbreviations can vary from institution to institution. Although the student will find many of the accepted variations listed in this appendix, other references may be needed to supplement this list.

| | | | | |
|---|---|---|---|---|
| EEG | electroencephalogram | | ICP | intracranial pressure |
| e.g. | for example | | ICU | intensive care unit |
| EGD | esophagogastroduodenoscopy | | i.e. | that is |
| ELISA | enzyme-linked immunosorbent assay | | IGT | impaired glucose tolerance |
| EMA | antiendomysial antibody | | IM | intramuscularly |
| EMG | electromyography | | inc | incontinent |
| ER | emergency room | | I&O (I/O) | intake and output |
| ERT | estrogen replacement therapy | | IV | intravenous |
| ESR | erythrocyte sedimentation rate | | IU | international unit |
| ESRD | end-stage renal disease | | J | joule |
| ESRF | end-stage renal failure | | K | potassium |
| F | Fahrenheit | | kcal | kilocalorie |
| FACSM | Fellow American College of Sports | | KCl | potassium chloride |
| | Medicine | | kg | kilogram |
| FBS | fasting blood sugar | | KS | Kaposi's sarcoma |
| FDA | Food and Drug Administration | | KUB | kidney, ureter, bladder |
| FEF | forced mid-expiratory flow | | L | liter |
| FEV | forced mid-expiratory volume | | LBM | lean body mass |
| FFA | free fatty acid | | lbs | pounds |
| FH | family history | | LCT | long chain triglyceride |
| FTT | failure to thrive | | LDH | lactic dehydrogenase |
| FUO | fever of unknown origin | | LES | lower esophageal sphincter |
| FVC | forced vital capacity | | LFT | liver function test |
| FX | fracture | | LIGS | low intermittent gastric suction |
| g | gram | | LLQ | lower left quadrant |
| GB | gallbladder | | LMP | last menstrual period |
| g/dL | grams per deciliter | | LOC | level of conciousness |
| GERD | gastroesophageal reflux disease | | LP | lumbar puncture |
| GI | gastrointestinal | | LUQ | lower upper quadrant |
| GM-CSF | granulocyte/macrophage colony | | lytes | electrolytes |
| | stimulating factor | | MAC | midarm circumference |
| GTF | glucose tolerance factor | | MAMC | midarm muscle circumference |
| GTT | glucose tolerance test | | MAOI | monoamine oxidase inhibitor |
| GVHD | graft versus host disease | | MCHC | mean corpuscular hemoglobin |
| h | hour | | | concentration |
| HAV | hepatitis A virus | | MCT | medium chain triglyceride |
| HBV | hepatitis B virus | | MCV | mean corpuscular volume |
| HbA$_{1c}$ | glycated hemoglobin | | mEq | milliequivalent |
| Hct | hematocrit | | mg | milligram |
| HC | head circumference | | Mg | magnesium |
| HCV | hepatitis C virus | | MI | myocardial infarction |
| HDL | high density lipoprotein | | mm | millimeter |
| HEENT | head, eyes, ears, nose, throat | | mmHg | millimeters of mercury |
| Hg | mercury | | MNT | medical nutrition therapy |
| Hgb | hemoglobin | | MODY | maturity onset diabetes of the young |
| HHNK | hyperosmolar hyperglycemic | | MOM | milk of magnesia |
| | nonketotic (syndrome) | | mOsm | milliosmol |
| HIV | human immunodeficiency virus | | MR | mitral regurgitation |
| HLA | human leukocyte antigen | | MRI | magnetic resonance imaging |
| HOB | head of bed | | MS | multiple sclerosis, morphine sulfate |
| H&P (HPI) | history and physical | | MVA | motor vehicle accident |
| HR | heart rate | | MVI | multiple vitamin infusion |
| HS or h.s. | hours of sleep | | N | nitrogen |
| HTN | hypertension | | NG | nasogastric |
| HX | history | | NH$_3$ | ammonia |
| IBD | inflammatory bowel disease | | NICU | neurointensive care unit, neonatal |
| IBS | irritable bowel syndrome | | | intensive care unit |
| IBW | ideal body weight | | NKA | no known allergies |
| ICF | intracranial fluid | | NKDA | no known drug allergies |

| | | | | |
|---|---|---|---|---|
| NPH | neutral protamine Hagedorn insulin | | RLQ | right lower quadrant |
| NPO | nothing by mouth | | R/O | rule out |
| NSAID | nonsteroidal antiinflammatory drug | | ROM | range of motion |
| NTG | nitroglycerin | | ROS | review of systems |
| N/V | nausea and vomiting | | RQ | respiratory quotient |
| $O_2$ | oxygen | | RR | respiratory rate |
| OA | osteoarthritis | | RUL | right upper lobe |
| OC | oral contraceptive | | RUQ | right upper quadrant |
| OHA | oral hypoglycemic agent | | Rx | take, prescribe, treat |
| OR | operating room | | s | without |
| ORIF | open reduction internal fixation | | SBO | small bowel obstruction |
| OT | occupational therapist | | SBS | short bowel syndrome |
| OTC | over the counter | | SGOT | serum glutamic oxaloacetic transaminase |
| $paco_2$ | partial pressure of dissolved carbon dioxide in arterial blood | | SGPT | serum glutamic pyruvic transaminase |
| $pao_2$ | partial pressure of dissolved oxygen in arterial blood | | SBGM | self blood glucose monitoring |
| pc | after meals | | SOB | shortness of breath |
| PCM | protein calorie malnutrition | | S/P | status post |
| PD | Parkinson's disease | | SQ | subcutaneous |
| PE | pulmonary embolus | | ss | half |
| PED | percutaneous endoscopic duodenostomy | | stat | immediately |
| PEEP | positive end expiratory pressure | | susp | suspension |
| PEG | percutaneous endoscopic gastrostomy | | T | temperature |
| PEM | protein energy malnutrition | | T, tbsp | tablespoon |
| PERRLA | pupils equal, round, and reactive to light and accommodation | | t, tsp | teaspoon |
| pH | hydrogen ion concentration | | T&A | tonsillectomy and adenoidectomy |
| PKU | phenylketonuria | | $T_3$ | triiodothyronine |
| PMN | polymorphonuclear | | $T_4$ | thyroxine |
| PN | parenteral nutrition | | TB | tuberculosis |
| PO | by mouth | | TEE | total energy expenditure |
| PPD | packs per day | | TF | tube feeding |
| PPN | peripheral parenteral nutrition | | TG | triglyceride |
| prn | may be repeated as necessary (*pro re nata*) | | TIA | transient ischemic attack |
| PT | patient, physical therapy, prothrombin time | | TIBC | total iron binding capacity |
| | | | tid | three times daily |
| | | | TKO | to keep open |
| PTA | prior to admission | | TLC | total lymphocyte count |
| PTT | prothromboplastin time | | TNM | tumor, node, metastasis |
| PUD | peptic ulcer disease | | TPN | total parenteral nutrition |
| PVC | premature ventricular contraction | | TSF | triceps skinfold |
| PVD | peripheral vascular disease | | TURP | transurethral resection of the prostate |
| q | every | | U | unit |
| qd | every day | | UA | urinalysis |
| qh | every hour | | UBW | usual body weight |
| qid | four times daily | | UL | tolerable upper intake level |
| qns | quantity not sufficient | | URI | upper respiratory intake |
| qod | every other day | | UTI | urinary tract infection |
| RA | rheumatoid arthritis | | UUN | urine urea nitrogen |
| RBC | red blood cell | | VLCD | very low calorie diet |
| RBW | reference body weight | | VOD | venous occlusive disease |
| RD | registered dietitian | | VS | vital signs |
| RDA | recommended dietary allowance | | w.a. | while awake |
| RDS | respiratory distress syndrome | | WBC | white blood cell |
| REE | resting energy expenditure | | WNL | within normal limits |
| RLL | right lower lobe | | wt | weight |
| | | | WW | whole wheat |
| | | | yo | year old |

# NORMAL VALUES FOR PHYSICAL EXAMINATION

*Vital Signs*

## Temperature
Rectal: C = 37.6°/F = 99.6°
Oral: C = 37°/F = 98.6° (± 10°)
Axilla: C = 37.4°/F = 97.6°

**Blood Pressure:** average 120/80 mmHg

## Heart Rate (beats per minute)

| Age | At rest awake | At rest asleep | Exercise or fever |
|---|---|---|---|
| Newborn | 100–180 | 80–160 | ≤ 220 |
| 1 week–3 months | 100–220 | 80–200 | ≤ 220 |
| 3 months–2 years | 80–150 | 70–120 | ≤ 200 |
| 2–10 years | 70–110 | 60–90 | ≤ 200 |
| 11 years–adult | 55–90 | 50–90 | ≤ 200 |

## Respiratory Rate (breaths per minute)

| Age | Respirations |
|---|---|
| Newborn | 35 |
| 1–11 months | 30 |
| 1–2 years | 25 |
| 3–4 years | 23 |
| 5–6 years | 21 |
| 7–8 years | 20 |
| 10–11 years | 19 |
| 12–13 years | 19 |
| 14–15 years | 18 |
| 16–17 years | 17 |
| 17–18 years | 16–18 |
| Adult | 12–20 |

**Cardiac Exam:** carotid pulses equal in rate, rhythm and strength; normal heart sounds; no murmurs present

*HEENT* Exam (head, eyes, ears, nose, throat)

*Mouth:* pink, moist, symmetrical; mucosa pink, soft, moist, smooth

*Gums:* pink, smooth, moist; may have patchy pigmentation

*Teeth:* smooth, white, shiny

*Tongue:* medium red or pink, smooth with free mobility, top surface slightly rough

*Eyes:* pupils equal, round, reactive to light and accommodation

*Ears:* tympanic membrane taut, translucent, pearly gray; auricle smooth without lesions; meatus not swollen or occluded; cerumen dry (tan/light yellow) or moist (dark yellow/brown)

*Nose:* external nose symmetrical, nontender without discharge; mucosa pink; septum at the midline

*Pharynx:* mucosa pink and smooth

*Neck:* thyroid gland, lymph nodes not easily palpable or enlarged

**Lungs:** chest contour symmetrical; spine straight without lateral deviation; no bulging or active movement within the intercostal spaces during breathing; respirations clear to auscultation and percussion

**Peripheral Vascular:** normal pulse graded at 3+, which indicates that pulse is easy to palpate and not easily obliterated; pulses equal bilaterally and symmetrically

**Neurological:** normal orientation to people, place, time, with appropriate response and concentration

**Skin:** warm and dry to touch; should lift easily and return back to original position indicating normal turgor and elasticity

**Abdomen:** umbilicus flat or concave positioned midway between xyphoid process and symphysis pubis; bowel motility notes normal air and fluid movement every 5–15 seconds; graded as normal, audible, absent, hyperactive, or hypoactive

# Appendix C

# DIETARY REFERENCE INTAKES

## 1997–2001 Recommended Dietary Allowances (RDA) and Adequate Intakes (AI)

| Age (yr) | Thiamin RDA (mg/day) | Riboflavin RDA (mg/day) | Niacin RDA (mg/day)[a] | Biotin AI (µg/day) | Pantothenic acid AI (mg/day) | Vitamin B6 RDA (mg/day) | Folate RDA (µg/day)[b] | Vitamin B12 RDA (µg/day) | Choline AI (mg/day) | Vitamin C RDA (mg/day) | Vitamin A RDA (µg/day)[c] | Vitamin D AI (µg/day)[d] |
|---|---|---|---|---|---|---|---|---|---|---|---|---|
| **Infants** | | | | | | | | | | | | |
| 0–0.5 | 0.2 | 0.3 | 2 | 5 | 1.7 | 0.1 | 65 | 0.4 | 125 | 40 | 400 | 5 |
| 0.5–1 | 0.3 | 0.4 | 4 | 6 | 1.8 | 0.3 | 80 | 0.5 | 150 | 50 | 500 | 5 |
| **Children** | | | | | | | | | | | | |
| 1–3 | 0.5 | 0.5 | 6 | 8 | 2 | 0.5 | 150 | 0.9 | 200 | 15 | 300 | 5 |
| 4–8 | 0.6 | 0.6 | 8 | 12 | 3 | 0.6 | 200 | 1.2 | 250 | 25 | 400 | 5 |
| **Males** | | | | | | | | | | | | |
| 9–13 | 0.9 | 0.9 | 12 | 20 | 4 | 1.0 | 300 | 1.8 | 375 | 45 | 600 | 5 |
| 14–18 | 1.2 | 1.3 | 16 | 25 | 5 | 1.3 | 400 | 2.4 | 550 | 75 | 900 | 5 |
| 19–30 | 1.2 | 1.3 | 16 | 30 | 5 | 1.3 | 400 | 2.4 | 550 | 90 | 900 | 5 |
| 31–50 | 1.2 | 1.3 | 16 | 30 | 5 | 1.3 | 400 | 2.4 | 550 | 90 | 900 | 5 |
| 51–70 | 1.2 | 1.3 | 16 | 30 | 5 | 1.7 | 400 | 2.4 | 550 | 90 | 900 | 10 |
| >70 | 1.2 | 1.3 | 16 | 30 | 5 | 1.7 | 400 | 2.4 | 550 | 90 | 900 | 15 |
| **Females** | | | | | | | | | | | | |
| 9–13 | 0.9 | 0.9 | 12 | 20 | 4 | 1.0 | 300 | 1.8 | 375 | 45 | 600 | 5 |
| 14–18 | 1.0 | 1.0 | 14 | 25 | 5 | 1.2 | 400 | 2.4 | 400 | 65 | 700 | 5 |
| 19–30 | 1.1 | 1.1 | 14 | 30 | 5 | 1.3 | 400 | 2.4 | 425 | 75 | 700 | 5 |
| 31–50 | 1.1 | 1.1 | 14 | 30 | 5 | 1.3 | 400 | 2.4 | 425 | 75 | 700 | 5 |
| 51–70 | 1.1 | 1.1 | 14 | 30 | 5 | 1.5 | 400 | 2.4 | 425 | 75 | 700 | 10 |
| >70 | 1.1 | 1.1 | 14 | 30 | 5 | 1.5 | 400 | 2.4 | 425 | 75 | 700 | 15 |
| **Pregnancy** | | | | | | | | | | | | |
| ≤18 | 1.4 | 1.4 | 18 | 30 | 6 | 1.9 | 600 | 2.6 | 450 | 80 | 750 | 5 |
| 19–30 | 1.4 | 1.4 | 18 | 30 | 6 | 1.9 | 600 | 2.6 | 450 | 85 | 770 | 5 |
| 31–50 | 1.4 | 1.4 | 18 | 30 | 6 | 1.9 | 600 | 2.6 | 450 | 85 | 770 | 5 |
| **Lactation** | | | | | | | | | | | | |
| ≤18 | 1.4 | 1.6 | 17 | 35 | 7 | 2.0 | 500 | 2.8 | 550 | 115 | 1200 | 5 |
| 19–30 | 1.4 | 1.6 | 17 | 35 | 7 | 2.0 | 500 | 2.8 | 550 | 120 | 1300 | 5 |
| 31–50 | 1.4 | 1.6 | 17 | 35 | 7 | 2.0 | 500 | 2.8 | 550 | 120 | 1300 | 5 |

## 1997–2001 Tolerable Upper Intake Levels (UL)

| Age (yr) | Niacin (mg/day)[a] | Vitamin B6 (mg/day) | Folate (µg/day)[a] | Choline (mg/day) | Vitamin C (mg/day) | Vitamin A (µg/day)[b] | Vitamin D (µg/day) | Vitamin E (mg/day)[c] | Calcium (mg/day) | Phosphorus (mg/day) | Magnesium (mg/day)[d] | Iron (mg/day) |
|---|---|---|---|---|---|---|---|---|---|---|---|---|
| **Infants** | | | | | | | | | | | | |
| 0–0.5 | — | — | — | — | — | 600 | 25 | — | — | — | — | 40 |
| 0.5–1 | — | — | — | — | — | 600 | 25 | — | — | — | — | 40 |
| **Children** | | | | | | | | | | | | |
| 1–3 | 10 | 30 | 300 | 1000 | 400 | 600 | 50 | 200 | 2500 | 3000 | 65 | 40 |
| 4–8 | 15 | 40 | 400 | 1000 | 650 | 900 | 50 | 300 | 2500 | 3000 | 110 | 40 |
| 9–13 | 20 | 60 | 600 | 2000 | 1200 | 1700 | 50 | 600 | 2500 | 4000 | 350 | 40 |
| **Adolescents** | | | | | | | | | | | | |
| 14–18 | 30 | 80 | 800 | 3000 | 1800 | 2800 | 50 | 800 | 2500 | 4000 | 350 | 45 |
| **Adults** | | | | | | | | | | | | |
| 19–70 | 35 | 100 | 1000 | 3500 | 2000 | 3000 | 50 | 1000 | 2500 | 4000 | 350 | 45 |
| >70 | 35 | 100 | 1000 | 3500 | 2000 | 3000 | 50 | 1000 | 2500 | 3000 | 350 | 45 |
| **Pregnancy** | | | | | | | | | | | | |
| ≤18 | 30 | 80 | 800 | 3000 | 1800 | 2800 | 50 | 800 | 2500 | 3500 | 350 | 45 |
| 19–50 | 35 | 100 | 1000 | 3500 | 2000 | 3000 | 50 | 1000 | 2500 | 3500 | 350 | 45 |
| **Lactation** | | | | | | | | | | | | |
| ≤18 | 30 | 80 | 800 | 3000 | 1800 | 2800 | 50 | 800 | 2500 | 4000 | 350 | 45 |
| 19–50 | 35 | 100 | 1000 | 3500 | 2000 | 3000 | 50 | 1000 | 2500 | 4000 | 350 | 45 |

*Note:* For all nutrients, values for infants are AI.

[a] Niacin recommendations are expressed as niacin equivalents (NE), except for recommendations for infants younger than 6 months, which are expressed as preformed niacin.

[b] Folate recommendations are expressed as dietary folate equivalents (DFE).

[c] Vitamin A recommendations are expressed as retinol activity equivalents (RAE).

[d] Vitamin D recommendations are expressed as cholecalciferol and assume an absence of adequate exposure to sunlight.

[e] Vitamin E recommendations are expressed as α-tocopherol.

*Source:* Adapted with permission from the *Dietary Reference Intakes* series, National Academy Press. Copyright 1997, 1998, 2000, 2001, by the National Academy of Sciences. Courtesy of the National Academy Press, Washington, D.C.

| Vitamins | | | Minerals | | | | | | | | | | |
|---|---|---|---|---|---|---|---|---|---|---|---|---|---|
| Vitamin E RDA (mg/day)[e] | Vitamin K AI (μg/day) | Calcium AI (mg/day) | Phosphorus RDA (mg/day) | Magnesium RDA (mg/day) | Iron RDA (mg/day) | Zinc RDA (mg/day) | Iodine RDA (μg/day) | Selenium RDA (μg/day) | Copper RDA (μg/day) | Manganese AI (mg/day) | Fluoride AI (mg/day) | Chromium AI (μg/day) | Molybdenum RDA (μg/day) |
| 4 | 2.0 | 210 | 100 | 30 | 0.27 | 2 | 110 | 15 | 200 | 0.003 | 0.01 | 0.2 | 2 |
| 5 | 2.5 | 270 | 275 | 75 | 11 | 3 | 130 | 20 | 220 | 0.6 | 0.5 | 5.5 | 3 |
| 6 | 30 | 500 | 460 | 80 | 7 | 3 | 90 | 20 | 340 | 1.2 | 0.7 | 11 | 17 |
| 7 | 55 | 800 | 500 | 130 | 10 | 5 | 90 | 30 | 440 | 1.5 | 1.0 | 15 | 22 |
| 11 | 60 | 1300 | 1250 | 240 | 8 | 8 | 120 | 40 | 700 | 1.9 | 2 | 25 | 34 |
| 15 | 75 | 1300 | 1250 | 410 | 11 | 11 | 150 | 55 | 890 | 2.2 | 3 | 35 | 43 |
| 15 | 120 | 1000 | 700 | 400 | 8 | 11 | 150 | 55 | 900 | 2.3 | 4 | 35 | 45 |
| 15 | 120 | 1000 | 700 | 420 | 8 | 11 | 150 | 55 | 900 | 2.3 | 4 | 35 | 45 |
| 15 | 120 | 1200 | 700 | 420 | 8 | 11 | 150 | 55 | 900 | 2.3 | 4 | 30 | 45 |
| 15 | 120 | 1200 | 700 | 420 | 8 | 11 | 150 | 55 | 900 | 2.3 | 4 | 30 | 45 |
| 11 | 60 | 1300 | 1250 | 240 | 8 | 8 | 120 | 40 | 700 | 1.6 | 2 | 21 | 34 |
| 15 | 75 | 1300 | 1250 | 360 | 15 | 9 | 150 | 55 | 890 | 1.6 | 3 | 24 | 43 |
| 15 | 90 | 1000 | 700 | 310 | 18 | 8 | 150 | 55 | 900 | 1.8 | 3 | 25 | 45 |
| 15 | 90 | 1000 | 700 | 320 | 18 | 8 | 150 | 55 | 900 | 1.8 | 3 | 25 | 45 |
| 15 | 90 | 1200 | 700 | 320 | 8 | 8 | 150 | 55 | 900 | 1.8 | 3 | 20 | 45 |
| 15 | 90 | 1200 | 700 | 320 | 8 | 8 | 150 | 55 | 900 | 1.8 | 3 | 20 | 45 |
| 15 | 75 | 1300 | 1250 | 400 | 27 | 13 | 220 | 60 | 1000 | 2.0 | 3 | 29 | 50 |
| 15 | 90 | 1000 | 700 | 350 | 27 | 11 | 220 | 60 | 1000 | 2.0 | 3 | 30 | 50 |
| 15 | 90 | 1000 | 700 | 360 | 27 | 11 | 220 | 60 | 1000 | 2.0 | 3 | 30 | 50 |
| 19 | 75 | 1300 | 1250 | 360 | 10 | 14 | 290 | 70 | 1300 | 2.6 | 3 | 44 | 50 |
| 19 | 90 | 1000 | 700 | 310 | 9 | 12 | 290 | 70 | 1300 | 2.6 | 3 | 45 | 50 |
| 19 | 90 | 1000 | 700 | 320 | 9 | 12 | 290 | 70 | 1300 | 2.6 | 3 | 45 | 50 |

| Minerals | | | | | | | | | |
|---|---|---|---|---|---|---|---|---|---|
| Zinc (mg/day) | Iodine (μg/day) | Selenium (μg/day) | Copper (μg/day) | Manganese (mg/day) | Fluoride (mg/day) | Molybdenum (μg/day) | Boron (mg/day) | Nickel (mg/day) | Vanadium (mg/day) |
| 4 | — | 45 | — | — | 0.7 | — | — | — | — |
| 5 | — | 60 | — | — | 0.9 | — | — | — | — |
| 7 | 200 | 90 | 1000 | 2 | 1.3 | 300 | 3 | 0.2 | — |
| 12 | 300 | 150 | 3000 | 3 | 2.2 | 600 | 6 | 0.3 | — |
| 23 | 600 | 280 | 5000 | 6 | 10 | 1100 | 11 | 0.6 | — |
| 34 | 900 | 400 | 8000 | 9 | 10 | 1700 | 17 | 1.0 | — |
| 40 | 1100 | 400 | 10,000 | 11 | 10 | 2000 | 20 | 1.0 | 1.8 |
| 40 | 1100 | 400 | 10,000 | 11 | 10 | 2000 | 20 | 1.0 | 1.8 |
| 34 | 900 | 400 | 8000 | 9 | 10 | 1700 | 17 | 1.0 | — |
| 40 | 1100 | 400 | 10,000 | 11 | 10 | 2000 | 20 | 1.0 | — |
| 34 | 900 | 400 | 8000 | 9 | 10 | 1700 | 17 | 1.0 | — |
| 40 | 1100 | 400 | 10,000 | 11 | 10 | 2000 | 20 | 1.0 | — |

[a] The UL for niacin and folate apply to synthetic forms obtained from supplements, fortified foods, or a combination of the two.

[b] The UL for vitamin A applies to the preformed vitamin only.

[c] The UL for vitamin E applies to any form of supplemental α-tocopherol, fortified foods, or a combination of the two.

[d] The UL for magnesium applies to synthetic forms obtained from supplements or drugs only.

Note: An Upper Limit was not established for vitamins and minerals not listed and for those age groups, listed with a dash(—) because of a lack of data, not because these nutrients are safe to consume at any level of intake. All nutrients can have adverse effects when intakes are excessive.

*Source:* Adapted with permission from the *Dietary Reference Intakes* series, National Academy Press. Copyright 1997, 1998, 2000, 2001, by the National Academy of Sciences. Courtesy of the National Academy Press, Washington, D.C.

...e: *National Academy of Sciences Recommended Dietary Allowances.* Washington, D.C.: National Academy Press, ... Copyright 2000 by the National Academy of Sciences. Reprinted courtesy of the National Academy Press, ...ington, D.C.

# CDC GROWTH CHARTS: UNITED STATES

*Weight-for-age percentiles: Boys, birth to 36 months*

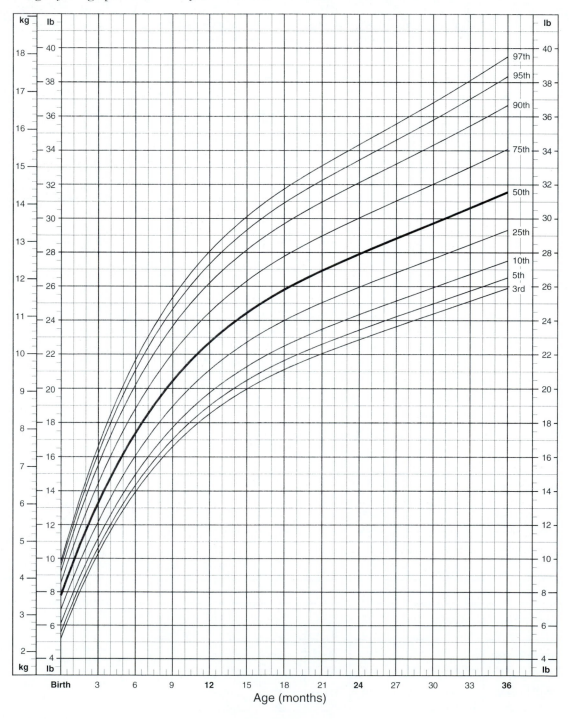

*Source:* Centers for Disease Control and Prevention. National Center for Health Statistics. CDC Growth Charts: United States. 2000. Available at http://www.cdc.gov/nchs/about/major/nhanes/growthcharts/charts.htm. Accessed August 1, 2001.

*Weight-for-length percentiles: Girls, birth to 36 months*

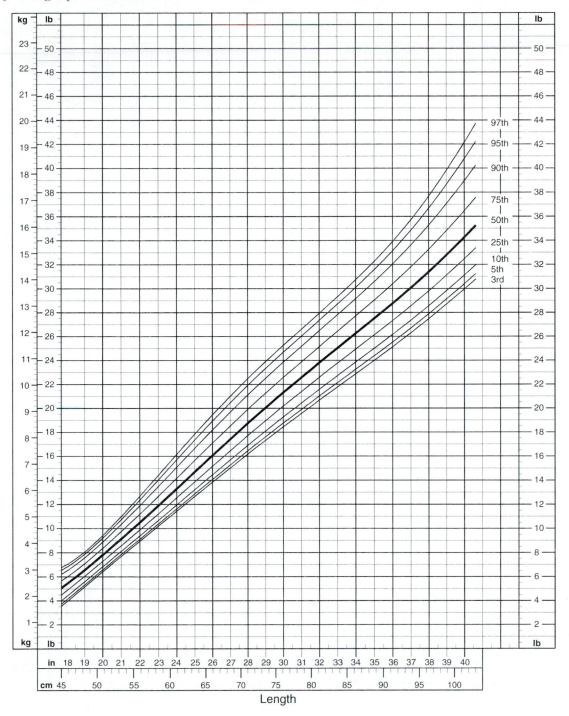

Revised and corrected June 8, 2000.
*Source:* Centers for Disease Control and Prevention. National Center for Health Statistics. CDC Growth Charts: United States. 2000. Available at http://www.cdc.gov/nchs/about/major/nhanes/growthcharts/charts.htm. Accessed August 1, 2001.

*Weight-for-age percentiles: Boys, 2 to 20 years*

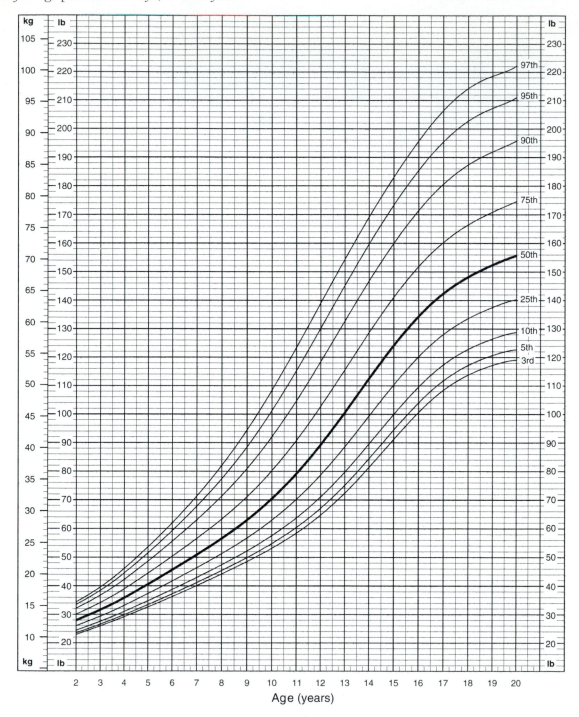

Age (years)

*Source:* Centers for Disease Control and Prevention. National Center for Health Statistics. CDC Growth Charts: United States. 2000. Available at http://www.cdc.gov/nchs/about/major/nhanes/growthcharts/charts.htm. Accessed August 1, 2001.

*Weight-for-stature percentiles: Girls*

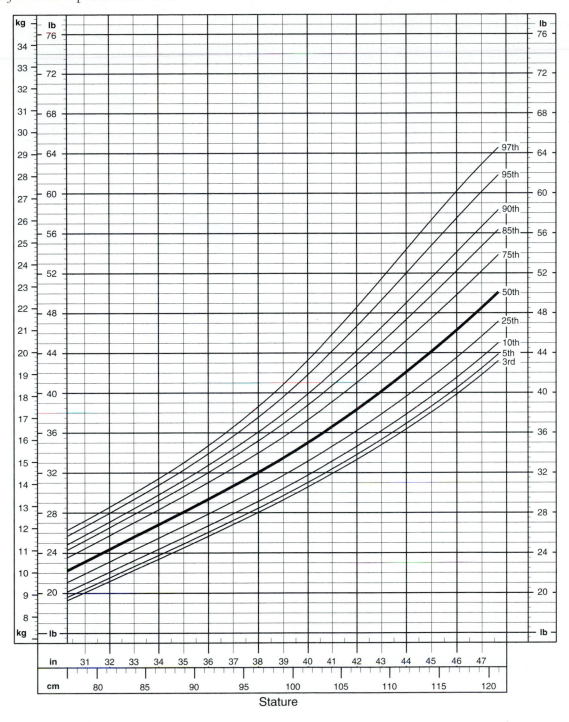

Stature

Revised and corrected November 21, 2000.
*Source:* Centers for Disease Control and Prevention. National Center for Health Statistics. CDC Growth Charts: United States. 2000. Available at
http://www.cdc.gov/nchs/about/major/nhanes/growthcharts/charts.htm. Accessed August 1, 2001.

# PRENATAL WEIGHT GAIN GRID

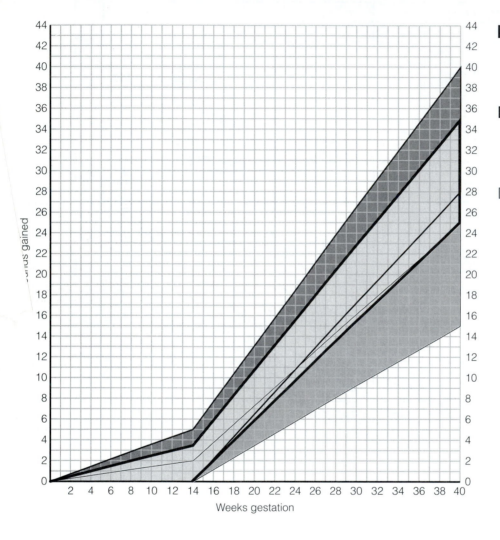

Normal-weight women should gain about 3 ½ pounds in the first trimester and just under 1 pound/week thereafter, achieving a total gain of 25 to 35 pounds by term.

Underweight women should gain about 5 pounds in the first trimester and just over 1 pound/week thereafter, achieving a total gain of 28 to 40 pounds by term.

Overweight women should gain about 2 pounds in the first trimester and ⅔ pound/week thereafter, achieving a total gain of 15 to 25 pounds.

Weeks gestation

*Source:* Whitney EN, Rolfes SR. *Understanding Nutrition,* 9th ed. Belmont, CA: Wadsworth, © 2002.

# Appendix F

# MEDICAL RECORD DOCUMENTATION FORM

*Progress Notes*

# DIABETIC MEDICAL NUTRITION THERAPY CALCULATION FORM

*Meal Plan*

|  |  | **Grams** | **Percent** |
|---|---|---|---|
| Patient's name _____ Date: _____ | Carbohydrate | _____ | _____ |
|  | Protein | _____ | _____ |
| Dietitian: _____ Phone: _____ | Fat | _____ | _____ |
|  | Calories | _____ | _____ |

| Time | Number of Exchanges/Choices | Menu Ideas | Menu Ideas |
|---|---|---|---|
|  | _____ Carbohydrate group<br>_____ Starch<br>_____ Fruit<br>_____ Milk _____<br>_____ Meat group _____<br>_____ Fat group _____ |  |  |
|  | _____ _____<br>_____ _____ |  |  |
|  | _____ Carbohydrate group<br>_____ Starch<br>_____ Fruit<br>_____ Milk _____<br>_____ Vegetables<br>_____ Meat group _____<br>_____ Fat group _____ |  |  |
|  | _____ _____<br>_____ _____<br>_____ _____ |  |  |
|  | _____ Carbohydrate group<br>_____ Starch<br>_____ Fruit<br>_____ Milk _____<br>_____ Vegetables<br>_____ Meat group _____<br>_____ Fat group _____ |  |  |
|  | _____ _____<br>_____ _____<br>_____ _____ |  |  |

## Appendix H
# RENAL MEDICAL NUTRITION THERAPY CALCULATION FORM

Patient's name: _____

Date: _____

Your dietitian is: _____

Telephone number: _____

_____ grams protein

_____ calories

_____ milligrams phosphorus

_____ milligrams sodium

### *Your Daily Meal Plan*

**Breakfast** | | | **Sample Menu**
--- | --- | --- | ---
Milk | _____ | choices | _____
Nondairy milk substitute | _____ | choices | _____
Meat | _____ | choices | _____
Starch | _____ | choices | _____
Fruit | _____ | choices | _____
Fat | _____ | choices | _____
High-calorie | _____ | choices | _____
Salt | _____ | choices | _____

**Snack** | | | **Sample Menu**
--- | --- | --- | ---
| _____ | choices | _____
| _____ | choices | _____

**Lunch** | | | **Sample Menu**
--- | --- | --- | ---
Milk | _____ | choices | _____
Nondairy milk substitute | _____ | choices | _____
Meat | _____ | choices | _____
Starch | _____ | choices | _____
Vegetable | _____ | choices | _____
Fruit | _____ | choices | _____
Fat | _____ | choices | _____
High-calorie | _____ | choices | _____
Salt | _____ | choices | _____

**Snack** | | | **Sample Menu**
--- | --- | --- | ---
| _____ | choices | _____
| _____ | choices | _____

**Dinner** | | | **Sample Menu**
--- | --- | --- | ---
Milk | _____ | choices | _____
Nondairy milk substitute | _____ | choices | _____
Meat | _____ | choices | _____
Starch | _____ | choices | _____
Vegetable | _____ | choices | _____
Fruit | _____ | choices | _____
Fat | _____ | choices | _____
High-calorie | _____ | choices | _____
Salt | _____ | choices | _____

**Snack** | | | **Sample Menu**
--- | --- | --- | ---
| _____ | choices | _____
| _____ | choices | _____

*Source:* Allen JC, Watters, C. *Lactations: Physiology, Nutrition & Breastfeeding,* 49–102, 1983. Reprinted by permission of Kluwer Academy.

# SUBJECTIVE GLOBAL ASSESSMENT FORM

## PG-SGA Scoring Guide

**Note: PG-SGA is also available in several languages for non–English-speaking clients and caregivers.**

The Patient-Generated Subjective Global Assessment (PG-SGA) provides a comprehensive evaluation of nutritional status level, which can then be used to determine the level of medical nutrition therapy required. The tool includes prognostic components of client history (amount and pattern of weight loss, qualitative assessment of nutritional intake, and standard performance status scales) and clinical history (nutrition impact symptoms, disease process, metabolic stress, and physical examination). Serial assessments using the PG-SGA are necessary in cancer patients to monitor any changes in nutritional status, as there is high risk for nutrition deterioration in this population. The PG-SGA scoring is based on the following parameters.

The first four boxes of the scored PG-SGA are filled out by the client, who provides a current history of weight change, food intake, symptoms, and functional capacity. The check-off format enables clients to be more forthcoming about symptoms that adversely impact intake and quality of life and that are not often thought of in a nutritional context by clinicians. After the client completes the first four boxes, the dietetics professional, doctor, nurse, or other therapist trained in PG-SGA completes the lower section.

Scoring is based on a scale from 0 to 4 points, ranging from no nutritional impact to mild, moderate, severe, and potentially life threatening. The points are determined by adding the checked off points in parentheses on the form, as well as from Boxes 1–4.

| | |
|---|---|
| Box 1: | the point score for the weight loss during the past month if available (or the past 6 months if this is the only information available) plus the points for what happened to the weight during the past 2 weeks. |
| Box 2: | the highest point category checked off by the client. |
| Box 3: | the additive score, for all symptoms checked off by the client. |
| Box 4: | the highest point category checked off by the client. |
| Disease section: | one point for each diagnosis identified in Box 2. |
| Metabolic section: | a score based on metabolic stressors identified in Box 3. |
| Physical section: | a score based on the physical assessment; refer to Box 4. |

Once each of these evaluations is made, the trained clinician proficient in nutrition physical assessment determines a global physical scoring (well-nourished or moderately or severely malnourished) using criteria outlined in Box 5. Triaging nutrition intervention is then determined using the information provided in Box 6.

*Source:* Reprinted with permission from Ottery FD, Kasenic S, DeBolt S, Roger K. Volunteer network accrues >1900 patients in 6 months to validate standardized nutritional triage. Abstract 282. Meeting of the American Society of Clinical Oncology, 1998.

# Scored Patient-Generated Subjective Global Assessment (PG-SGA)

| Patient ID Information |
| --- |

## History

### 1. Weight:

In summary of my current and recent weight:

I currently weigh about _____ pounds
I am about _____ feet _____ inches tall

One month ago I weighed about _____ pounds
Six months ago I weighed about _____ pounds

During the past two weeks my weight has:
☐ decreased    ☐ not changed    ☐ increased

### 2. Food Intake: As compared to my normal, I would rate my food intake during the past month as:

☐ unchanged
☐ more than usual
☐ less than usual
   I am now taking:
   ☐ normal food but less than normal
   ☐ little solid food
   ☐ only liquids
   ☐ only nutritional supplements
   ☐ very little of anything
   ☐ only tube feedings or only nutrition by vein

### 3. Symptoms: I have had the following problems that have kept me from eating enough during the past two weeks (check all that apply):

☐ no problem eating
☐ no appetite, just did not feel like eating
☐ nausea              ☐ vomiting
☐ constipation        ☐ diarrhea
☐ mouth sores         ☐ dry mouth
☐ things taste funny or have no taste  ☐ smells bother me
☐ problems swallowing ☐ feel full quickly
☐ pain; where? _____
☐ other * _____
   * Examples: depression, money, or dental problems

### 4. Activities and Function: Over the past month, I would generally rate my activity as:

☐ normal with no limitations
☐ not my normal self, but able to be up and about with fairly normal activities
☐ not feeling up to most things, but in bed or chair less than half the day
☐ able to do little activity and spend most of the day in bed or chair
☐ pretty much bedridden, rarely out of bed

**Additive Score of the Boxes 1–4** ☐ A

## The remainder of this form will be completed by your doctor, nurse, or therapist. Thank you.

### 5. Disease and its relation to nutritional requirements

All relevant diagnoses (specify) _____
Primary disease stage (circle if known or appropriate)    I   II   III   IV   Other _____
Age _____

Numerical score from Box 2 ☐

### 6. Metabolic demand

☐ no stress    ☐ low stress    ☐ moderate stress    ☐ high stress

Numerical score from Box 3 ☐

### 7. Physical

Numerical score from Box 4 ☐

### Global Assessment

☐ Well-nourished or anabolic (SGA-A)
☐ Moderate or suspected malnutrition (SGA-B)
☐ Severely malnourished (SGA-C)

**Total numerical score of Boxes A+B+C+D** ☐

*(See triage recommendations below)*

Clinician Signature _____ RD RN PA MD DO Other _____    Date _____

**Nutritional Triage Recommendations:** Additive score is used to define specific nutritional interventions including patient and family education, symptom management including pharmacologic intervention, and appropriate nutrient intervention (food, nutritional supplements, enteral, or parenteral triage). First line nutrition intervention includes optimal symptom management.

**0–1**    No intervention required at this time. Reassessment on routine and regular basis during treatment.

**2–3**    Patient and family education by dietitian, nurse, or other clinician with pharmacologic intervention as indicated by symptom survey (Box 3) and laboratory values as appropriate.

**4–8**    Requires intervention by dietitian, in conjunction with nurse or physician as indicated by symptoms survey (Box 3).

**≥9**    Indicates a critical need for improved symptom management and/or nutrient intervention options.

# PHENYLKETONURIA NUTRITION RESOURCES

*Phenylketonuria Diet Calculation Guide*

Date: __7__ / __17__ / _____

Name: _____ Laura Mayberry _____

Birthdate: __7__ / __8__ / _____     Age: _____ **9 days** _____

Length/Height: _____ 18.5 _____ (cm or (in))     Weight: _____ 5.8 _____ (kg or (lb))

| Medical Food Mixture | Amount | ___ (mg) | ___ (mg) | Phe (mg) | Tyr (mg) | Protein (g) | Energy (kcal) |
|---|---|---|---|---|---|---|---|
| Similac w/Fe | 15 g | ___ | ___ | 68.9 | 64 | 1.58 | 7.69 |
| Phenex 1 | 52 g | ___ | ___ | 0 | 780 | 7.8 | 249.6 |
| _____ | ___ g/Tbsp | ___ | ___ | ___ | ___ | ___ | ___ |
| _____ | ___ g/Tbsp | ___ | ___ | ___ | ___ | ___ | ___ |
| _____ | ___ mL | ___ | ___ | ___ | ___ | ___ | ___ |
| _____ | ___ mL | ___ | ___ | ___ | ___ | ___ | ___ |

(_____ mg/mL)

Add water to make _____ mL ( ____ 16 ____ fl oz).     20 kcal/oz

| Food List | Servings | | | | | | |
|---|---|---|---|---|---|---|---|
| Breads/Cereals | ___ | ___ | ___ | ___ | ___ | ___ | ___ |
| Fats | ___ | ___ | ___ | ___ | ___ | ___ | ___ |
| Fruits | ___ | ___ | ___ | ___ | ___ | ___ | ___ |
| Vegetables | ___ | ___ | ___ | ___ | ___ | ___ | ___ |
| Free Foods A | ___ | ___ | ___ | ___ | ___ | ___ | ___ |
| Free Foods B | ___ | ___ | ___ | ___ | ___ | ___ | ___ |
| Total per day | | ___ | ___ | 68.9 | 844 | 9.38 | 326.5 |
| Total per kg | | ___ | ___ | 25.8 | 316.6 | 3.5 | 122.5 |

**Comments:**

_L. Bernstein, MS, RD, FADA_
Nutritionist

*Recommended Daily Nutrients Intakes (Ranges) for Infants with PKU*

| Infants | Phe mg/kg | Tyr mg/kg | Protein g/kg | Energy kcal/kg | Fluid mL/kg |
|---|---|---|---|---|---|
| 0 to <3 mo | 25–70 | 300–350 | 3.00–3.50 | 120 (95–145) | 135–160 |
| 3 to <6 mo | 20–45 | 300–350 | 3.00–3.50 | 120 (95–145) | 130–160 |
| 6 to <9 mo | 15–35 | 250–300 | 2.50–3.00 | 110 (80–135) | 120–145 |
| 9 to <12 mo | 10–35 | 250–300 | 2.50–3.00 | 105 (80–135) | 120–135 |

*Source:* Adapted with permission from Acosta B. *Ross Metabolic Formula Nutrition Support Protocols.* Columbus, OH: Ross Laboratories, 1993.

*Nutrient Composition of Similac Powder with Iron and Phenex 1*

| | Similac powder per 100 grams | Phenex 1 per 100 grams |
|---|---|---|
| Energy, kcal | 513 | 480 |
| Protein equiv., g | 10.56 | 15 |
| Phenylalanine, mg | 460 | 0 |
| Tyrosine, mg | 427 | 1500 |

*Source:* Adapted with permission from Acosta B. *Ross Metabolic Formula Nutrition Support Protocols.* Columbus, OH: Ross Laboratories, 1993.

# INDEX